Climate from tree rings

This book is based largely on material presented
at the Second International Workshop on Global
Dendroclimatology, Norwich, 1980. This event was
supported by the World Meteorological Organization
jointly with the United Nations Environment
Programme and by the Scientific Affairs Division
of the North Atlantic Treaty Organization.

Climate from tree rings

Edited by

M.K.HUGHES
Department of Biology, Liverpool Polytechnic

P.M.KELLY
Climatic Research Unit, University of East Anglia

J.R.PILCHER
Palaeoecology Laboratory, Queen's University of Belfast

V.C.LAMARCHE JR
Laboratory of Tree-Ring Research, University of Arizona

CAMBRIDGE UNIVERSITY PRESS
Cambridge
London   New York   New Rochelle
Melbourne   Sydney

CAMBRIDGE UNIVERSITY PRESS
Cambridge, New York, Melbourne, Madrid, Cape Town, Singapore, São Paulo, Delhi

Cambridge University Press
The Edinburgh Building, Cambridge CB2 8RU, UK

Published in the United States of America by Cambridge University Press, New York

www.cambridge.org
Information on this title: www.cambridge.org/9780521113069

First published 1982
This digitally printed version 2009

A catalogue record for this publication is available from the British Library

Library of Congress Catalogue Card Number: 81-17056

ISBN 978-0-521-24291-2 hardback
ISBN 978-0-521-11306-9 paperback

CONTENTS

There is an urgent need for greater understanding of
the nature and causes of the fluctuations in climate that
have occurred since the most recent ice age. In a world
faced with a growing population and accelerating use of
energy, water, and food resources, the demand for estimates of climatic change over the next few decades is
pressing. This demand cannot be met at present because our
knowledge of the course and causes of past climates is too
limited. The instrumental records of climate variables
such as temperature, pressure, and rainfall are sparse
over much of the globe before the beginning of the 20th
century. A few extend back two or three hundred years but
these are mainly restricted to the North Atlantic sector.
In order to gain some understanding of climatic variations
on timescales of up to a few decades, climate records for
much of the globe over several centuries are needed.
Information on large-scale climatic change during the
early- and pre-instrumental periods may only be established by the use of historical (documentary) and proxy
climate records (Barry et al., 1979). The term "proxy
climate records" describes dateable evidence of a
biological or geological phenomenon whose condition is at
least in part determined by climate at the time of its
formation.

Tree growth provides a variety of measurements which
can be used as proxy climate records. As records which may
be assigned accurately to a particular year and place,
tree growth occupies a special position in the catalogue
of proxy climate information alongside annually layered
lake sediments, some ice cores, and certain historical
data. Trees are found over a large part of the Earth's
landmass and although not all trees form dateable annual
layers from which climate information may be decoded, many
do. Among these there is a wide variety of responses of
growth to climate, both on the basis of inherited
capabilities and of their expression under a given set of
site conditions. Consequently, tree rings record responses
to a wider range of climate variables over a larger part
of the Earth than any other kind of annually-dated proxy
record. Climatically responsive trees forming annual
growth rings may be found in many regions of the world. A
global network of several centuries length is conceivable,
although not yet established. Those not familiar with the
field of dendroclimatology will find an excellent introduction and a glossary in Fritts (1976).

There have been striking advances in the field of
dendroclimatology during the past 20 years. Tree-ring data
have been used to derive various climate variables for
certain localities and have been used in a handful of
cases to provide estimates of spatial patterns of climate
anomalies for larger regions. Despite these successes, the
full potential of tree rings as sources of proxy climate
data is far from realised. Although the reconstruction of
global climate must be an international exercise, the
related sciences of dendrochronology and dendroclimatology
have developed rather parochially. Dendrochronology was
concerned with climate right at its conception in the
early decades of the present century. In fact, much of the
research reported at the Second International Workshop on
Global Dendroclimatology in 1980 was foreseen in essence
by A.E.Douglass (University of Arizona) as early as 1914.
The spur to the development of dendrochronology came,
however, not from climatology but from archaeology. The
science developed as an archaeological dating tool both in
the southwestern United States and in Europe. From an
archaeological standpoint, these two areas are far apart
and basic techniques developed along very different lines
with little contact between researchers. In recent years,
as the frontiers of the science have extended into new
geographical areas, experience in these regions has led to
further modification and refinement of dendroclimatic
techniques. But climate does not respect political, cultural, or methodological barriers.

The Second International Workshop on Global Dendroclimatology was held at the University of East Anglia,
Norwich, United Kingdom, from the 7th to 11th July, 1980.
This was not a conference but a working meeting of invited
scientists selected on the basis of their proven willingness to engage in collaborative research in dendroclimatology. The First International Workshop on Dendroclimatology took place at the University of Arizona in 1974.
That workshop achieved such success in establishing
international collaboration in the following six years
that scientists from 12 countries and all inhabited
continents attended the meeting in 1980. At a planning
meeting held in July 1979, a group from a number of these
countries agreed that a second workshop should be held in
order to provide an international and interdisciplinary
forum in which methodological differences could be aired
constructively in order to provide the basis of understanding, if not agreement, that is necessary for collaborative effort.

The specific aims of the Second International Workshop on Global Dendroclimatology, as set out in the pros-
pectus issued to participants, were as follows:

a) to review basic concepts and techniques to provide a
   common basis for discussion;
b) to consider what kinds of information available from
   tree rings would be of value to climatologists,
   hydrologists, historians, economists, and others
   working with palaeoclimatic proxy data; in doing so,
   the properties of the original data and their consequences for the types of information that may be
   derived must be reviewed;
c) to evaluate critically the past and potential cont-

ribution of tree-ring data to palaeoclimatic and
related studies;

d) to exchange information on analytical techniques and
results as well as on recent technical developments;
particular emphasis should be placed on those techniques essential to the acquisition of data appropriate to dendroclimatology;

e) to review systematically the existing geographical
and temporal extent of appropriate tree-ring collections, of data analysis, and of applications; and

f) to prepare a review of the current status of dendroclimatology and a set of recommendations for the
development of dendroclimatic research on a global
scale.

In order to achieve these ends, the Workshop was organised
in three phases. In the first phase, invited contributors
prepared their papers which were then sent to discussants
for written comment. These papers were circulated to all
participants prior to the meeting, greatly facilitating
discussion during the Workshop sessions. The second phase
was the meeting itself, which consisted of sessions and
working group periods. The participants entered into the
spirit of discussion and collaboration fully making the
meeting itself a very positive experience. The third phase
has been the communication of the results of the Workshop
to the scientific community and to decision-makers. The
detailed findings of the Workshop have been presented in
the "Report and Recommendations" (Hughes et al., 1980),
which is available from Dr.J.R.Pilcher (Queen´s University
of Belfast). This volume is based largely on the proceedings of the Workshop.

The Workshop was an expensive enterprise. It was
generously supported by the World Meteorological Organization jointly with the United Nations Environment
Programme and by the Scientific Affairs Division of the
North Atlantic Treaty Organization. The U.S. National
Science Foundation provided travel funds for many of the
participants from the United States.

The Editors acknowledge permission to reprint
material granted by the American Meteorological Society
(Figure 2.16), Quaternary Research, the University of
Washington (Kutzbach & Guetter, Chapter 2), Lehrmittelverlag des Kantons Zürich (Bräker, Figures 4.14 and 4.15),
and the American Water Resources Association (Holmes,
Stockton & LaMarche, Chapter 5). They also thank all who
contributed to the Workshop and to this volume and the
staff of the Climatic Research Unit who assisted with the
production of this book (particularly B.Harris, A.Andrup,
S.Napleton, and S.Boland). Finally, the invaluable work
done by our Assistant Editor, D.A.Campbell, is gratefully
acknowledged.

The contributors to this volume wish to thank the
following individuals and organisations: R.G.Barry,
T.Barnett, P.Berthet, T.Bird, W.R.Boggess, J.Carlson,
L.G.Drew, G.Farmer, I.M.Flaschka, H.Fluhler, J.D.Good-
ridge, M.Heeb, N.Huckstep, B.Keller, Dr.Kirchhofer,
D.A.Larson, D.Lee, A.M.Marek, R.Mortimer, M.J.Sieber,
T.Starr, M.A.Thompson, L.D.Ulan, B.Watson, T.Webb III, the
Californian Department of Water Resources, C.S.I.R.O.
Division of Atmospheric Physics, Fonds de la Recherche
Fondementale Collective de la Belgique, Institut pour
l´Encouragement de la Recherche Scientifique dans
l´Industrie et l´Agriculture, the U.S. National Science
Foundation (Grants ATM 74-23041, ATM 77-19216, and ATM
79-19223), and the U.S. Office of Naval Research (Grant
N00014-77-G-0074).

CONTRIBUTORS

R.W.Aniol: Institut für Ur und Frühgeschichte, Universität Köln, Weyertal 125, D5000 Köln 41, Federal Republic of Germany

M.G.L.Baillie: Palaeoecology Laboratory, Queen's University, Belfast BT7 1NN, Northern Ireland

Z.Bednarz: Akademia Rolnicza w Krakówie, Instytut Hodowli Lasu, 31-024 Kraków, ul.św. Marka 37, Poland

A.L.Berger: Institut d'Astronomie et de Géophysique, Université Catholique de Louvain, Chemin du Cyclotron 2, B-1348 Louvain-la-Neuve, Belgium

T.T.Bitvinskas: Lietuvos TSR Mokslu Akademija, 232600 Vilnius MTP-1, Lenino pr.3, Lithuanian SSR, U.S.S.R.; also LTSR MA Botanikos Institutas Dendroklimatochronologine Laboratorija, 22300 Kaunas, Laisves aleja 53, Lithuanian SSR, U.S.S.R.

J.A.Boninsegna: Instituto Argentino de Nivologia y Glaciologia, Casilla de Correo 330, 5500 Mendoza, Argentina

O.U.Bräker: Eidgenössische Anstalt für das Forstliche Versuchswesen, CH-8903 Birmensdorf ZH, Switzerland

D.W.Brett: Botany Department, Bedford College, Regent's Park, London NW1 4NS, U.K.

L.B.Brubaker: College of Forest Resources, University of Washington, Seattle, Washington 98195, U.S.A.

D.A.Campbell: Laboratory of Tree-Ring Research, University of Arizona, Tucson, Arizona 85721, U.S.A.; then Climatic Research Unit, School of Environmental Sciences, University of East Anglia, Norwich NR4 7TJ, U.K.; now Burwood Downs, Yetman, NSW 2409, Australia

L.E.Conkey: Laboratory of Tree-Ring Research, University of Arizona, Tucson, Arizona 85721, U.S.A.

E.R.Cook: Lamont-Doherty Geological Observatory, Columbia University, Palisades, New York 10964, U.S.A.

J.P.Cropper: Laboratory of Tree-Ring Research, University of Arizona, Tucson, Arizona 85721, U.S.A.

P.W.Dunwiddie: Quaternary Research Center, University of Washington, Seattle, Washington 98195, U.S.A.

T.G.J.Dyer: Geography Department, University of Witwatersrand, Johannesburg, South Africa

D.Eckstein: Universität Hamburg, Institut für Holzbiologie, Leuschnerstrasse 91, D-2050 Hamburg 80, Federal Republic of Germany

E.Frisse: Universität Hamburg, Institut für Holzbiologie, Leuschnerstrasse 91, D-2050 Hamburg 80, Federal Republic of Germany

H.C.Fritts: Laboratory of Tree-Ring Research, University of Arizona, Tucson, Arizona 85721, U.S.A.

H.Garfinkel: College of Forest Resources, University of Washington, Seattle, Washington 98195, U.S.A.

G.A.Gordon: Laboratory of Tree-Ring Research, University of Arizona, Tucson , Arizona 85721, U.S.A.; now Institute for Quaternary Studies, Boardman Hall, University of Maine, Orono, Maine 04469, U.S.A.

B.M.Gray: Climatic Research Unit, School of Environmental Sciences, University of East Anglia, Norwich NR4 7TJ, U.K.

D.A.Graybill: Laboratory of Tree-Ring Research, University of Arizona, Tucson, Arizona 85721, U.S.A.

P.J.Guetter: Center for Climatic Research, Institute of Environmental Studies, University of Wisconsin, Madison, Wisconsin 53706, U.S.A.

J.Guiot: Institut d'Astronomie et de Géophysique, Université Catholique de Louvain, Chemin du Cyclotron 2, B-1348 Louvain-la-Neuve, Belgium; now Laboratoire de Botanique, Historique et Palynologie, Université d'Aix Marseilles III, Rue Henri Poincaré, 13397 Marseille Cedex 4, France

R.L.Holmes: Laboratory of Tree-Ring Research, University of Arizona, Tucson, Arizona 85721, U.S.A.

M.K.Hughes: Biology Department, Liverpool Polytechnic, Byrom Street, Liverpool L3 3AF, U.K.

J.H.Hunt: Laboratory of Tree-Ring Research, University of Arizona, Tucson, Arizona 85721, U.S.A.

G.C.Jacoby: Lamont-Doherty Geological Observatory, Columbia University, Palisades, New York 10964, U.S.A.

P.M.Kelly: Climatic Research Unit, School of Environmental Sciences, University of East Anglia, Norwich NR4 7TJ, U.K.

K.C.Kuivinen: Department of Geography, University of Nebraska, Lincoln, Nebraska 68588, U.S.A.

J.E.Kutzbach: Center for Climatic Research, Institute of Environmental Studies, University of Wisconsin 53706, U.S.A.

V.C.LaMarche Jr: Laboratory of Tree-Ring Research, University of Arizona, Tucson, Arizona 85721, U.S.A.

Lin Zhenyao: Institute of Geography, Academia Sinica, Peking, Peoples Republic of China

M.P.Lawson: Department of Geography, University of Nebraska, Lincoln, Nebraska 68588, U.S.A.

G.R.Lofgren: Laboratory of Tree-Ring Research, University of Arizona, Tucson, Arizona 85721, U.S.A.

A.Long: Laboratory of Isotope Geochemistry, Department of Geosciences, University of Arizona, Tucson, Arizona 85721, U.S.A.

S.J.Milsom: Biology Department, Liverpool Polytechnic, Byrom Street, Liverpool L3 3AF, U.K.

A.V.Munaut: Laboratoire Palynologie et Phytosociologie, Université Catholique de Louvain, 4 Place Croix du Sud, 1348 Louvain-la-Neuve, Belgium

J.Ogden: Department of Botany, University of Auckland, Private Bag, Auckland, New Zealand

R.L.Phipps: Tree-Ring Laboratory, Water Resources Division, U.S. Geological Survey, Mailstop 461, Reston, Virginia 22092, U.S.A.

J.R.Pilcher: Palaeoecology Laboratory, Queen's University, Belfast BT7 1NN, Northern Ireland

A.B.Pittock: CSIRO, Atmospheric Physics Division, P.O.Box 77, Mordialloc, Victoria 3195, Australia

M.J.Salinger: Climatic Research Unit, School of Environmental Sciences, University of East Anglia, Norwich NR4 7TJ, U.K.

B.Schmidt: Institut für Ur und Frühgeschichte, Universität Köln, Weyertal 125, D5000 Köln 41, Federal Republic of Germany

F.H.Schweingruber: Eidgenössische Anstalt für das Forstliche Versuchswesen, CH-8903 Birmensdorf ZH, Switzerland

F.Serre-Bachet: Laboratoire de Botanique, Historique et Palynologie, Université d'Aix Marseille III, Rue Henri Poincaré, 13397 Marseille Cedex 4, France

C.W.Stockton: Laboratory of Tree-Ring Research, University of Arizona, Tucson, Arizona 85721, U.S.A.

T.M.L.Wigley: Climatic Research Unit, School of Environmental Sciences, University of East Anglia, Norwich NR4 7TJ, U.K.

Wu Xiangding: Institute of Geography, Academia Sinica, Peking, Peoples Republic of China

Zheng Sizhong: Institute of Geography, Academia Sinica, Peking, Peoples Republic of China

INTRODUCTION
The Editors

Where trees form annual growth layers, there exists the likelihood that the characteristics of those layers reflect the conditions under which they were formed. Differences in annual growth layers, which are seen as tree rings, may be parallel in many trees within a region indicating that some common set of external factors is influencing growth. Such similarities in growth variation may be strong and spatially extensive. Where this is true, it is reasonable to assume that the external agents forcing the pattern of variability common to trees in a region relate to climate. There are no other environmental factors likely to act on the same range in the space, time, and frequency domains. It should be possible, therefore, to extract a record of the climate variables recorded in the rings of wood formed in the past. This is the basic assumption of dendroclimatology.

Many tree species in the temperate regions show patterns of common year-to-year variability in one or more measure describing the state of the tree ring. This phenomenon has been exploited by scientists in two main fields: dendrochronology (tree-ring dating of wood found in archaeological, geological, or other contexts) and dendroclimatology (the use of tree rings as proxy climate indicators). Both fields depend heavily on the identification and verification of patterns of common year-to-year variability in many wood samples from a site or region. A variety of methods for establishing calendrical control have been used, and a priori criteria have been developed for assessing whether material from a given tree ring variable, species, site, or region is likely to yield a strong enough common pattern of variability for either dating or dendroclimatic use. For dendroclimatology, a range of techniques has been used to treat the absolutely dated time series of tree-ring measurements in such a way that climate-related information (signal) is retained with minimum distortion, whilst the non-climatic variation (noise) is removed as far as possible.

This chapter is concerned with the acquisition and preparation of tree-ring data for the specific purpose of reconstructing past climate and related phenomena. Some indication is given of promising new approaches. LaMarche outlines a general sampling strategy for extending the existing tree-ring data base into a new geographical area. He demonstrates that some generally applicable principles and concepts may be used to guide sample selection from the initial identification of a promising region through to the choice of individual trees and radii. He draws on his great experience in exploiting new regions for dendroclimatology in order to assess the generality of these concepts and emphasises the need for constant re-assessment. Cook takes up this theme, with the particular case of the relative importance of site and species selection in securing good proxy records of climate. Whilst in many cases the climate-growth response is peculiar to a particular site-habitat-species complex, there are some species, such as Tsuga canadensis Carr. in eastern North America, which show a characteristic response. There is a tantalising hint in Baillie's discussion of correlations between chronologies in different parts of Ireland that Quercus may behave similarly.

Reference has been made to the influence of past conditions on the state of annual growth layers. A number of different measures have been used to describe this, most commonly the width of the annual ring (the annual radial increment). In recent years, other physical and chemical variables have been studied in the context of dendroclimatology. Particular progress has been made in the use of X-ray densitometry to reveal intra- and inter-annual variations in wood density. Density measurements integrate changes in proportions of cell types and their relative dimensions. Schweingruber describes the methodological basis of this approach. The results of a remarkably successful reconstruction of past temperature from ring-width and density records of coniferous trees from Swiss sites have been reported elsewhere (Schweingruber et al., 1978a). These variables may also be recorded directly as described by Eckstein and Frisse for Quercus in Europe. They report indications of a significant climate signal in early wood vessel size in Quercus. Oaks have been shown to be a difficult, but not impossible, material for X-ray analysis (Milsom, 1979).

Of all the possible chemical descriptions of the state of the annual growth layer, most attention has been paid to variations in the stable isotopes of hydrogen, carbon, and oxygen. Long reviews the basis of this approach. He shows that the climatic interpretation of stable isotope variations in tree rings is at the learning stage, and stresses the dangers of making naive assumptions of a simple relationship between isotopic parameters and particular climate variables. Whilst it is possible that general trends may be inferred from stable isotope data from tree rings, more precise palaeoclimatic inferences will require a more thorough understanding of the systems involved, necessitating carefully designed growth experiments and suitable structured field measurements. This need for a fuller understanding of the physiological and biochemical mechanisms involved is emphasised by Wigley, who also discusses the study of carbon isotope measurements in tree rings.

Regardless of which tree-ring variables are measured, the measurement procedures are designed to yield a set of absolutely dated time series that are records of the environmental conditions experienced by each tree in past years. Graybill describes the most widely used procedures for enhancing the climate signal by methods intended to remove growth trends and biologically induced changes. Series of individual radii and trees are combined to give the maximum common signal. Descriptive sample statistics are then calculated and should be used to evaluate samples, to characterise time series, and to assess the relationships between annual ring variables. These methods have been developed with ring-width measurements in mind and may require further modification for use with other variables and sets of variables. Whilst standardising by division of the actual value by the calculated growth curve value may be most appropriate for ring-width, density data may only require the computation of differences since the variance of raw density data appears not to vary through time. Hughes discusses various assumptions that are made in the course of preparing tree-ring samples for dendroclimatic analysis.

Dendrochronologists are familiar with the phenomenon of _pointer_ years described by Aniol and Schmidt. In such years, a very high proportion of all trees at a site or in a region show a change of ring width or other variable in the same sense. Whilst the proportion of years showing such behaviour increases with sample size, some years remain as a record of disparate tree response. Pointer years potentially record climatic phenomena which are coherent over a large area. This has not yet been fully exploited.

Many of the statistical techniques described in this chapter have been developed for use on material from sites where trees are obviously limited by heat or moisture deficit at one or more times of the year. In general, these trees grow in open-canopy forests as distinct from the more closed canopies of mesic forests. In the regions into which dendroclimatology is presently expanding, such as Tasmania, eastern North America, and much of Europe, strong common patterns of year-to-year variability (good cross-dating in dendrochronological terms) are being found in trees from closed-canopy forests. In these cases, factors related to stand and population dynamics and disturbance as well as age-trend may influence ring width and other variables. The variations in the tree-ring series that result from such factors are likely to be of medium to low frequency and thus may be confused with medium- to low-frequency climatic change. It is of considerable importance that methods of standardisation that are capable of removing as much of this non-climatic variation as possible be developed. It is likely that suitable algorithms will start from a consideration of the common and of the unshared variance in several frequency bands for the whole data set of cores and trees. An attempt can then be made to remove unshared variance in order to find the best available expression of the common variance in all frequency bands.

In addition to its value in preparing chronologies from living trees for dendroclimatic analysis, such a method of standardisation would be needed for the pretreatment of the many short series to be included in long composite chronologies. A further problem in the compilation of long composite chronologies concerns the kinds of adjustment to the time-series properties of the series to be merged that may be made safely without inadvertently distorting the low-frequency climate signal.

Methods of establishing precise calendrical control are shared by dendrochronology and dendroclimatology, but major differences exist in other areas. In particular, specific criteria are being developed in dendroclimatology for the selection of suitable regions, sites, species, and trees to achieve the clearest climate signal. Similarly, techniques of measurement and analysis have been, and are still being, developed which are designed specifically to describe, extract, and strengthen that signal without distortion. The particular method chosen at each stage depends on both the tree-ring variable or variables being measured and the climate information that the investigator hopes to extract. Consequently, whilst many of the methods described have very general application, it is not possible to prescribe a procedure that would be appropriate in every case. Indeed, it is important that new approaches be investigated in order that the considerable scientific resources presented to dendroclimatology by as yet unused material be tapped.

SAMPLING STRATEGIES
V.C.LaMarche Jr.
INTRODUCTION

The use of tree rings to study past climate on regional, hemispheric, and global scales requires an expansion of the existing data base to include new habitats, new tree species, and new regions. In order to use the limited available resources with maximum effectiveness, it is necessary to concentrate our efforts in those areas and on those taxa that will yield long, accurately dated, and climatically indicative series of tree-ring data, and that will provide proxy climate records of areas, variables, and seasons of particular interest to palaeoclimatologists. The quantitative evaluation of palaeoclimatic data networks (Kutzbach & Guetter, this volume) is beyond the scope of this report. Rather, I will concentrate on approaches to tree-ring sampling and analysis based on some important dendro-chronological and dendroclimatic concepts that can influence the development of a sampling programme. Because many of these concepts were strongly influenced by early experience with trees of only a few species in a restricted geographic area and occupying a limited range of habitats, it is also worth examining their applicability in situations far different from those in which they were originally conceived. Therefore, some of the general observations presented here are based in part on newly published or unpublished results from recent work in the Southern Hemisphere (Ogden, 1978a,b; Dunwiddie, 1979; LaMarche et al., 1979a,b,c,d,e; Dunwiddie & LaMarche, 1980a,b) and in the Pacific Northwest (Brubaker, 1980) and in the eastern United States (Cook & Jacoby, 1979). See also Chapters 3 and 4 of this volume.

Fritts (1976) has presented many of the underlying principles and concepts of dendrochronology and dendroclimatology, and discussed them in the context of dendroclimatic problems. Table 1.1 lists several of the most critical, and depicts their approximate importance at different stages of a decision-making process ranging in scale from the selection of a promising region for study down to the determination of which trees and which radii to sample on a given site. Although some of these concepts (and some decisions) are clearly inter-related, a general approach of this kind can be very useful in developing a research strategy for exploratory dendroclimatic studies in a new area.

ANNUAL RING CHARACTERISTICS

The fundamental concept in dendrochronology is that of the annual ring. For our purposes, it can be described as a growth band in the xylem of a tree or shrub which has anatomically definable boundaries and is formed during a single period of cambial activity that occurs only once per year. Although one can conceive of situations in which trees regularly and dependably form exactly two or exactly three (or more) distinct growth bands each year, this is rarely, if ever, found to be the case in practice. The dependability of the annual ring is essential for the development of accurately dated tree-ring records. Accuracy of dating is, in turn, critically important for most dendroclimatic applications because the calibration of tree-ring data with concurrent meteorological observations requires that the two data sets be matched exactly in time.

The concept of the annual ring is epitomised by the growth bands seen in transverse section in the wood of many gymnosperms in temperate regions (Fritts, 1976, p.13). They lay down wood cells only during a limited period of cambial activity during the warmer part of the year and are dormant in winter. Microscopic examination shows an abrupt discontinuity between the small, thick-walled tracheid cells formed late in one growing season and the large, thin-walled cells representing the first flush of growth at the beginning of the next. The wood of angiosperms is anatomically more complex but may display annual ring structure clearly defined by similar discontinuities in tracheid size, or by changes in the size or frequency of vessels, density of parenchyma, a terminal parenchyma layer, or some combination of these features (Lilly, 1977). At the other end of the spectrum are many trees and shrubs - perhaps the majority on a global basis - that display only a crude growth banding, often of a quasi-annual nature (Mucha, 1979), and lack objectively definable annual rings. In such situations, close study may yield approximate age information useful for inferences about age structure, fire frequency, competitive status, or other parameters, but does not provide a starting point for dendroclimatic research.

Although angiosperms (notably _Quercus_ sp.) have been used extensively for dating purposes and, to a much lesser degree, for dendroclimatic analysis and palaeoclimatic reconstruction, the vast majority of tree-ring applications to the study of past climate have been based on coniferous gymnosperms. Examination of samples from a wide variety of conifers suggests that there are genetically linked clusters of growth-ring characteristics, which may be modified by local environmental characteristics in some taxa. Thus, one may be able to identify target areas on the basis of a knowledge of the local flora and the general climatic regime and to focus on the most promising species and sites with little prior information on the dendrochronological potential of a region. Similar

Table 1.1: Influence of concepts on sampling strategy.

STRATEGY	ANNUAL RING	LIMITING FACTORS	CLIMATIC RESPONSIVENESS	SENSITIVITY AND CROSS-DATING	CIRCUIT UNIFORMITY	REPETITION
Regional Selection	XXX	XXX	X			
Species Selection	XXX	X	XXX	XXX	X	
Site Selection	X	XXX	XXX	XXX	X	
Sample Size Estimation			X	X	XXX	XXX
Tree and Radius Selection					XXX	XXX

Strong influence XXX, Some influence X, Little or no influence Blank.

considerations seem to apply to angiosperms, but experience is more limited. For many angiosperms, new
criteria for both dating and analysis will have to be
developed in order to study and interpret fluctuations in
ring width or other parameters in palaeoclimatic terms.

The two main problems in the dating of the growth
rings in conifers are those involving intra-annual bands
("false rings") and diffuse annual ring boundaries. A very
high frequency of locally absent ("missing") rings may
preclude dating of a particular sample, tree, or sometimes
an entire site. However, in most cases, careful use of
cross-dating techniques (Stokes & Smiley, 1968) can
overcome this problem and yield a sufficient number of
accurately dated radial tree-ring series for further
dendroclimatic analysis, particularly if the sample size
is large and represents trees of a range of ages in a
variety of microhabitats within the site. Table 1.2
illustrates the relationship between taxonomic status and
growth-ring characteristics in a number of conifers. It is
clearly oversimplified. For example, some small and
geographically restricted genera, such as Athrotaxis of
Australia, show few problems of any kind (Ogden, 1978b).
Other, more diverse and widespread genera, such as Pinus,
include species that rarely form false rings as well as
species in which the frequency is so great as to preclude
dating.

The relationship between the taxon and the clarity
of the annual ring boundary is more consistent and appears
to be largely determined at the family level. Some of the
Cupressaceae, many of the Podocarpaceae, and all of
Araucariaceae may produce ring sequences in which the
annual ring boundary is diffuse. That is, boundaries in
which the true latewood cells are not pronouncedly smaller
than in the earlywood, and in which there is a gradational
transition to the earlywood of the following ring
(Dunwiddie & LaMarche, 1980b). This characteristic does

not seem to be strongly related to local environment or
regional climate because it is found in trees of cool
temperate latitudes as well as the subtropics. Moreover,
diffuse ring sequences may occur only on some sites, in
some trees, on some radii, or for a short distance
(reflecting time) on a single radius, more or less at
random. In contrast, there is clear evidence of
environmental control of the incidence of intra-annual
latewood bands. For example, it has been observed that the
frequency of false rings in species such as Juniperus
scopulorum and Pinus ponderosa increases from north to
south in western United States. This is presumably related
at least in part to the increasing frequency of effective
rainfall events in summer, inducing a resumption of the
formation of earlywood-type cells after latewood formation
had already begun. Similar climatically determined trends
are seen in the Southern Hemisphere genera Callitris,
Widdringtonia, and Austrocedrus (LaMarche et al.,
1979a,b,c,d,e) where the lowest incidence of false rings
is observed in areas of Mediterranean-type climate, with a
pronounced winter rainfall maxmum and very dry summers.
In general, a propensity to form false rings, where local
environmental conditions permit, seems characteristic of
most of the Cupressaceae and of many species of Pinus.

LIMITING FACTORS AND CLIMATIC RESPONSIVENESS

The concepts of limiting factors, climatic
responsiveness, and "sensitivity" can have a critical
influence on sampling strategies (Table 1.1) but they are
inter-related in complex ways. Furthermore, criteria
applicable in some areas may be invalid or even misleading
in others. The consequences of the law of limiting
factors, as developed by Fritts (1976) in the context of
dendroclimatology, can be paraphrased as follows: a
tree-ring series (defined by ring width, density, or other
parameters) will contain a climate signal only if climatic

Table 1.2: Ring characteristics of some major coniferous taxa.

INCREASING FREQUENCY OF DIFFUSE ANNUAL RING BOUNDARIES → → →

INCREASING
FREQUENCY OF
INTRA-ANNUAL
BANDS

	TAXODIACEAE	PINACEAE	CUPRESSACEAE	PODOCARPACEAE	ARAUCARIACEAE
↓		Picea			
↓	Athrotaxis	Abies	Fitzroya	Phyllocladus	
↓	Sequoia	Larix	Pilgerodendron	Saxegothaea	
	Sequoiadendron	Pseudotsuga		Podocarpus	
		Tsuga		Dacrydium	
					Araucaria
		Cedrus			
			Thuja		Agathis
	Taxodium				
			Libocedrus		
			Callocedrus		
		Pinus	Austrocedrus		
			Juniperus		
			Cupressus		
			Widdringtonia		
			Callitris		

factors have been directly or indirectly limiting to
processes within the tree. Such trees can be described as
"climatically responsive". This is the critical concept
underlying the use of tree-ring data to make inferences
about past climate. The traditional argument follows, that
the most responsive trees will be found near their
climatically determined limits of distribution, where
climatic factors are most frequently limiting. The classic
examples are trees at ecotonal boundaries such as the
lower forest border regions, and the subalpine and
subarctic treelines.

   Although the basic validity of the limiting-factors
concept cannot be questioned, it is clear that useful
chronological and palaeoclimatic information is contained
in the ring widths of many trees growing in habitats where
the limiting effects of factors such as temperature or
precipitation may not be so obvious. Examples include the
well-established and widespread application of cross-
dating using oak from mesic sites in western Europe
(Pilcher, 1973). In North America, successful climate
reconstructions have recently been made based on ring-
width data from relatively moist areas such as the Pacific
Northwest (Brubaker, 1980) and the northeast (Conkey,
1979a; Cook & Jacoby, 1979; Conkey, this volume; Cook,
this volume). Unpublished results from the Southern Hemisphere show that a strong tree-growth response to
temperature exists in trees of several species in the
mesic forests of New Zealand, and statistically
significant temperature and runoff reconstructions have
been made, using data from trees in similar habitats in
Tasmania (Campbell, 1980, this volume; LaMarche & Pittock,
this volume).

SENSITIVITY AND CROSS-DATING

   The concept of "sensitivity" has long been associated with cross-dating and chronology development, on
the one hand, and with climatic responsiveness, on the
other. However, "sensitivity" is basically a statistical
concept, describing one of the properties of a time
series. As used by Schulman (1956), it is the average
absolute mean first difference of a ring-width or
ring-width index series; high mean sensitivity describes a
series with large relative year-to-year differences in
ring width. However, high mean sensitivity is not
necessary for development of accurately dated ring-width
chronologies, nor does low mean sensitivity necessarily
imply that a strong climate signal is absent. For example,
long chronologies have been developed for mesic-forest
species in genera such as Athrotaxis, Librocedrus, and
Dacrydium (Ogden, 1978b; LaMarche et al., 1979a,b,c,d,e),
where the consistently narrow (or "critical") rings may
occur only a half-dozen times per century. Accurate dating
is attainable because locally absent rings are very rare.

   Similarly, high mean sensitivity is a sufficient,
but not necessary, indicator of climatic responsiveness.

While the two parameters seem to be closely linked in
trees near the arid forest margins, trees in other kinds
of habitats can yield palaeoclimatically significant
ring-width records characterised by low mean sensitivity.
The best example is provided by trees whose growth is
limited to a large extent by low warm season temperature,
such as at the upper (LaMarche & Stockton, 1974) and
northern treelines. Their growth records typically show
low mean sensitivity and a high degree of persistence, as
measured by the first order autocorrelation coefficient.
It is not fully understood but this persistence or serial
dependence may be related in part to the retention of
foliage for several years or more in many of these species
(LaMarche, 1974b), which has the effect of smoothing out
year-to-year temperature differences while enhancing lower
frequency trends and fluctuations. Conversely, high mean
sensitivity may be found to characterise tree-ring series
in ecological situations far removed from climatically
determined boundaries. For example, most species in the
Southern Hemisphere genus Phyllocladus yield ring-width
series with very high mean sensitivity and low first order
autocorrelation (LaMarche et al., 1979a,b,c,d,e). Yet,
these trees are found in very moist temperate forests over
a broad altitudinal range.

   Although new criteria may be applicable in some
areas, the more traditional concepts can still provide
very valuable guidelines in regional evaluation and in
species and site selection. The drought-responsive
conifers near the arid lower-forest border in dry areas
have their analogues in many mesic forest regions, where
local mesoclimatic or edaphic conditions lead to more
frequent drought stress because of relatively low
rainfall, higher air temperatures, greater insolation, or
low soil moisture retention. North American examples
include sites in the "rain-shadow" area of the northeastern Olympic Peninsula of Washington State and in the
Catskill Mountains of New York (Cook & Jacoby, 1979). In
Europe, certain Larix sites in the Alps (Serre, 1978a) may
also fall in this category. However, different criteria
may apply to selection of sites in the relatively moist,
closed-canopy forests covering much of the Pacific
Northwest, eastern North America, and western Europe, as
well as southern South America, southern South Africa,
Tasmania, New Zealand, and other regions. For example,
Brubaker (this volume) chooses mesic-forest sites mainly
on the basis of maximum tree age and a history of minimal
disturbance.

TREE AND RADIUS SELECTION

   Obtaining increment cores or cross-sections is
obviously the final step in a tree-ring sampling programme. Once reconnaissance work has led to the identification of suitable species and desirable site
characteristics in a target area, decisions must be made
about which trees to sample, and where to sample them, and

about sample size requirements. Clearly, sampling the
oldest trees in a stand will yield the longest tree-ring
records, although heart-rot, fire hollows, or other
problems may reduce the length of record attainable (in
special circumstances, records from stumps, logs, remnants
or wood from artefacts or structures may permit chronology
extension). Where a high frequency of missing rings is
likely to cause dating problems, a larger sample is
required and trees of a broader range of age classes and
microhabitats should be represented to ensure calendrical
accuracy of the resulting chronology.

Criteria for identification of old trees depend very
much on the habitat involved. In general, trees attain
maximum ages at two environmental extremes. At one extreme
are the dominant conifers in mesic, closed-canopy forests
that are growing under what appear to be optimal conditions in a relatively stable environment. Examples
include Sequoia, Sequoiadendron, Thuja, Chamaecyparis, and
Pseudotsuga of the coastal and mountain coniferous forests
of western North America (Franklin & Dyerness, 1973) and
Fitzroya, Libocedrus, Podocarpus, Dacrydium, Athrotaxis,
and Araucaria from the temperate mesic forests of the
Southern Hemisphere (LaMarche et al., 1979a,b,c,d,e). The
oldest trees in stands of this type are usually the
largest, have heavy upper branches, and may show some
crown dieback. The stem often shows evidence of a spiral
grain, and is normally fully bark-covered except where
fire-scarring has occurred. In contrast, the oldest trees
in semi-arid regions and many subalpine areas are found on
what appear to be the most inhospitable sites (Schulman,
1954; LaMarche, 1969). In arid regions, old trees are
typically found on rock ledges, exposed knolls and ridges
and on rocky slopes. Often occurring in pure stands, the
oldest trees are widely separated, short in stature, and
may show pronounced spiral grain and extensive crown
dieback. In some families, such as the Cupressaceae and
Pinaceae, trees of many genera adopt a strip-bark growth
habit with advanced age. A feature common to old trees in
all habitats is a low rate of radial growth during at
least the last few centuries of the tree's life.

Sampling variability is probably the most important
determinant of sample size requirements. The greater the
differences within and between trees, of parameters such
as relative ring width, wood density, or isotopic composition, the larger the sample must be in order to obtain
estimates of the "true" or "population" value for a
particular year at the desired level of precision.
Although quantitative methods have been developed for
sample size estimation based on results of analysis of
variance (Fritts, 1976) or other analyses, the necessary
statistics are normally available only for well-studied
regions. For exploratory work in new areas, pragmatism
dictates large sample sizes, particularly where the
climate signal is concentrated in the lower frequency
ranges, or where disturbance may be an important factor.

Decisions can also be made in the field based on
examination of the external form of the trees and the
internal ring structure shown in increment cores or
cross-sections. A lobate growth pattern, for example, is
reflected in poor circuit uniformity, which means that
more radii must be sampled to characterise the growth
record of an individual tree. Alternatively, if lobate
growth is associated with pronounced fluting or buttressing near the base of the stem, an attempt might be made to
sample at higher levels in the tree.

TREE-RING DATA NETWORKS

A tree-ring sampling programme is normally carried
out in the context of some broader chronological, ecological, or climatological problem. When the dendroclimatology of a region is known sufficiently well, it is
often possible to assemble a set of tree-ring data designed to answer specific questions or to generate proxy
values for a hydrometeorological variable of particular
interest. Such data sets can sometimes be selected from an
existing data base but more often their development
requires additional sampling of different species, in
different habitats, or in different areas.

An important element in the design of a data network
is a knowledge of the climate response characteristics for
a tree-ring parameter as a function of species, substrate,
altitude, climatic zone, or some other determining factor.
Table 1.3 shows schematically some possible approaches to
climate reconstruction in relation to climate response
diversity and the spatial distribution of data sites.
Further discussion of these concepts, with examples, is
presented elsewhere (LaMarche, 1978).

COMMENT

E.R.Cook

It is a pleasure to see that sampling strategies for
dendroclimatology are finally being examined and
formulated from a much needed global perspective. My own
experience in the eastern deciduous forests of North
America suggests that virtually no consistent set of site
selection criteria exists for that region as a whole
(Cook, this volume). This observation agrees with
LaMarche's experience in the Southern Hemisphere. It thus
seems likely that many of the site selection criteria
developed from tree-ring research in western North America
are quite unique to that region and to the lower-forest
border environments that abound there. For example, very
old, climatically sensitive eastern hemlock, Tsuga
canadensis Carr., can be found on talus slopes, in moist
ravines, and on poorly drained, low-relief sites with
excellent soil development. This was suggested years ago
by Lutz (1944) who studied tree-ring sequences of
swamp-grown hemlock and eastern white pine, Pinus strobus
L., in the State of Connecticut. Eastern hemlock, in
particular, has such a shallow root system that it can be

"drought sensitive" even where the local water table is less than a metre below the surface. The dendroclimatic usefulness of some species is more site dependent however. For example, pitch pine, Pinus rigida Mill., shows considerable ring-width sensitivity over a wide range of sites, but longevity is rarely found except on the more xeric, open-canopy sites. This circumstance is often found because pitch pine is very shade intolerant and cannot compete for light with more vigorous broad-leaved decid- uous trees. The thrust of this discussion is that species selection may be more important than site selection in some cases, and the dendroclimatologist should know the tree species in terms of each one's environmental require- ments and tolerances.

With the expansion of the global data base, field guidelines for the most efficient selection of sites and species in particular regions may be developed. In eastern North America, Europe, and other parts of the world where civilisation has a major impact on forest ecosystems, the development of such guidelines will be more difficult. Sites are more likely to be disturbed and native vegetation to be removed. The dendrochronologist may be forced to depend on old-growth stands of whatever dateable species are available. In this sense, the only pre- requisite that must be adhered to in developing a dendro- chronologically valid site is the presence of cross-dating to ensure absolute dating control. As much as cross-dating implies that a common set of climatically related limiting factors is influencing tree growth, tree-ring chronologies developed from a criterion of maximum age site selection should be useful for dendroclimatic studies.

COMMENT

M.G.L.Baillie

We have recently constructed seven oak site chrono- logies from various parts of Ireland. Each chronology was constructed identically by producing an index master using only the polynomial option (Graybill, this volume). These chronologies show very variable amounts of chronology variance accounted for by climate in response functions (5% to 64%). In no way could this have been deduced by observation or predicted in advance from site charact- eristics.

What we can observe is that, for the common period 1850 to 1969, the correlations between these chronologies show no clear distance dependence, that is, the corr- elation values and the distances between the sites are not themselves correlated. This implies that each site is receiving the same basic signal modified only by site conditions. If this is true, then rather than using any one of these site chronologies in dendroclimatology a better approach might be to use only one Irish chronology (the Irish site chronology!) which would be the mean of the individual site chronologies. In support of this, the overall master chronology constructed from these seven site chronologies shows very considerable similarity to the original 30-tree, generalised chronology from the north of Ireland. If this overall Irish chronology contains more climate information than the individual sites, then the whole question of site chronologies in our temperate area must be seriously reviewed, species by species.

If generalised chronologies can be shown to have

Table 1.3: Influence of climate-growth response characteristics on strategies for dendroclimatic reconstruction.

CLIMATE RESPONSE CHARACTERISTICS

SPATIAL DISTRIBUTION OF TREE-RING SITES	RESPONSE DIVERSITY			RESPONSE UNIFORMITY
	DIFFERENT DIRECTIONS	DIFFERENT SEASONS	DIFFERENT VARIABLES	SAME DIRECTION SAME SEASON SAME VARIABLES
Point	Independent verification of climate reconstructions	Lengthening of seasonal climate interval reconstructed	Reconstruction of joint local climate anomalies	Local recon- struction and testing for non-climate growth effects
Transect	or	and	Reconstruction of joint meridional or zonal anomalies	Reconstruction of individual meridional or zonal anomalies
Array	Improvement of climate calibration and reconstruction	Improvement of seasonal resolution of past climate	Reconstruction of joint regional anomaly patterns	Reconstruction of individual regional anomaly patterns

some suitable application it would open the way for us to
overcome the fundamental problem of our short modern
chronologies, that of stepping back in time to historic
timbers of unknown site and provenance. If genotype were
shown to be of greater relevance than phenotype in
characterising tree response to climate, a major obstacle
to the use of composite chronologies as proxy records of
climate would be removed.

MEASUREMENT OF DENSITOMETRIC PROPERTIES OF WOOD

F.H.Schweingruber

Translation: M.J.Seiber

INTRODUCTION

It is well known that wide and narrow annual rings
and high and low densities occur in wood. Measurement of
ring width usually presents no problems, whereas the
determination of the different densities within one annual
ring was for a long time fraught with difficulties. Until
about 10 years ago, therefore, research work concentrated
on the measurement of ring width and cambial activity.
Little attention was paid to cell wall thickness and
density as an expression of the physiological activity of
living cells. In 1965, a group established by the International Union of Forest Research Organisations investigated the problems of density measurement. Phillips
(1965) summarised the technical knowledge at that time.

The development of density measurement methods has
opened up a new field in wood technology, wood biology,
and dendroclimatology. The term "density" is taken to mean
the density (weight/volume) of untreated wood (according
to DIN 52182) measured at a given moisture content, in
this case 8%. Densitometric measurement actually records
radiographic densities. These, however, can be converted
into volumetric-gravimetric densities (Schweingruber et
al., 1978a).

METHODS FOR MEASUREMENT OF WOOD DENSITIES

Volumetric-gravimetric methods. Until the beginning
of the 1960s, most wood densities were measured by weight
and volume. This is still the case today in wood technology, but this method is less suitable for wood biology
and dendroclimatology, which use smaller samples whose
volume is more difficult to determine.

Photometric methods.

a) Light transmission. Using light-transmission factors
   of microsections, Green (1965) and Müller-Stoll
   (1947, 1949) were able to construct transmission
   curves. As every cell in a microsection is visible
   and the resolving power of microscope optical
   systems is high, this method is useful in wood
   biology, but the difficulty of converting optical
   transmission curves into wood density curves limits
   its application in dendroclimatology.

b) Measurement of cell wall and cell type proportions.
   If the density of the cell wall is taken as
   constant, average densities and detailed density
   curves can be obtained by measuring the proportion
   of cell walls and cell types per unit area (Liese &
   Meyer-Uhlenried, 1957; Knigge & Schulz, 1961;
   Eckstein et al., 1977). Electronic image analysers
   can now scan relatively large sample areas and
   register selected values. This method will achieve
   importance in all branches of wood research in the
   near future (Sell, 1978).

Mechanical methods.

a) Hardness. The hardness probe was developed by
   Mayer-Wegelin (1950). Radial cell rows in transverse
   wood sections are probed with constant impact force.
   The depth to which the needle penetrates is a
   function of the wood density.

b) Tensile strength. By using microtension machines,
   tensile strength diagrams can be constructed which
   closely reproduce the density sequences of annual
   rings (Kloot in Trendelenburg & Meyer-Wegelin,
   1955). Neither of these methods is much used today
   as calibration is difficult.

Radiation methods.

a) Beta-particles. Cameron et al. (1959) discovered the
   connection between wood density and absorption of
   beta-particles by irradiated wooden bodies. This
   method had potential but was not without dangers.
   It is no longer used, having been replaced by the
   X-ray method.

b) X-rays. This method was developed by Polge (1963,
   1966) and improved by Parker (1971) and Lenz et al.
   (1976). It is applicable in practically all fields
   of wood research and since the establishment of
   several X-ray laboratories in the last 10 years,
   appropriate fields of application, potentials, and
   limits have become more defined.

Applications. In wood density research, two areas of
investigation are today in the foreground: the effect of
ecological factors on density, and the connection between
the density relations of wood and its industrial uses.
These problems are being investigated by different
disciplines. Wood technologists are especially interested
in the industrial uses of wood as a raw material.
Relationships are being sought between chemical, physical
and mechanical properties and the potential applications
of wood in the energy, construction (solid wood, particle
boards, veneers), and paper fields. Foresters are
interested in forest yield. The weight of wood, as well as
its volume, is becoming increasingly important.
Therefore, density relations are being studied with
reference to the following: (1) the choice of suitable

taxa (species, provenance), for example, those with long
fibres for the afforestation of relatively large areas;
(2) the relationship between tree type, soil, and wood
production; (3) the effect of forestry activities, such as
thinning and pruning on wood production; and (4) the
recognition of influences harmful to trees and wood
production, for example, leaf loss through insects, fungi,
salts, and gases.

More than 10 laboratories are working in the technical forestry field. Historical scientists are seeking to
provide an absolute scale for their chronologies, which
are only relative and based on typology. Using the X-ray
technique, wood samples from boreal and temperate regions,
especially conifer wood, can be dated. Few datings have
yet been commissioned. There are no special archaeological
or historical X-ray laboratories. Recently, interest in
climate change has stimulated dendroclimatologists to
search more intensively for the relationship between
weather, climate, site, and growth rings. By measuring
densities it is possible to obtain substantial climate
information from trees in boreal and temperate zones.
Such research has only just begun; there are only a few
laboratories working on this problem.

THE X-RAY METHOD

Basics. The X-ray method has proved of great worth
in dendroclimatology, but it must be emphasised that
satisfactory results can only be obtained by the careful
and competent handling of numerous technical details.
Great attention must be paid to the technical development
and maintenance of the apparatus. The following main steps
are the same for all radiodensitometric analyses: choice
and treatment of sample before radiation; radiation; film
development; densitometric analysis; and evaluation.

From the technical viewpoint, any wood in good
condition can be used for X-ray dendroclimatological
investigations. In decomposed woods, from archaeological,
moor or lakeshore settlements, for example, only growth-
ring width can be evaluated. It is difficult to obtain
ecologically meaningful density curves from hardwoods for
anatomical reasons.

At the Swiss Federal Institute of Forestry Research
(Birmensdorf), the following procedures are used:
a) Technical steps. These are detailed in Table 1.4.
b) Computer processing of raw data. After the
   adjustment and graphical representation of the
   original data via the computer comes the question of
   further analysis, which necessitates a reduction in
   the mass of data. This can be done by: (1) the
   selection of meaningful density values; (2) the
   choice and processing of chronologies from suitable
   sites; and (3) the carefully directed application of
   simple statistical procedures.
c) Time consumption for collection of data. Present
   experience in our laboratory shows that a practised

technician can analyse 20 0.3 to 2mm sample cores of
150 annual rings each in a week. This includes
correction and synchronisation of the data, but not
final evaluation or the sporadic necessity for
repairs to the apparatus.

At other laboratories, different types of apparatus have
been put into use according to structure and financial
means. The main technical alternatives are given below:
a) Radiation. (1) The samples, resting on the X-ray
   film, are placed on a travelling stage which moves a
   few centimetres under the X-ray source. (2) The
   samples are kept stationary about two metres under
   the X-ray source. This process presumably gives more
   constant results, as it eliminates the problem of
   constant transport rate of the sample carrier and,
   because of the long exposure time, the importance of
   emission fluctuations in the X-ray tube.
b) Film development. Fully and half automatic developing machines are used; occasionally the films are
   developed by hand.
c) Densitometer. Various models are suitable.
d) Data conversion. (1) All data registered by the
   densitometer are directly transferred to the computer, which, by mathematical operations, converts
   the optical density values to wood density values
   and picks out the pre-selected ones, for example,
   ring width, maximum density. A sophisticated
   computer service is needed. (2) All data registered
   by the densitometer are drawn by a line plotter and
   also through a microprocessor which makes the data
   conversions and extracts the pre-selected values.
   These alone are then fed into the computer.

Organisation and costs of an X-ray laboratory. These are
dependent on the efficiency of the laboratory. A special-
ist in each of the following fields is needed: technology
and electronics; ecology; computer processing and
statistics. A radiation-proof cubicle, a relatively
dust-free room for the densitometer, and access to a large
computer are also necessary. A financial outlay of
sFr.150,000 (£35,000 sterling) is to be reckoned with
(1980 prices). This includes X-ray apparatus, densitometer, puncher, line plotter, and microprocessor, but not
film-developing equipment or accommodation and salary
costs.

APPLICATION OF THE X-RAY TECHNIQUE IN DENDRO-
CLIMATOLOGY

Present knowledge. On the basis of previous investigations in Canada (Parker & Henoch, 1971), France (Huber,
1976), and Switzerland (Schweingruber & Schär, 1976), the
following important points have been established. There is
a multiple inter-relationship between width and density
within one annual ring. Ecological information is
contained in the density as well as in the width of growth

Table 1.4: Methodology for radiographic-densitometric measurements of wood density. After Schweingruber et al. (1978a).

PHASES OF PROJECT	PRINCIPAL PROBLEMS OR SOURCES OF ERROR	REMEDIAL ACTIONS TO ELIMINATE OR REDUCE ERROR
Collection of samples:		
Phytosociological-pedological site analyses		Phytosociological description of site (after Braun-Blanquet) and of soil profiles
Sampling with increment borer	Poor orientation of core in a vertical direction	Use of borer guide
Transport of samples		Use of lightweight water-permeable core cases
Protection of tree from fungal infections		Use of hygrophilic fungicidal paste in core hole
Intermediate storage:		
	Resin dries in cores and affects density when prepared	Storage in 80% alcohol
Preparation of samples:		
Processing of the cylindrical cores into parallel-faced bore laths	Round cores are poorly suited to densitometric evaluation since axial diameter varies from zero to 5mm	Round cores are worked into flat laths of 1.25mm thickness; use of glueing apparatus and a spring tension-adjustment mechanism
		Orientation of cores using a microscope; sawing of round core into parallel-sided bore-lath using precision double-bladed circular saw
Labelling of bore-laths		Samples are labelled on front side with india ink
Acclimatisation of cores	Differing amounts of cell wall moisture falsify the density diagrams	Acclimatisation of cores in 55% ambient humidity (wood humidity 10%)
Preparation of radiograph:		
Covering film with sample	Film damaged by improper handling, resulting in undesirable dark blotches	Use of a sample carrier
Choice of film	Not all films available on the market are suitable for quantitative analysis	Choice of an evenly layered, fine-grained film
Irradiation	Various (see Polge, 1966)	Irradiation of sample from a distance of 2.5m; determination of optimal exposure; determination of optimal irradiation surface
Film development	Non-homogeneous development or darkening of film over entire surface	Development of film in a high-speed automatic developer
Densitometry:		
Use of Joyce-Loebel Microdensitometer		
Linearisation of optical density with wood density		Use of step wedge and continuous wedge
Choice of form and size of light-measuring slit	Density results depend not only on ring anatomy but also on measuring slit	Size = 0.03 x 0.08mm
Coordination of light-measuring slit with orientation of rings	Rings do not run parallel to long axis of light-measuring unit	Use of a simultaneous control device for light-measuring slit and compensation prism on the micro-densitometer by means of a servo-mechanism

rings. Cell wall growth in late wood (maximum density) is mainly limited, in boreal and subalpine regions, by summer temperatures. Cambial activity, on the other hand, depends on factors affecting the local habitat. In arid zones, cell wall growth and cambial division rate are limited by precipitation. Maximum density curves from different tree species synchronise well with each other. Maximum density curves are valid over a wide geographical area. Sets of chronologies for density and ring-width variables at a group of sites can be used to produce reconstructions of high quality.

Aims. The main question is: how is cambial activity (ring width) related to cell wall growth (density) and how are these influenced by environmental factors? See Figure 1.1. The question can be divided into three parts:

   a) Origins: studies of the relationship between ecological factors, site, and the formation of cells and cell walls;

   b) Region of validity: analysis of annual ring sequences in trees from comparable sites in different regions, and comparison of the results with weather data; and

   c) Historical development of climate: reconstruction of climate with the aid of fossil wood in regions which are comparable with reference to trees.

Material for investigation. In order to answer these questions, it is necessary to choose sample material with discrimination. For growth measurements and studies on annual ring sequences, trees should be chosen from sites whose topography, ecology, pedology, and plant sociology are known. For the determination of the geographical region of validity of single growth-ring parameters, old, standing trees in comparable sites should be chosen. A network of sampling sites in subalpine and boreal regions of the Northern and Southern Hemispheres makes comparisons of maximum densities over wide areas possible, and provides information about temperature fluctuations in the last 300 years. For the reconstruction of climate during the Late Glacial and Holocene periods, material can be obtained from moraines, alluvia, moors at the alpine and northern forest lines, former forest areas and alluvia in arid and semi-arid coniferous zones, sediments from permafrost zones, especially on shore terraces, and marine deposits of the Holocene.

   X-ray investigation of oak and beech is, for anatomical reasons, difficult. In addition, climatological interpretation is hindered by their origin in temperate zones, and often by their state of preservation. There is a mass of dendrochronological material now available from Europe, but it is mainly based on oak. Coniferous material exists in many parts of the world which may yield density values of use as proxy climate records. Although more difficult, it may well be that there are angiosperms that would merit such study.

THE WOOD DENSITOMETRY WORKING GROUP

   The International Union of Forestry Research Organisations, Division V, created the Wood Densitometry Working Group in April 1980. They publish the _Wood Microdensitometry Bulletin._ The first issue appeared in

Table 1.4: Continued.

PHASES OF PROJECT	PRINCIPAL PROBLEMS OR SOURCES OF ERROR	REMEDIAL ACTIONS TO ELIMINATE OR REDUCE ERROR
Comparison of absorptive properties of cellulose acetate and wood	Cellulose acetate and wood do not absorb X-rays to the same degree	Comparison of gravimetric-volumentric density with radiographic density of cellulose and wood
Data acquisition:		
Analogue data	Maximum extension of micro-densitometer recorder's advance mechanism is insufficient	Use of data collecting and control mechanism for continuous, periodic, and selective recording
Checking and correction	Errors occur for both technical and anatomical reasons	Correction and cross-dating by computer programs and printouts

NOTE:

It is essential to verify that the values given by the data processor are recorded correctly. This is accomplished in four steps:

   1. Information from the punched tape is transferred to magnetic tape. Both data sets are drawn for comparison;

   2. The sets of data are compared by a computer program which recognises and flags discrepancies and errors in the printout. The various ring parameters are then put into a standard sequence;

   3. The errors are corrected by hand by comparing the printout with the density diagram and by checks incorporated in the subsequent cross-dating procedure;

   4. The data are transformed into a coordinate system, and the data are printed as a time sequence. Connecting lines are hand drawn for plot comparisons to locate missing rings and abnormal density values.

July 1980. The aim of the newsletter is "to promote and
enhance communications between densitometry laboratories,
and to provide a medium through which ideas, problems,
findings etc., can be exchanged. The Newsletter will be
aimed particularly at those involved and interested in
wood densitometry using radiation techniques with alpha,
beta or gamma sources." The Editor is Dr.J.A.Evertsen,
Institute for Industrial Research and Standards, Forest
Products Department, Ballymun Road, Dublin 9, Ireland.

THE INFLUENCE OF TEMPERATURE AND PRECIPITATION
ON VESSEL AREA AND RING WIDTH OF OAK AND BEECH
D.Eckstein & E.Frisse

Severe changes in the environmental conditions of a
tree are well documented in the annual increment of the
wood. Until now the analysis of tree rings has mostly been
based on the width of the increment (Fritts, 1976) and,
more recently, also on its density. The research programme
discussed here additionally takes into account the
differences in the anatomical structure of the xylem. In a
preliminary study on deciduous trees, the vessel system
turned out to yield the most promising variables. The
investigation dealt with two oak trees and eight beech
trees from four different sites in Germany and Austria
(Frisse, 1977). Ring widths were measured in the usual
manner. Time series were also established for the vessel
area, each value for one year being the mean of 40 single
vessel areas measured by means of a semi-automatic image
analyser connected on-line to a computer. The data
processing was done mainly at the Computer Center of the
University of Arizona in Tucson using the methods of
Fritts (1976) and his group.

The ring-width sequences have a serial correlation
of 0.4 to 0.7 and a mean sensitivity of 0.17 to 0.32,
whereas the sequences of vessel area, especially for oak,
show a serial correlation of nearly zero, but also a very
low mean sensitivity of 0.07 to 0.14. The main results of

Figure 1.1: The influence of environmental factors
on ring characteristics.

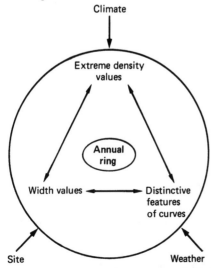

the response function analysis for all four sites are
listed in Table 1.5. In summary, the year-to-year
variation of vessel area is, with one exception, more
climatically influenced than ring-width variation; the
precipitation factor playing a dominant role.

On the basis of these results, the Schleswig site
was chosen for a reconstruction of climate conditions.
From this site, four different time series are available:
vessel area and ring width for oak and beech. The response
functions are shown in Figure 1.2. During the calibration
period from 1910 to 1967, precipitation in spring (March
to June) and temperature in winter (November to April)
could be determined to 39% and 37%, respectively. This is
demonstrated in Figure 1.3. In spite of this promising
quality of calibration, the reconstruction of climate for
these two seasons for the independent period from 1885 to
1909 appears to be rather vague (Figure 1.4). Obviously,
it will be necessary to increase the number of trees as a
basis for wood anatomical time series.

Figure 1.2: Response functions of vessel area and
ring width of oak and beech from the Schleswig site.

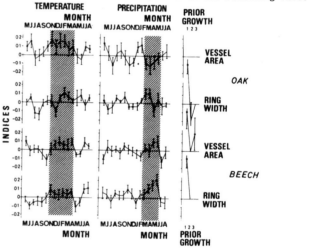

Figure 1.3: Calibration of winter temperature (top)
and spring precipitation (bottom) for the period
from 1910 to 1967 (actual values = thick lines).

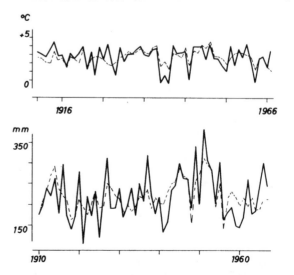

STABLE ISOTOPES IN TREE RINGS

A.Long

INTRODUCTION

As water evaporates, condenses or freezes, the proportions of the isotopic molecules, $H_2^{16}O$, $HD^{16}O$, $H_2^{18}O$, first entering the new phase are not the same as in the old phase. Vapour in equilibrium with liquid will contain measurably less of the isotopically heavier molecules than the liquid contains. This phenomenon is known as isotope fractionation. In equilibrium fractionation, the extent of the effect is temperature dependent. Thus, the isotope ratio of precipitation is a function of temperature.

Calcium carbonate crystallised under equilibrium conditions bears an oxygen isotope ratio dependent on the oxygen isotope ratio of the water from which it crystallised and its temperature of crystallisation. This phenomenon is the basis for oxygen isotope palaeothermometry in ocean sediments. Carbon also exhibits temperature-dependent fractionation of its stable isotopes ^{13}C and ^{12}C. Wood is composed almost entirely of carbon, hydrogen, and oxygen, each with measureable stable isotopes whose ratios are known to be temperature dependent in other systems. Tree rings in many instances are assignable to specific years or even months of growth. Therefore, it is not surprising that recent interest in climate prediction, short-time-scale palaeoclimatology, and climate monitoring for pre-instrumental times has stimulated research into the temperature signal, if any, contained in stable isotope ratios in tree rings.

H.C. Urey, the acknowledged father of isotope geochemistry, suggested in 1947 that such a signal was a possibility: "It seems probable that plant carbon comp-ounds synthesised at different temperatures may contain varying amounts of ^{13}C" (Urey, 1947). The uncertainty, as Urey recognised, arises from the fact that wood manufacture is not an inorganic process treatable by equilibrium chemical thermodynamic calculations. Rather it is a system of very complex biological reactions utilising both equilibrium and kinetic processes, which may or may not have temperature coefficients. Urey thus suggested the empirical approach to the question of a temperature signal in tree-ring isotope variations. Bigeleisen (1949) pointed out that isotope fractionation effects in pure kinetic reactions, except for those involving hydrogen isotopes, are essentially independent of temperature. Therefore, even temperature dependence is not assured. Temperature dependence in biological processes is further complicated by the fact that photosynthesis is itself controlled by temperature.

DOES A TEMPERATURE SIGNAL EXIST IN THE STABLE ISOTOPES OF TREE RINGS?

Empirical studies. Despite a general agreement that the processes which affect the isotope ratios in organic matter are complex and best approached empirically, attempts at theoretical equilibrium prediction have been made. Libby (1972) calculated temperature coefficients expected for cellulose in "equilibrium" with oxygen, carbon dioxide and water (Table 1.6). Attempts to measure temperature coefficients fall into two categories: growth chamber experiments on relatively fast-growing, non-woody plants, and field measurements on leaves or tree rings compared with meteorological records. Some experimenters choose to measure isotopes in the entire specimen, others separate out a specific compound or class of compounds, usually cellulose, for isotopic analysis.

Techniques. The analysis of different chemical fractions in wood is discussed first. Analysis of stable isotopes in a specific compound, or class of compounds such as the cellulose group, has the advantage of dealing with data from a substance whose biosynthesis and time of origin are known better than other constituents of tree rings. Several studies have concluded that some lignification

Figure 1.4: Reconstruction of winter temperature (left) and spring precipitation (right), 1885 to 1909 (actual values = thick lines).

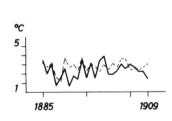
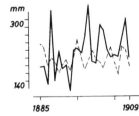

Table 1.5: Summary of the results of the response function analyses.

SITE	SPECIES	VARIABLE	STEP	F-LEVEL	CLIMATE %	PRIOR GROWTH %
Schleswig	oak	vessel area	7	3.3	49.4	0.0
		ring width	5	2.1	24.5	14.2
	beech	vessel area	7	2.9	16.1	28.2
		ring width	8	3.6	38.3	25.5
Reinbek	beech	vessel area	5	4.9	40.0	19.4
		ring width	7	3.3	9.9	58.4
Glashütten	beech	vessel area	5	3.1	8.3	23.0
		ring width	4	3.3	5.8	37.6
Hochfilzen (Austria)	beech	vessel area	15	3.1	47.5	31.1
		ring width	7	1.1	17.7	49.1

occurs during heartwood formation (Cain & Suess, 1976; Long et al., 1976). Moreover, if the sample contains varying proportions of non-cellulose material and this non-cellulose material has isotope ratios different from cellulose, the non-climatic variance of the sample isotope ratios may be greater. In fact, $\delta^{13}C$ and δD values for cellulose, lignin, and resins from the same tree ring may be quite different (Epstein et al., 1976; Wilson & Grinsted, 1977a).

Although most analysts prefer working with cellulose or holocellulose, some maintain that the best signal comes through when whole wood composes the sample (Libby, personal communication, 18.11.1975). Chemical separations necessarily expose wood to possible isotopic alteration from solvents. Recent studies in the Laboratory of Isotope Geochemistry at the University of Arizona have shown, however, that when proper precautions are taken with cellulose, solvents have no significant effect on the sample's carbon or oxygen isotopic composition.

Second, post depositional hydrogen isotope exchange in cellulose is considered. In the glucose unit ring which makes up the cellulose polymer chain, half of the hydrogens are attached to carbon atoms and half to oxygen atoms. Those attached to oxygen are known to be easily exchangeable with hydrogen in water. Not only could these exchange during cellulose separation, but perhaps with sap water in the living tree. Consequently, only the C-H hydrogen is certain to be original biosynthesised hydrogen, whose values are obtainable either after replacement of O-H hydrogen with hydrogen of known isotope composition (Grinsted, 1977) or after replacement with $-NO_2$ (Epstein et al., 1976).

In principle, D/H results using these two procedures should give comparable results, or at least the patterns of variability should be the same. In practice, however, this has not yet been shown (Epstein & Yapp, 1976; Wilson & Grinsted, 1977a). A possible explanation for the difference may lie within the macromolecular structure of cellulose (Wilson, personal communication, 23.8.1978). Cellulose molecules are arranged in a tree more or less vertically to give maximum strength and resilience. Regions of the structure ordered enough to give X-ray patterns are called crystalline cellulose, other regions of the cellulose are essentially amorphous. The latter are likely to exchange O-H hydrogen during laboratory procedures more readily than the crystalline cellulose, which may retain "memory" of sap waters running at the time the cellulose was originally deposited in the ring. Wilson suggests that the difference between D/H by H_2O exchange and D/H by nitration may in itself be representative of the climate at the time the ring grew.

Finally, isotope measurement techniques for cellulose are described. Techniques for cellulose separation are standard (see, for example, Green, 1963) and need not be repeated here. Oxygen isotopic differences

between α-cellulose and holocellulose are not measurable (Gray & Thompson, personal communication, 22.5.1979). All ratio measurements are made on dual or triple collector isotope ratio mass spectrometers from either CO_2 or H_2 labelled with carbon, oxygen or hydrogen from the sample. Gases introduced into these mass spectrometers must be free of impurities for reproducible results. All chemical and physical steps in the procedure must be quantitative and non-fractionating. Precision and reproducibility of the final values are highly operator dependent.

a) Preparation of CO_2 for $\delta^{13}C$ analysis. The pre-dried cellulose sample weighing about 5mg is burned in the presence of O_2 and CuO in a recycling system or a flow-through system with helium as a carrier gas. Complete combustion and total recovery of dry CO_2 are of primary importance in obtaining high precision.

b) Preparation of CO_2 for $\delta^{18}O$ analysis. The cellulose must be pre-dried in a vacuum oven to ensure that no oxygen from absorbed H_2O appears in the CO_2. Three techniques are currently in use. No laboratory to my knowledge has compared all three, though all have been compared in pairs.

1) Vacuum pyrolysis (Hardcastle & Friedman, 1974; Ferhi et al., 1975). Heat the ~20mg cellulose sample to $1250^{\circ}C$ in a fused quartz tube and allow pyrolysis products to pass through hot diamonds. The products are CO_2, CO, and H_2. Pump away the H_2, react the CO to CO_2 + C in an electric discharge, and the combined CO_2 is the sample gas. Ferhi (personal communication, 19.8.1979) finds this technique highly reproducible compared to the $HgCl_2$ technique.

2) Vacuum pyrolysis (Thompson & Gray, 1977). Heat the cellulose sample in a nickel crucible at $1,125^{\circ}C$. Hydrogen diffuses out of nickel at these

Table 1.6: Theoretical equilibrium temperature coefficients for cellulose calculated by Libby (1972).

ISOTOPE RATIO	TEMPERATURE COEFFICIENT (‰/°C) *
$^{18}O/^{16}O$	+0.92 to +1.14
$^{13}C/^{12}C$	+0.36
D/H (O-H)	+0.4
D/H (C-H)	+2.0

* NOTE:

Conventionally, stable isotope ratios are reported in parts per thousand (‰) deviation from a common reference material, computed from

$$\delta\ (‰) = [Rx/Rref - 1]\ 10^3$$

where Rx and Rref are (moles heavy isotope)/(moles light isotope) in unknown and reference material, respectively. For oxygen and hydrogen, the reference is "SMOW", a water sample similar isotopically to average ocean water. For carbon, the reference is "PDB", a fossil similar isotopically to marine carbonates.

temperatures, which eliminates H_2O as a product. Convert CO to CO_2 and combine as above. Epstein et al. (1977) find this technique and the $HgCl_2$ technique both reproducible.

3) $HgCl_2$ reaction (Rittenberg & Ponticorvo, 1956, modified by Wilson in Grinsted, 1977). Heat cellulose with dry $HgCl_2$ at 400 to 565°C in a closed, evacuated quartz glass tube overnight. The products are CO_2, CO, HCl, Cl_2 and Hg. Absorb HCl and Cl_2 into dry quinoline, react CO to CO_2, and the combined CO_2 constitutes the sample. Grinsted (1977) and Burk (1979) find this technique quite reproducible.

c) Preparation of H_2 for δD analysis. Four different sample preparation procedures yield hydrogen from different compounds and parts of molecules and, therefore, may not be comparable. One group asserts that any treatment beyond oven drying loses isotope information. In this case, the dry sample is simply burned and the water of combustion reacted with hot uranium to produce H_2 for mass spectrometric analysis. Another group recognises that wood resins are isotopically light in hydrogen, thus these mobile substances must be removed before burning. Most now work with some form of cellulose and attempt to eliminate the effect of exchangeable hydrogen on the glucose unit. Nitration replaces essentially all of the O-H hydrogen, but H_2O exchange may affect only the more accessible hydrogen. Both may yield valid but different climate signals. Cellulose and cellulose nitrate must be kept dry and can be safely burned in a combustion line or closed tube with CuO, and the water of combustion reacted to H_2 as above for isotope analysis.

Laboratory measurements of $\delta^{13}C$ temperature coefficients in land plants. To date, only measurements on non-woody plants are published. Controlled environment growth chamber experiments on trees are cumbersome and time-consuming. Such experiments on Douglas fir are now in progress, however, at the Duke University Phytotron. It seems acceptable in the meantime to assume that all C_3-type plants manufacture glucose in the same manner, and that insignificant further fractionation occurs in the step from glucose to cellulose production and deposition.

a) $^{13}C/^{12}C$ fractionation in vivo. Table 1.7 summarises results of growth chamber experiments on plants. The experiments pose difficulties controlling $\delta^{13}C$ of the plant's environment because plants respire CO_2 whose $\delta^{13}C$ is about 20‰ lighter than that of the CO_2 withdrawn from the atmosphere. What these data have in common is that they are all negative and small compared to the relatively large positive values predicted by Libby (1972). Not shown in the table is the variability of the individual plant specimens. Rigorous statistical treatment of the data would probably not reveal significant trends.

b) $^{13}C/^{12}C$ fractionation in vitro. Three experiments have examined fractionation by CO_2-fixing enzymes isolated from leaves. This step fractionates carbon the most and is the likely candidate for the location of the temperature effect. Their results are listed in Table 1.8. Here again no deciding vote is cast.

Field measurements of $\delta^{13}C$ temperature coefficients in tree rings. Acknowledging that the pathway of carbon from the atmosphere to tree-ring cellulose is complex and that the times and conditions under which plants do and do not carry out these processes may vary with species, soil, and light conditions, inter alia, many researchers have opted for the empirical approach. Several examples of this type of study are shown in Table 1.9. Authors do not state variance in terms of temperature predictability from measured $\delta^{13}C$ values, however. Considering this and the differences in their temperature coefficients, the only possible conclusion from this is that not all variables are accounted for.

One of the complications of tree-ring $\delta^{13}C$ work is that even for cellulose the $\delta^{13}C$ varies significantly around the circumference of a single ring or groups of rings (Mazany et al., 1978; Tans, 1978). The relative

Table 1.7: Carbon isotope temperature coefficients measured in growth chamber experiments.

PLANT TYPE	MATERIAL	TEMPERATURE COEFFICIENT (‰ per °C)	REFERENCE
Xanthium strumarium	leaf	−1 (non-linear or large variance)	Smith et al. (1976)
pea rape barley	whole plant	−0.13 −0.29 −0.25	Smith et al. (1973)
wheat	leaf (cellulose?)	−0.4	Schmidt et al. (1978)
eleven different C_3 species	whole plant	−0.0125 (6 sp.) −0.046 (5 sp.)	Troughton & Card (1975)

ring-to-ring variations, however, seem to be similar for different radial transects (Mazany et al., 1978). In order to determine if the actual δ^{13}C value contains palaeoclimate information beyond mere relative changes (that is, if it can produce numerical climate inferences), future work must employ either the entire circumference of the tree ring, or representative radial samples.

Field measurements of δ^{13}C in leaf cellulose with temperature and precipitation. Arnold (1979) measured the δ^{13}C in cellulose isolated from living juniper twigs collected around the perimeters of several trees at each site for nine different meteorological stations in Arizona. He compared an average δ^{13}C for each site with average spring temperature and precipitation and found an excellent correlation with a function of both parameters. Based on the high correlation coefficient obtained in this study, the δ^{13}C of any juniper site in Arizona should be predictable from growing season temperature and precipitation measurements with an accuracy of about +/-0.1‰ but temperature alone gave a predictability of only about +/-0.5‰. From Arnold's study we can gain optimism that δ^{13}C in trees does indeed bear a climate signal but the signal is more complex than most workers have until now assumed.

Laboratory measurements of the influence of temperature and other environmental variables on the δ^{18}O in plant cellulose. Recent papers by Ferhi et al. (1977) and Ferhi & Letolle (1977, 1979) have demonstrated that the δ^{18}O value of plant cellulose is governed by the δ^{18}O value of water taken up by plant roots, the δ^{18}O in atmospheric water vapour, and the conditions of evapo-transpiration that prevail at the leaf surfaces. They found that δ^{18}O values in cellulose were independent of temperature. The following equation explains their experimental data (Ferhi & Letolle, 1979)

$$\delta^{18}O(cellulose) = 0.38[(1-h)\delta_i + \varepsilon_e + \varepsilon_k + h(\delta_v - \varepsilon_k)] + 0.56\,\delta^{18}O(CO_2) + 0.06\,\delta^{18}O(O_2) \qquad (1.1)$$

where h is the relative humidity; δ_i, δ_v, are δ^{18}O values of soil water and atmospheric water vapour, respectively; ε_e and ε_k are equilibrium and kinetic isotopic enrichment factors for water vapour to water liquid; and CO_2 and O_2 refer to δ-values of these gases in the environmental atmosphere. In the same paper, these authors showed how δ^{18}O in plant cellulose collected from contrasting climates (Sweden to the Ivory Coast) adhere qualitatively to this equation.

Field measurements of the influence of temperature and other environmental factors on the δ^{18}O in tree-ring wood and cellulose. Table 1.10 summarises the work of those who found empirical relationships between temperature and δ^{18}O in tree rings. Contained in these coefficients is the temperature effect on δ^{18}O of precipitation, which, depending on location, ranges from about 0.6‰ per °C down to near zero. This array of values, though confusing, cannot necessarily be written off as contradictory, because of the different species, localities, and methods of correlation. A more fundamental understanding of the physical and biochemical processes is necessary.

Table 1.8: Carbon isotope temperature coefficients measured for enzymatic reaction.

ENZYME (PLANT TYPE)	TEMP. COEF. (‰ per °C)	REFERENCE
RuDP-C (sorghum)	+1.2	Whelan et al. (1973)
RuDP-C (soybean)	−0.22	Christeller et al. (1976)
PEP-C (wheat and maize)	−0.03	Schmidt et al. (1978)

Table 1.9: ^{13}C/^{12}C temperature coefficients for naturally grown trees.

TREE TYPE	LOCATION	δ^{13}C CORRELATED WITH	TEMP. COEF.	REFERENCE
Poplar	Norway England	Running mean average spring temperature	−0.1	Lerman (1974)
Monterey Pine	New Zealand	Mean summer temp.	+0.2	Wilson & Grinsted (1977b)
King Billy Pine	Tasmania	Running mean Feb. max. temp.	+0.24 to +0.48	Pearman et al. (1976)
Elm	Massachusetts	Mean summer temp.	−0.7	Farmer (1979)
Oak	Germany	Running mean average winter temp. from stations in three countries	+2.0 to +2.7	Libby & Pandolfi (1974)

Epstein et al. (1977) compared $\delta^{18}O$ and δD in terrestrial plant cellulose with the inferred $\delta^{18}O$ and δD of their environmental water. From this comparison they concluded, as did Ferhi & Letolle (1979), that about two-thirds of the cellulose oxygen is derived from atmospheric CO_2, but unlike Ferhi & Letolle, that the remaining third comes directly from environmental (soil) water. Epstein et al. (1977) also point out that this relationship is complicated by evapo-transpiration effects, which they and others have shown to occur in leaves, and by the possibility that atmospheric CO_2 may exchange isotopically with water in leaf cells. Indeed DeNiro & Epstein (1979) grew wheat in an environment of ^{18}O enriched CO_2 and found that ^{18}O of cellulose is determined essentially by the $\delta^{18}O$ in plant (leaf) water, and that the CO_2 involved in cellulose synthesis exchanges oxygen with plant water. As did Ferhi & Letolle, Burk (1979) has also demonstrated the applicability of a Craig & Gordon (1965) type model for $\delta^{18}O$ on naturally grown trees. Burk's excellent quantitive model agreement using latitudinally or altitudinally distributed sites is extremely encouraging for the future of isotope dendroclimatology. His empirical equation gives a one-to-one relationship between $\delta^{18}O$ of leaf water and source water. This apparent contradiction with the results of Ferhi must be resolved experimentally. Thus, a question remains about the factor in oxygen pathways during photosynthesis that prevents a one-to-one relationship between $\delta^{18}O$ of leaf water and $\delta^{18}O$ of cellulose. Some evidence exists that the rôle of molecular oxygen may be more than trivial (Andrews et al., 1971; Berry et al., 1978). Since atmospheric O_2 does not exchange readily with water, this may be the constant source required to limit the influence of leaf water on the $\delta^{18}O$ of plant cellulose.

Field experiments on the relationships between δD in tree rings and climate. The first work published on this topic was that of Schiegl (1970, 1974), who was also the first to recognise the applicability of Craig & Gordon's (1965) evaporative basin type isotope fractionation processes in wood production. Thus, for trees grown in humid climates,

such as _Picea_ from southern Germany (see Table 1.11), Schiegl (1974) found that D/H in whole wood (less resins) from tree rings recorded annual D/H in atmospheric water vapour and precipitation. The apparent temperature effect he found was approximately that predicted from the temperature effect on δD of precipitation. He also found that decreased humidity increased the D/H in tree rings, as predicted by the Craig & Gordon (1965) equation. Table 1.11 summarises empirical temperature effects of δD in tree rings and shows three disparate conclusions. These may not be incompatible because they represent three different techniques possibly analysing three different chemical groups of hydrogen. Any humidity effect which exists in the data is perhaps obscured by the fact that δD in plant cellulose responds more dramatically to environ-mental isotope differences than $\delta^{18}O$ (in plant cellulose) but less dramatically than $\delta^{18}O$ to changes in humidity, considering measurement precision.

SIGNIFICANCE OF EMPIRICAL DATA,
PROBLEMS, PROSPECTS, AND FUTURE WORK

A very tempting generalisation from the studies quoted above is that they verify theoretical predictions about kinetic chemical reactions: except for hydrogen, stable isotopes have little or no pure temperature coefficient. Apparent temperature effects arise either coincidentally or because temperature is correlated with another causative factor such as δD and $\delta^{18}O$ of prec-ipitation, or relative humidity. Species effects probably exist but are likely to be actually ecological or physio-logical. The multiplicity of variables in the field precludes the necessary experimental controls. Diff-erences in stable isotope values, especially in carbon isotopes among C_3-type plants, do occur, but cannot be explained by biochemical differences. Stomatal and leaf structure differences may bias isotope ratios but are not likely to affect relative values. Some trees shed leaves and lie dormant while others can store photosynthates during winter. Different ecological zones have different δ-values in precipitation and $\delta^{18}C$ in the local

Table 1.10: $^{18}O/^{16}O$ temperature coefficients measured in naturally grown tree rings.

SPECIES (MATERIAL)	LOCALITY	CORRELATED WITH	TEMP. COEF. (‰ per $^{\circ}C$)	REFERENCE
Quercus petraea (whole wood)	Spessart Mts. (Germany)	Average ann. temp. of England	+18.3	Libby et al. (1976)
Abies alba (whole wood)	Bavaria (Germany)	Average ann. and spring temps.	+0.7	Libby et al. (1976)
Cryptomeria japonica (whole wood)	1350m elev. (Japan)	Averages annual: from coastal spring: station	+4.0 +1.3	Libby et al. (1976)
Picea glauca (α-cellulose)	Edmonton (Alberta)	Mean annual temp.	+1.3 ±0.1	Gray & Thompson (1976)
Pinus radiata (cellulose)	Hamilton (N.Z.)	Seasonal variations	~+0.3	Wilson & Grinsted (1976)
Picea glauca	W. coast N. America	Average annual temp. –latitude transect	+0.2	Burk (1979)

environment. Atmospheric CO_2 $\delta^{13}C$ varies throughout the season or even during the day at a single location. Different species have different stomatal responses to temperature, light intensity, and water stress. These factors alone could produce "species" effects or even apparent temperature and precipitation effects. The best correlation of $\delta^{13}C$ with climate is that of Arnold (1979). Future work will decide if his excellent correlations with leaf cellulose also hold up for tree-ring cellulose in time sequences for trees from a single site.

Such are the problems with empirical results. With so many uncontrolled variables, one cannot be sure which are primarily affecting the observed results. Because evidence from controlled growth experiments on oxygen is theoretically justifiable, we feel closer to understanding oxygen and hydrogen than carbon isotope responses in plants. Both oxygen and C-H hydrogen in cellulose appear to monitor δ-values of changes in local precipitation, modified by a humidity factor. Promising studies by Schiegl, by Epstein and his students, and by Burk should be followed up by more ring sequence studies, especially in humid inland areas near meteorological stations. Both $\delta^{18}O$ and δD (C-H) on cellulose may allow reconstruction of both temperature and humidity sequence (Long et al., 1979). It now seems possible that if $\delta^{13}C$ is a function of growing season temperature and precipitation and $\delta^{18}O$ and δD are functions of annual temperature and humidity relatable to precipitation, then annual temperature, seasonal temperatures, and precipitation values may be retrievable from the isotopic measurements. This could be possible only with better understanding of how the isotopes behave in trees.

The essential bridge between the field experiment and theory is the highly controlled growth chamber experiment, wherein not only the temperature, humidity, photoperiod and light intensity are controlled but also

the ppm and $\delta^{13}C$ of CO_2, $\delta^{18}O$ of O_2, and $\delta^{18}O$ and δD of liquid and vapour H_2O. All three isotope ratios in ring cellulose, leaf cellulose, and photosynthetic reaction intermediates from the plants should be measured. Measurements of $\delta^{18}O$ and δD in stem and leaf sap are also necessary. All measurements must be taken on an appropriate time sequence. These experiments are yet to be done on woody plants and must be done to give us necessary understanding of the controlling mechanisms and their theoretical operation. They will not be simple or cheap, but I am optimistic that the investment will not only prove worth while to fundamental science but will also be applicable to the solution of environmental problems now facing our world community.

This field is still in the juvenile stage of development and climatic interpretations from single ring sequence isotope measurements should be considered as speculative. However, results so far have shown that stable isotopes in tree rings really do respond to climatic influences but that in single locations the climatic variation is sufficiently slight to render the extraction of a climate signal difficult. Further, this means that it is hazardous to interpret the data in terms of temperature isolated from relative humidity or storm track effects. With careful site selection, sample replication using multiple trees and network sampling, some of the noise will, no doubt, turn into information such as the migration of storm tracks with changing climate.

OXYGEN-18, CARBON-13, AND CARBON-14 IN TREE RINGS
T.M.L.Wigley

INTRODUCTION

Long (this volume) has provided a balanced review of stable isotopic measurements in tree rings and their interpretation. Here, some of the more important recent

Table 1.11: D/H temperature coefficients measured in naturally-grown tree rings.

SPECIES (MATERIAL)	LOCALITY	CORRELATED WITH	TEMP. COEF. (% per °C)	REFERENCE
Picea sp. (extracted wood)	Southern Germany	annual temp.	∿+6	Schiegl (1974)
Quercus petraea	Spessart Mts. (Germany)	winter temps. of England	∿+140	Libby et al. (1976)
Many species (C-H hydrogen in nitrocellulose)	latitude transect through western N. America	local precip.	Constant offset with respect to precip. (∿+5)	Epstein, Yapp & Hall (1976)
Eastern conifers (C-H hydrogen in cellulose nitrate)	New Paltz (New York)	local precip.	Constant offset with respect to precip. (∿+5)	J.White (personal communication)
Pinus radiata (C-H hydrogen in cellulose by equilibration technique)	Hamilton (New Zealand)	local seasonal temperatures	∿-5	Wilson & Grinsted (1975)

results for the stable isotopes ^{18}O and ^{13}C, and for the unstable isotope ^{14}C, are reviewed briefly.

Whenever isotopic values show variations, there is some possibility that these may reflect some climatic influence. For any one isotope, however, there may be differences in response between species, between trees of the same species, between different chemical constituents or between different locations in a single tree. Long (this volume) has stressed the importance of using the same chemical constituent when comparing the results of different workers, and some early analyses using whole wood instead of identical chemical fractions are suspect. Isotopic variations are known to occur around the circumference of a single ring, but provided different radii show parallel variations, this need not be a problem.

Chemical constituent and circumferential differences apart, isotopic variations in tree rings arise in the following ways:

    a) through variations in external parameters, atmospheric ^{14}C or ^{13}C, precipitation ^{18}O, and so on; this constitutes the input to the system;

    b) through physical fractionation processes which may be directly temperature-dependent, indirectly temperature-dependent (for example, via relative humidity which may be related statistically to temperature), or independent of temperature;

    c) through equilibrium or kinetic chemical fractionation processes; and/or

    d) through purely arithmetical effects, as outlined by Wigley et al. (1978), resulting from the combined effect of isotopic variations within a ring and variations in ring width.

Most studies to date have been either purely statistical, or based on semi-empirical models of the processes thought to control isotopic composition in the tree. This latter approach is to be preferred, but still requires calibration using measured data. Ideally, the same rigorous model calibration and verification criteria which are used in tree-ring width studies should be applied to isotopic studies. Because a much greater effort per data point is required in isotopic studies, few have been able to afford the "luxury" of statistical rigour.

The field of isotope dendroclimatology is still in its infancy. Long's contribution to this volume is one of the few reviews available. A more general review of isotope palaeoclimatology concentrating on the past 5,000 years, and covering biological materials such as tree rings and peat deposits, has been published by Gray (1981). A number of important papers are included in the <u>Proceedings of the International Meeting on Stable Isotopes in Tree-Ring Research</u>, May 22-25, 1979 (Jacoby, 1980b). For a general review of isotopes in palaeoclimatology, the reader is referred to Fritz & Fontes (1980).

OXYGEN ISOTOPES

As Long (this volume) has pointed out, the $\delta^{18}O$ in plant cellulose comes from four sources: soil water and atmospheric water vapour (both related to precipitation $\delta^{18}O$), atmospheric carbon dioxide, and atmospheric oxygen. The semi-empirical formula of Ferhi & Letolle (1979) given by Long apportions these contributions. The first term, the soil water/water vapour term,

$$(1-h)\delta_i + \varepsilon_e + \varepsilon_k + h(\delta_v - \varepsilon_k) \qquad (1.2)$$

is the familiar Craig-Gordon expression (Craig & Gordon, 1965; Schiegl, 1970, 1974) which has been shown experimentally to determine the $\delta^{18}O$ of leaf water (Förstel, 1978, 1979).

Burk (1979) and Burk & Stuiver (1981) have demonstrated that the above relationship can be used to explain the variations of tree-ring cellulose $\delta^{18}O$ both latitudinally (from La Jolla, California, to Fairbanks, Alaska) and altitudinally (up Mt. Rainier). They suggest that leaf water and leaf CO_2 are in isotopic equilibrium and that the cellulose-CO_2 fractionation is essentially constant (see also Gray, 1981). This allows Equation (1.2) to be used directly. This approach differs somewhat from that of Ferhi & Letolle (1979) and further work is required to reconcile the two. Burk & Stuiver's results indicate that $\delta^{18}O$ in tree-ring cellulose is directly related to precipitation $\delta^{18}O$ with the difference being determined by growing season mean temperature and relative humidity. Note that both δ_i and δ_v in Equation (1.2) are related to precipitation $\delta^{18}O$ (Förstel, 1978, 1979), although these relationships require further clarification, and this explains why a statistically simple link with precipitation $\delta^{18}O$ can be obtained. Furthermore, spatial variations in precipitation $\delta^{18}O$ are closely correlated with mean annual temperature. Relative humidity, h, which also appears in Equation (1.2), is often correlated with temperature, and this may explain why, in some cases, significant correlations have been found between tree-ring cellulose $\delta^{18}O$ and temperature (Gray & Thompson, 1976; Gray, 1981).

Although promising for climate reconstruction, much work still needs to be carried out. The most convincing studies today involve spatial correlations rather than correlations over time at a single site. Even in the spatial studies, the presence of relative humidity as a possibly important variable adds another complicating dimension to the reconstruction problem. Difficulties are magnified when one considers single-site reconstructions. Even if, for example, one were able to reconstruct precipitation $\delta^{18}O$ values at a site, this is not an easy parameter to interpret in terms of climate. Although spatial variations in $\delta^{18}O$ reflect spatial variations in temperature, this relationship does not carry over to a

single site on short timescales. Year-to-year variations
in precipitation $\delta^{18}O$ show no correlation with temperature
(Siegenthaler & Oeschger, 1980; I.A.E.A., 1981). For tree
rings, Burk & Stuiver could find no correlation between
annual cellulose $\delta^{18}O$ values and temperature at a single
site. This contrasts with the results of Gray & Thompson
(1976), but these authors used 5-year blocks of tree-ring
material, and it may be that time-averaging simplifies the
relationship between cellulose $\delta^{18}O$ and climate. There is
some evidence that this is so for precipitation.

On long timescales (millennia), precipitation $\delta^{18}O$
and temperature appear to be well correlated, as evidenced
by the long timescale changes in the Camp Century ice core
$\delta^{18}O$ (Dansgaard et al., 1971). There is even crude evidence
of correlations on the century timescale (Dansgaard et
al., 1975), although these are not statistically convincing because of the lack of a reliable independent
climate record for comparison. The same lack of climate
data makes it difficult to test whether there is a link
between ice-core (and hence precipitation) $\delta^{18}O$ and
temperature on the decadal timescale (no quantitative
comparison has been published); and precipitation isotope
measurements have been made for too short a time to test
for any relationship using these data. There is, therefore, an unfortunate gap in our understanding of
single-site precipitation $\delta^{18}O$ fluctuations on short
timescales which needs to be filled before tree-ring
cellulose $\delta^{18}O$ data can be used confidently for climate
reconstruction. Possibly tree-ring $\delta^{18}O$ data can be used to
reveal the details of rainfall $\delta^{18}O$-climate links.

CARBON-13

There is some controversy over whether ^{13}C in tree
rings is determined primarily by climatic effects or by
changes in atmospheric ^{13}C. Since the beginning of the
industrial revolution, Man has added to atmospheric CO_2
levels by burning fossil fuels and, possibly, by deforestation and changing land-use. The ambient atmospheric value
for $\delta^{13}C$ is around $-7‰$. Forests and fossil fuels have $\delta^{13}C$ of
around $-25‰$. Hence, Man-made inputs to atmospheric CO_2
should lower its isotope ratio. Keeling et al. (1979,
1980) have measured a reduction from $-6.69‰$ in 1956 to
$-7.34‰$ in 1974.

Freyer (1979a,b; 1980) has analysed $\delta^{13}C$ for a number
of trees from sites on the east coast of North America and
the west coast of Europe and finds a general decrease
between 1845 and 1970 of approximately $2‰$. He attributes
this largely to atmospheric $\delta^{13}C$ changes. Not all authors
obtain the same results, but Freyer has compiled the
largest amount of data. Recently, Francey (1981) has
published results from seven Tasmanian trees. Six of these
show no trend in $\delta^{13}C$ over the past 100 years, while the
seventh shows a decreasing trend. In an earlier work,
Pearman et al. (1976) suggested that temperature was the
main controlling variable for the isotopic variations in
these Tasmanian trees, and Francey interprets his results

as indicating that there is no significant atmospheric ^{13}C
effect.

Francey's and Freyer's results are apparently in
conflict. However, it is possible that, because of the
complexity of the processes which cause $\delta^{13}C$ changes in
tree-ring cellulose, the relative importance of
atmospheric and climatic factors may vary considerably
from tree to tree, species to species, and even with
location and tree age. Most recently, Freyer (1981) has
analysed five Scots pines from northern Sweden covering
the past 500 years and states that these show both
changing atmospheric $\delta^{13}C$ and major climate events such as
the Little Ice Age and the early 20th century warming.
After about 1850 he finds an overlying steady decrease in
the mean $\delta^{13}C$ for these five trees amounting to
approximately $2‰$, which he attributes to both fossil fuel
burning and deforestation. Prior to 1850 only climate
effects are noticeable. Just why the post-industrial
atmospheric $\delta^{13}C$ effect is not evident in Francey's trees
is uncertain, but it could be that these trees have
experienced, by chance, a compensating climatic effect, or
that they are somehow less sensitive to atmospheric $\delta^{13}C$
changes. Further research, and the development of models
similar to those in use in recent $\delta^{18}O$ studies, is required
to resolve these differences.

CARBON-14

Carbon-14 dates from tree rings, when compared with
dendrochronological dates, can be used to determine past
levels in atmospheric ^{14}C. The variations show short-term
(decades to centuries) changes superimposed on a long-term
trend which is largely due to changes in the geomagnetic
field intensity. These short-term fluctuations have been
convincingly linked to changes in solar activity (Stuiver
& Quay, 1980), and have been correlated with certain
evidence of climatic change (de Vries, 1958; Damon, 1968;
Suess, 1968). This suggests a possible link between solar
activity, climate, and atmospheric ^{14}C which has been
exploited by Eddy (1977). The argument used is this:
increased solar activity produces higher temperatures, and
also causes a reduction in atmospheric ^{14}C levels. Thus,
atmospheric ^{14}C, as measured by tree rings, may be a useful
proxy climate variable. There is, however, an important
gap in this argument; the (demonstrated) solar activity
changes which produce atmospheric ^{14}C variations are not
the same sort of solar changes which might lead to a
change in the Earth's heat budget.

The main support for the ^{14}C-climate link is
circumstantial and not based on a proven physical process
nor on a rigorous statistical correlation. Support rests
heavily on the rough coincidence in timing of the Maunder
minimum (a period of reduced sunspot activity, but not
necessarily any change in solar luminosity), the Little
Ice Age climax, and a large change in atmospheric ^{14}C.
Stuiver (1980) has compared climate data over the past

1,000 years (spanning the Little Ice Age) with atmospheric ^{14}C data inferred from tree rings. Most of the data he has examined show no correlation. The most significant result involves the least reliable of his climate records, that based on an unreliable compilation of historical data from the Russian Plain. Williams et al. (1981) have carried out a detailed comparison between atmospheric ^{14}C data and a large number of proxy records from high latitudes of the Northern Hemisphere over the past 4,000 years. When the data are considered in total, there is no significant link. A few isolated places (the Alaska-Yukon area and Scandinavia) show good agreement when the ^{14}C record is lagged 50 years behind the climate record, but, since 20 comparisons are made in all, two "successes" might well occur by chance.

In contrast to these negative results, Harvey (1980) claims that there is a broad parallel between atmospheric ^{14}C variations and climate over the past 7,500 years. This study considers a much greater amount of data than similar earlier work by Eddy (1977), but, unfortunately, the author's conclusions are based on a visual comparison of an incomplete ^{14}C record with a climate record which is often poorly dated and sometimes of uncertain reliability. The comparison is, therefore, unconvincing.

The main difficulties in trying to establish a ^{14}C-climate link are that the atmospheric ^{14}C record itself is incomplete (although work is in progress to produce a detailed record extending over the past 8,000 years) and that the climate data used for comparisons are generally indirect, local, and poorly dated. Furthermore, even if atmospheric ^{14}C does reflect changes in solar luminosity, the climate response is liable to be quite complex, with different regions responding in different ways and possibly at different times. The link must, at present, be considered "not proven". However, the same uncertainties which preclude a rigorous proof, also preclude a rigorous negation of the ^{14}C-climate hypothesis. There are enough encouraging aspects to warrant a strong continued research effort.

CHRONOLOGY DEVELOPMENT AND ANALYSIS
D.A.Graybill
INTRODUCTION
In any new and rapidly developing science it is important to experiment with analytical methods that will permit comparable and replicable research results. One of the most basic steps in dendroclimatology is the analysis of ring-width variability in developing single-site tree-ring chronologies. These chronologies are the primary data for many purposes. The analytical approaches used in chronology development are, therefore, important in terms of our potential ability to compare and generalise the results of similar research in different geographical areas.

The purpose of this section is to present a summary of basic analytical strategies for chronology building. The main emphasis will be upon the procedures of their implementation with a series of recently revised computer programs (Graybill, 1979a). They were designed to refine or replace older programs that were furnished by the Laboratory of Tree-Ring Research (University of Arizona). Program RWLIST replaces RWLST and is used for data inspection. Programs INDEX and SUMAC replace INDXA. INDEX is used for standardisation of ring-width series to develop ring-width indices while SUMAC summarises series of indices and performs analyses of variance and cross-correlation.

In some cases, the ideas and goals of analysis that are embodied in the programs, such as standardisation via curve fitting, stem from the early days of tree-ring research when these procedures were accomplished manually. The first comprehensive computerised procedures for chronology development and statistical description were developed by Fritts (1963) in the program that came to be known as INDXA. A number of documented additions to it were later made (Fritts et al., 1969) as well as a host of undocumented changes as needs dictated. The program was widely disseminated to other researchers who have made or suggested useful modifications. By the late 1970s, it became apparent that the program needed major revision for increased efficiency with changes in the FORTRAN language, for greater flexibility in use and for keeping it adaptable to an increasing variety of computing environments. The new programs were designed accordingly. They are now fully operational and available from our laboratory, along with a detailed operating manual (Graybill, 1979b). A brief technical description of them is provided later.

GOALS OF ANALYSIS
The overriding goal of the series of analyses described here is to develop chronologies that maximise the macroclimatic response of the individual specimens that are used and that minimise other non-climatic variation. For the purposes of this discussion, it is useful to conceive of tree-ring chronologies as time series with certain frequency domain properties. This domain is composed of four different kinds of signals that are recognisable and are germane to the goals of analysis. For any individual specimen let

$$R(t) = C + B + D + E \qquad (1.3)$$

where $R(t)$ is the measured ring width in year t; C is the macroclimatic signal common to trees at a site; B is the biological growth curve as a function of increasing tree age; D is the tree disturbance signal that may be: $D1$, unique to a single specimen or tree and due to random events that affected its growth or $D2$, common to most or all specimens due to fire, insect damage, or other disturbance that affected an entire site; and E is the random growth signal unique to each specimen.

In order to maximise the climate signal, it is necessary to recognise and remove or control the others. Program RWLIST is an aid in the recognition of the tree disturbance signal, D, and the shape of the biological growth curve, B. Program INDEX is used to remove these two signals. Attempts to minimise the random growth signal, E, are made by averaging the indices of an adequate number of specimens with program SUMAC. The statistics it provides for each summary, as well as the analysis of variance and cross-correlation procedures, are useful guides in determining how much of the macroclimatic signal common to the data set remains in the final chronology.

Before considering these procedures in greater detail, it is important to stress that this is not a "black box" approach. The design of the analyses now requires more interaction with the data and more review of decisions by the investigator than was common with the older programs.

THE BASIC DATA SET

It is assumed at this point that the basic units of analysis are ring-width records of specimens from a single site that were collected, dated and measured by standardised procedures such as those described by Stokes & Smiley (1968). It is also assumed that the samples were taken from trees in a limited area with similar slope, exposure, and elevation so they might reasonably be expected to have a common macroclimatic signal. Determination of the actual number of trees per site and number of samples per tree that are required to obtain a "reasonable" amount of common climate signal will initially be a matter for experimentation. The analysis of variance discussion below will treat this topic in greater detail.

PRELIMINARY DATA INSPECTION

The first goals in the analysis are:

a) to check the ring-width data for preparation errors;

b) to attempt to determine if any specimen has growth anomalies that are suggestive of the tree dist- urbance signal, D, noted above;

c) to develop an understanding of the central tend- encies and range of variability of certain stat- istical properties of the data set; and

d) to decide on the type of curve-fitting procedure for each specimen for subsequent standardisation.

The accomplishment of these goals is facilitated by the program RWLIST. It first checks the data for a variety of errors that may have occurred in the process of developing computer-readable records of ring widths. The data are listed for final proof-reading and a limited number of descriptive statistics that have commonly been used to characterise ring-width series are also printed. These are the mean ring width, mean sensitivity, standard deviation,

and first order autocorrelation coefficient (Douglass, 1936; Fritts, 1976). The first two values are usually of greatest interest in terms of the focus here. The latter two were included in this program to facilitate other dendroclimatic research where ring widths instead of indices are being utilised (for example, LaMarche, 1974a). Given experience with the range of variability in the first two values for a given species or area they can be useful aids in determining which of the specimens from a site might best be used or rejected. In general, fast-growing trees with relatively high mean ring width and relatively low mean sensitivity would be those least limited by climatic factors and, hence, the least useful for dendroclimatic purposes. Inspection of those statistics across the entire data set can also be a useful exercise. Specimens with exceptionally different values from the rest may become apparent. These may represent actual statistical outliers relative to the more central tendencies of the data set, and warrant removal from further analysis.

The final output for each specimen is a printer plot of 20-year averages shifted by 10-year increments. Figure 1.5 presents an example of this. The plots of these averages have generally been found to be satisfactory in smoothing out the high frequency variation often seen in annual ring-width plots that tends to obscure longer trends. The 20-year period is arbitrary but has been suitable for most of our data. For other data sets with relatively low variance or mean sensitivity in the component series it might be useful to use a shorter period for averaging. Visual inspection and comparison of the growth characteristics of specimens across all plots permits decisions to be made about the use of all or portions of certain specimens and the course of succeeding analyses. In some cases specimens will exhibit surges or depressions in growth that are not common to others in similar time periods. This may often be an extreme case of the tree disturbance signal (D1) that is due to some unusual non-climatic event. The general practice has been to exclude this type of specimen from further analysis. The next decisions to be made are about the curve-fitting procedure. These are discussed after the options are presented.

STANDARDISATION

The primary purpose of standardisation is to remove non-climatic signals in a series that may include either a biological growth trend (B), tree disturbance signals (D1 and/or D2), or both. This is done by fitting an appropr- iate curve to a ring-width series R(t), and calculating a new time series, Index(t), as follows

$$\text{Index}(t) = R(t)/Y(t) \qquad (1.4)$$

where Y(t) is the expected yearly growth determined from

curve-fitting. This has the effect of scaling the variance so that it is more homogeneous than is normally seen in ring-width series. It also scales the mean of each series to about 1.0 and reduces the first order autocorrelation due to trend (Fritts, 1976, p.266).

The curve-fitting options that are presently available in program INDEX are the same as those found in the older INDXA but some of the computational aspects and the printed output differ considerably. The options are:

0 - Fit a negative exponential. If this is not feasible, fit a straight line with zero or negative slope.

1 - Fit a horizontal line through the mean.

2 - Fit a straight line of any slope.

3 - Fit a negative exponential. If this is not feasible, fit a straight line with any slope.

4 - Fit an orthogonal polynomial.

With the selection of option 0 or 3 the program first attempts to fit a negative exponential curve of the form

$$Y(t) = ae^{-bt} + k \qquad (1.5)$$

where $Y(t)$ is the expected growth at a given year t, e is the base of natural logarithms and a, b, and k vary from series to series. If the value of b is found to be negative or if a and k are negative, a straight line of the specified slope is fitted. More details on this are found in Fritts et al. (1969) and Fritts (1976). This form of curve, as well as option 2, has been most commonly used for North American conifers growing on arid sites with relatively low stand density.

The horizontal line fitted through the mean of a series (option 1) is commonly used for cases where the investigator has some confidence that no discernible growth trend is present. This would be more appropriate than a straight line of any slope (option 2) because it would help preserve any climatic trends that were equal to or longer than the component series. This type of growth is apparent in the outer sectors of long lived species such as bristlecone pine, Pinus longaeva and Pinus aristata, and limber pine, Pinus flexilis. The 8,680yr bristlecone chronology developed by Ferguson (1969), and later updated, is composed of numerous segments first standardised in this fashion.

The polynomial curve-fitting option is sometimes more appropriate than others. A great variety of non-climatic factors may exert their influence on tree growth, resulting in series with complex characteristics that are not suitably described by negative exponential or linear models. In general these complex series are more common as stand density increases, as the mesic nature of the site increases, and with deciduous species.

The orthogonal polynomial used in INDEX fits a univariate regression curve of the familiar form

$$Y(t) = a_0 + a_1 x + a_2 x^2 + \ldots + a_m x^m \qquad (1.6)$$

where $Y(t)$ is the expected growth for a given year, a_0 is a constant, a_1, $a_2 \ldots a_m$ are regression coefficients, and x is the ring width in year t. The technique follows that of Forsythe (1957). One of the major problems in using this approach is how to determine m, the final degree of the fitted model, given no a priori idea of what it should be. The procedure that was adopted makes decisions based on changes in variance due to regression. For any step m: SSR

Figure 1.5: Plot from RWLIST.

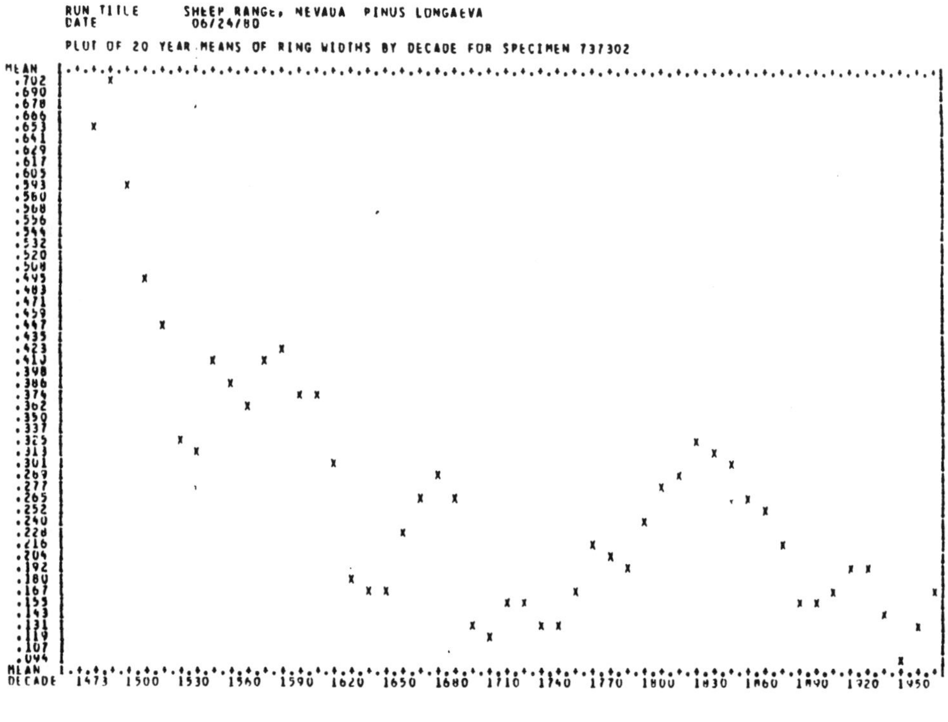

is the sum of squares due to the current orthogonal
polynomial; TSS is the corrected total sum of squares; and
TEST is SSR/TSS. The program checks to see at each degree
m if the value of TEST is greater than or equal to an
arbitrarily preset value. If this is true then degree m+1
and m+2 are developed. If the value of TEST at both of
these steps is less than the preset value then degree m is
the final one accepted. Otherwise, m is incremented by 1
and the procedure is repeated.

The preset value is now a variable under user
control in program INDEX. In the older INDXA this was
fixed at 0.05. Setting this at a lower value has the
effect of forcing more coefficients into the equation or
fitting the curve more closely to trends in the data.
This option should be used with extreme caution. If the
polynomial is forced to a relatively high degree and
closely fits most trends in a series then it will
obviously be acting as a high-pass filter. This will
remove not only the growth trend but may also remove
climatic trends that are shorter than the length of the
series. Depending on the series in question, it is
conceivable that all variance except that with high
frequency occurrence of one to a few years could be
removed.

Decisions about the type of curve to be fitted to a
series are not always simple ones. They need to be made
with as much information as possible about stand history,
site characteristics, and actual specimen characteristics
to hand. A common dilemma is how to decide whether certain
growth characteristics in complex series represent
climatic or non-climatic factors. For example, in Figure
1.5, is the growth surge in the early 1800s related to
climate or to other factors? If it is climatic, then an
exponential should be fitted; if it is not, then a
polynomial curve would be appropriate. A visual comparison
of that plot with those of other specimens from the site
indicated that this period of increased growth was common
to most series. A review of notes on the site indicated no
evidence of logging activity, fire, and so on, that might
have affected the growth of most trees. A negative
exponential was chosen.

After due consideration of the growth characteristics of all specimens from this site, program INDEX was
executed. The results of curve-fitting were then closely
inspected before moving to further analysis. This is
another important stage of decision-making that has often
been overlooked in the past. In part this was due to the
design of program INDXA which did not permit one to stop
at this point. To facilitate further processing the
current program can write the results of standardisation
to a peripheral unit such as a disc where they can be
saved and later retrieved by program SUMAC. The curve-
fitting procedure results in a list of the indices and the
statistics that describe them. The results have been made
more graphic by adding the option of a printer plot of

observed and expected values. This is plotted with the
same 20-year averages and 10-year overlap period used in
RWLIST. Examples of these for the specimen used in Figure
1.5 are provided in Figures 1.6 and 1.7.

Inspection and comparison of the statistics across a
data set may permit one to identify outliers or those with
some type of undesirable characteristic. In general, those
that exhibit relatively divergent values from others in
terms of exceptionally high first order autocorrelation,
low mean sensitivity, low variance, a very high or low
mean index value, or combinations thereof might be subject
to removal. Inspection of the plots might permit isolation
of the reason for some divergent statistical characteristics in terms of a poor curve fit. In some cases the
first decision about curve-fitting options may have been
inappropriate and the standardisation is either not
possible with current options or needs to be tried with a
different option.

The curve fitting procedures in INDEX have been
found suitable for a great variety of data but it is
recognised that no single set of procedures is universally
applicable. As different procedures appear to be required
they could be added to the present repertoire. For
example, given a series with especially complex growth
characteristics, there are other approximating functions
that might be preferable to the orthogonal polynomial,
such as the cubic spline (Reinsch, 1967). The use of this
technique for standardising tree-ring series is being
investigated by researchers at Lamont-Doherty Geological
Observatory (Cook & Peters, 1979; Peters & Cook, 1979).

SUMMARIES OF INDICES

The process of summarising or averaging the indices
of several series has the desired goal of decreasing the
random error (E) associated with each series, resulting in
a final chronology with a relatively high ratio of climatic signal to non-climatic noise. This kind of chronology
development is done by program SUMAC with either a Type 1
or Type 2 analysis. In a Type 1 analysis, the program
reads the specimen indices and associated parameters from
intermediate disc storage following a final execution of
program INDEX. The Type 2 analysis is designed for
situations where the job flow is not continuous with
INDEX. The input is of indices that may have been produced
by INDEX or INDXA but have been stored as physical cards
or card images. Following input operations, the two types
of analyses are identical.

A typical data set for these analyses is composed of
replicated series of two or more samples per tree from a
single site. Three types of summaries are normally
developed. The program reads a single control record and
is designed to order the summaries as described below,
decreasing some of the tedium and reducing mistakes common
to use of the INDXA program. The first set of summaries
are for each tree. The second set includes summaries of

the replicated samples across trees, for example, representing opposing cardinal directions of growth. The third type of summary is the average of indices from all replicated sets of specimens and is the one that might then potentially be used for dendroclimatic purposes. An example of the printed output of this type of summary is provided in Figure 1.8. The first set of statistics below the list of indices describes the entire chronology. Below this the second set of statistics describes the chronology for a period that all component series have in common. The data from this period are used later in analysis of variance (ANOVA) and in cross-correlation. The three statistics that have commonly been used in partial evaluation of a chronology are the mean sensitivity, standard deviation, and first order auto-correlation coefficient (Fritts & Shatz, 1975). In general a "good" chronology for climatic investigation is characterised by relatively high values of the first two of those statistics and a relatively low value of the third. Further evaluation is discussed in the ANOVA and cross-correlation sections.

In the current execution mode, a fourth type of summary is sometimes optionally produced. The data set may have originally included a few extraneous specimens from unreplicated samples. These can be arranged with all others to develop a chronology that is used only for cross-dating or other purposes. In addition to the type of example just described the program will now accept data

sets that have only one specimen per tree. In this case there would only be one final summary produced although ANOVA and cross-correlation can also be executed.

While the current example focussed on single-site chronology development, the program has the capability of developing a multiple-site chronology also with a Type 1 or Type 2 analysis. All that is required is a constant number of trees per site and specimens per tree in order that the succeeding ANOVA can execute properly.

A final type of chronology development procedure in SUMAC is designed to average indices for any number of diverse purposes. For example, site summaries and tree summaries developed at different times may be used. Other applications might involve averaging numerous overlapping archaeological and/or modern specimens for various research purposes (Baillie, 1977a; Dean & Robinson, 1978).

ANALYSIS OF VARIANCE (ANOVA)

The primary purpose of this analysis is to estimate the relative magnitude or importance of the major sources of variance in the common period of an averaged chronology. It should be noted that the ANOVA results are not used for hypothesis testing because various assumptions of the model cannot always be met (Sokal & Rohlf, 1969, pp.367-403). Instead the results are viewed as useful descriptive statistics that provide one of several bases for chronology evaluation and for refining future sampling strategies. A detailed inspection of the ANOVA comp-

Figure 1.6: Indices from INDEX.

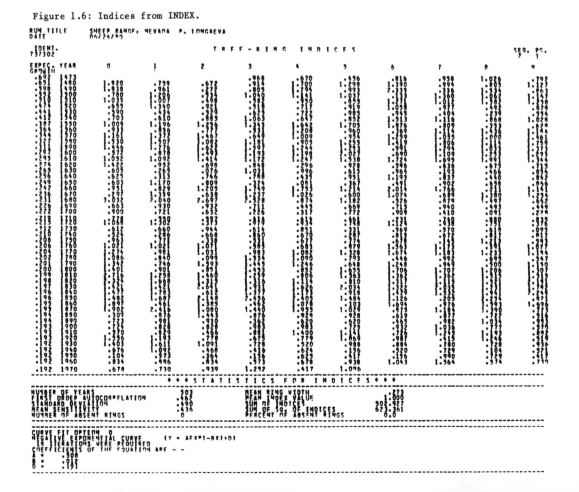

utational procedure and an example of its use have been
presented by Fritts (1976, pp.282-293). Further discussion
here concerns recent changes in program operations,
output, and use of the results.

The ANOVA section of SUMAC can be executed as an
option that follows the summary development or it can be
used as a separate program. The printed output shown in
Figure 1.9 now follows the tabular presentation of Fritts
(1976, p.288). A revision made by D.Duvick (Laboratory of
Tree-Ring Research, University of Arizona) will provide
information on the variance contributed by "subsamples".
These might be indices of two specimens from the same side
of a tree. Another useful revision of the program made by
J.P.Cropper (Laboratory of Tree-Ring Research, University
of Arizona) allows the ANOVA to be utilised when only one
specimen per tree is available for analysis.

When developing a single-site chronology there are
five potential sources of variance that can be described
in the program output. The relative contribution of each
of these is most readily described by the percentage
variance component seen in the far right hand column of
Figure 1.9. In this example, the subsampling option was
not utilised so only four types of variation are considered. The first, labelled "mean indices in total
chronology (Y)", is the percent variance held in common by
all specimens in the chronology. This is interpreted as
the macroclimatic component of the chronology that one is
attempting to maximise. This figure may exceed 80% with
chronologies from some arid sites but is generally lower
with deciduous species and more mesic sites. A recent
comparison of this figure for a 102-site grid of western
United States conifers and 20 selected sites with various
species from the eastern United States resulted in %Y

averages of 60 and 29 respectively (Dewitt & Ames, 1978,
p.12).

The second source of variation, "chronologies of
trees in groups (Y x T/G)", is due to variability or
differences between tree chronologies. This value tends to
be higher as the heterogeneity of site characteristics,
such as substrate, soil, slope, or drainage characteristics, increases. The third source of variation,
"chronologies of core classes (Y x C)", is due to
variability in the samples given the directional or other
defining criteria for this class. This value seldom
exceeds a few percent.The final source of variation,
"chronologies of cores with trees in groups (Y x C x
T/G)", is due to differences in the individual specimen
series. If there had been two or more sites in the
analysis the group categories (Y x C x G) and (Y x G)
would have been other sources of variation in the final
chronology.

The description of the use of ANOVA has focussed on
the entire common period of a chronology but other approaches can be taken. In a study of oak in North Wales, the
ANOVA portion of INDXA was used to derive statistics for
two different portions of the chronology, 1875 to 1900 and
1905 to 1930 (Hughes et al., 1978a). It had previously
been suggested that the earlier period was one of greater
variability in terms of the frequency of days with
westerly winds (Lamb, 1966). The %Y value for the earlier
period was 6% greater than that of the later one. The
greater climatic variability in the earlier period is
thought to have been recorded as a stronger similarity of
growth response between trees. The results of the ANOVA
procedures can also be used for the development of future
collection strategies. Given some experience with a

Figure 1.7: Plot of curve-fitting results from INDEX.

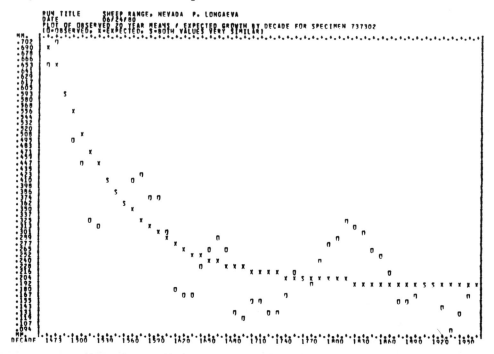

particular species or area where the %Y can be expected to be near a certain value, it is possible to estimate the number of samples that need to be collected for optimising the climatic signal (%Y) relative to non-climatic noise (100%-%Y). The ratio of signal to noise (SNR) in a final chronology varies directly with the number of trees (NT) that are included. Assuming one specimen per tree was collected this can be expressed as

$$SNR = NT \times [\%Y/(100\%-\%Y)] \qquad (1.7)$$

A graphic illustration of the relationships among these quantities is provided in Figure 1.10. It was developed by J.P.Cropper and H.C.Fritts (both of the Laboratory of Tree-Ring Research, University of Arizona) using data from the 102- and 20-site grids from western and eastern North America (DeWitt & Ames, 1978, p.14). A more concise statement of the number of trees required for any particular SNR can be developed from Equation 1.6 as

$$NT = SNR \times [(100\%-\%Y)/\%Y] \qquad (1.8)$$

An obvious problem now presents itself. What is a desirable or an acceptable signal-to-noise ratio? This is, quite simply, a topic for further research in any particular empirical setting. As appropriate data become available, it might be useful to design experiments that attempt to calibrate climatic variation with several different chronologies from one site incorporating successively larger numbers of trees. If site data are too restricted then simulation techniques might be profitably utilised (see Cropper, this volume). Reduction of error in a dendroclimatic chronology is always desirable but the benefits are necessarily tempered by specimen availability and collection costs.

CORRELATION ANALYSIS

The correlation options in SUMAC provide another basis for locating variability and examining the covariation that occurs in the component series of a chronology. It must be noted again that the correlation figures are descriptive statistics. Significance levels for them are not currently computed. The correlation procedures, like the ANOVA, can be used as an optional analysis when averaging series or as a separate program. The user has control over the series to be correlated, the number of lags of autocorrelation to be computed for each series, and the length of various subperiods for which correlation figures are of interest.

The first set of indices typically examined are those for the individual specimens while the second set is for tree chronologies. In each case, execution of the

Figure 1.8: Final chronology from SUMAC.

RUN TITLE SHEEP RANGE, NEVADA P. LONGAEVA
DATE 06/24/80

IDENT. 737510 S U M M A R Y O F T R E E - R I N G I N D I C E S SEQ. 42 PG. 1

SEQUENCE NUMBERS INCLUDED ARE 40 41

YEAR	INDICES 0	1	2	3	4	5	6	7	8	9	NUMBER OF CORES 0	1	2	3	4	5	6	7	8	9
1409										.606										1.
1410	.677	.814	.725	.477	1.070	.748	.366	.916	1.056	.552	1.	1.	1.	1.	1.	1.	1.	1.	1.	1.
1420	.899	.993	.949	1.183	.860	1.025	1.190	1.426	1.828	1.033	1.	1.	1.	1.	1.	1.	1.	1.	1.	1.
1430	1.106	.849	.792	.993	.873	1.066	1.169	.977	.802	.906	1.	1.	2.	2.	2.	2.	2.	2.	2.	2.
1440	.899	1.306	.708	1.129	.417	1.006	.736	1.056	.506	.827	2.	2.	2.	2.	2.	2.	2.	2.	2.	2.
1450	.500	.624	.907	1.155	.804	.688	.783	.577	.907	.578	2.	2.	2.	2.	2.	2.	2.	2.	2.	2.
1460	.365	1.195	1.187	1.261	.368	.900	1.134	1.138	.864	.896	2.	2.	2.	3.	3.	3.	3.	3.	3.	3.
1470	.913	.333	.829	1.150	.564	.581	1.106	1.094	1.268	1.196	3.	3.	3.	4.	4.	4.	4.	4.	4.	4.
1480	1.927	1.188	.958	1.050	1.716	1.506	1.616	1.504	1.071	1.289	4.	4.	4.	4.	4.	4.	4.	4.	4.	4.
1490	2.023	1.040	.520	1.372	1.982	.975	1.946	.451	2.127	1.063	4.	4.	4.	4.	4.	4.	4.	4.	4.	4.
1500	.419	1.215	.742	.822	.819	.629	.247	1.142	.979	1.039	5.	5.	5.	5.	5.	5.	5.	5.	5.	5.
1510	.662	1.031	.567	.727	.954	.806	1.191	.625	1.043	1.223	6.	6.	6.	6.	6.	6.	6.	6.	6.	6.
1520	.757	.872	.425	.184	.672	.537	1.047	.718	.468	.966	6.	6.	6.	6.	6.	6.	6.	6.	6.	6.
1530	.711	.478	.189	.594	.436	.875	.854	.604	.261	.805	6.	6.	6.	6.	6.	6.	6.	6.	6.	6.
1540	.697	.553	.775	.784	.208	.693	1.130	.575	.827	1.175	7.	7.	7.	7.	7.	7.	7.	7.	7.	7.
1550	1.040	1.056	.904	1.229	1.194	1.450	1.486	.928	.445	.764	9.	9.	9.	9.	9.	9.	9.	9.	9.	9.
1560	1.017	.877	1.043	.641	1.166	.998	.393	1.185	1.445	1.165	9.	9.	9.	9.	9.	9.	9.	9.	9.	9.
1570	1.035	1.738	1.854	.924	1.323	1.210	1.159	1.306	1.095	1.555	9.	9.	9.	9.	9.	9.	9.	9.	9.	9.
1580	1.416	1.487	1.106	1.299	.627	.406	1.326	1.549	1.188	1.489	11.	11.	11.	11.	11.	11.	11.	11.	11.	11.
1590	.932	1.072	1.362	1.417	1.068	1.318	1.111	1.370	1.049	1.156	11.	11.	11.	11.	11.	11.	11.	11.	11.	11.
1600	1.058	.859	1.280	.891	1.598	2.143	.621	1.350	1.408	1.287	11.	11.	11.	11.	11.	11.	11.	11.	12.	12.
1610	1.327	.921	1.091	.638	1.345	1.633	1.427	1.350	1.623	1.290	12.	13.	13.	13.	13.	13.	14.	14.	14.	14.
1620	1.424	1.134	.935	.953	.181	.970	.376	.953	.804	1.187	14.	14.	14.	14.	14.	14.	14.	14.	14.	14.
1630	.403	.438	.341	1.021	.863	.737	1.528	.607	.657	1.168	14.	14.	14.	14.	14.	14.	14.	14.	14.	14.
1640	1.183	.730	1.002	.986	.902	1.109	1.342	1.054	.762	1.234	14.	14.	14.	14.	14.	14.	15.	15.	15.	15.
1650	1.072	1.627	.822	.491	.478	.652	1.140	.980	1.011	.903	15.	15.	15.	15.	15.	15.	15.	15.	15.	17.
1660	1.236	1.169	1.179	1.012	1.236	.541	1.309	1.123	.203	.807	18.	18.	18.	18.	18.	18.	18.	18.	18.	18.
1670	.107	1.206	.832	1.034	.972	.549	.869	1.111	1.255	.839	18.	18.	18.	18.	18.	18.	18.	19.	19.	21.
1680	1.663	1.128	1.487	1.582	.761	.939	.150	1.026	1.253	1.322	22.	22.	22.	24.	24.	24.	24.	24.	24.	24.
1690	.789	.842	1.116	.733	.918	1.217	1.067	1.079	.614	1.027	24.	24.	24.	24.	24.	24.	24.	24.	24.	24.
1700	1.308	1.411	1.211	.353	1.003	1.692	1.136	.481	.313	.607	24.	24.	25.	25.	25.	25.	25.	25.	25.	25.
1710	1.042	.813	.950	.340	.868	.700	.266	.783	1.263	.875	25.	25.	25.	25.	25.	25.	25.	25.	25.	25.
1720	1.645	1.430	.339	1.236	1.628	1.585	1.865	1.468	1.339	.373	25.	25.	25.	25.	25.	26.	26.	26.	26.	26.
1730	.604	1.110	.930	.492	.992	.205	.755	.860	.951	.185	26.	26.	26.	26.	26.	26.	26.	26.	26.	26.
1740	.802	1.668	.267	1.286	.886	.615	1.171	1.096	.817	1.027	26.	26.	26.	26.	26.	26.	26.	26.	26.	26.
1750	1.028	.536	.089	.655	1.000	.302	.804	.172	.934	.745	26.	26.	26.	26.	26.	26.	26.	26.	26.	26.
1760	1.126	.932	1.012	.434	1.013	.576	1.625	1.224	.412	1.387	26.	26.	26.	26.	26.	26.	26.	26.	26.	26.
1770	1.475	1.421	1.161	1.054	1.261	1.427	1.804	.687	1.767	1.456	26.	26.	26.	26.	26.	26.	26.	26.	26.	26.
1780	.896	.897	.186	1.285	1.337	.362	.436	.935	.489	.931	26.	26.	26.	26.	26.	26.	26.	26.	26.	26.
1790	.377	1.018	1.328	1.175	.880	.455	1.002	.805	.612	.988	26.	26.	26.	26.	26.	26.	26.	26.	26.	26.
1800	.799	.764	1.253	1.361	.984	1.164	1.159	1.312	1.047	.892	26.	26.	26.	26.	26.	26.	26.	26.	26.	26.
1810	1.794	1.183	1.242	.273	.822	1.274	1.253	1.176	1.031	1.269	26.	26.	26.	26.	26.	26.	26.	26.	26.	26.
1820	.956	1.392	.658	.643	.824	1.046	1.089	.881	1.399	1.138	26.	26.	26.	26.	26.	26.	26.	26.	26.	26.
1830	1.309	1.428	2.223	1.668	1.001	1.564	.701	1.445	1.409	1.613	26.	26.	26.	26.	26.	26.	26.	26.	26.	26.
1840	1.218	1.401	.482	1.251	1.116	.402	1.187	.435	1.574	1.491	26.	26.	26.	26.	26.	26.	26.	26.	26.	26.
1850	1.210	1.039	1.533	1.482	1.520	1.016	.702	.112	2.185	.935	26.	26.	26.	26.	26.	26.	26.	26.	26.	26.
1860	1.573	.679	1.243	.189	.906	.382	1.223	1.065	1.629	1.309	26.	26.	26.	26.	26.	26.	26.	26.	26.	26.
1870	1.152	1.232	.466	.780	1.662	1.187	.962	.906	.810	.555	26.	26.	26.	26.	26.	26.	26.	26.	26.	26.
1880	.138	.598	.891	.345	.868	.927	.720	1.052	.996	.700	26.	26.	26.	26.	26.	26.	26.	26.	26.	26.
1890	.686	1.230	1.059	.355	.483	1.041	.525	1.155	1.183	1.215	26.	26.	26.	26.	26.	26.	26.	26.	26.	26.
1900	.635	.802	.501	.822	.425	1.158	1.224	1.306	1.266	1.116	26.	26.	26.	26.	26.	26.	26.	26.	26.	26.
1910	.881	1.728	1.305	1.551	2.206	1.700	1.028	1.700	1.518	1.576	26.	26.	26.	26.	26.	26.	26.	26.	26.	26.
1920	1.886	1.981	1.368	1.076	1.265	1.635	1.616	.721	1.456	.361	26.	26.	26.	26.	26.	26.	26.	26.	26.	26.
1930	1.545	1.488	1.493	.869	.172	1.179	.447	1.576	1.120	1.183	26.	26.	26.	26.	26.	26.	26.	26.	26.	26.
1940	.947	1.506	1.023	.678	1.004	.585	.821	.928	.611	.880	26.	26.	26.	26.	26.	26.	26.	26.	26.	26.
1950	.179	.453	.881	.246	.797	.729	.119	1.162	.891	.185	26.	26.	26.	26.	26.	26.	26.	26.	26.	26.
1960	.918	.833	.778	.412	.624	.975	.603	1.214	.774	1.340	26.	26.	26.	26.	26.	26.	26.	26.	26.	26.
1970	.810	.905	.583	1.367	.403	1.497					26.	26.	26.	26.	26.	26.				

* * * S T A T I S T I C S F O R S U M M A R Y * * *

TIME RANGE IS	1409-1975	NUMBER OF YEARS	567	STANDARD DEVIATION	.402
FIRST ORDER AUTOCORRELATION	.257	MEAN SENSITIVITY	.453	MEAN INDEX VALUE	.942
STANDARD ERROR	.086	SUM OF INDICES	562.396	SUM OF SQ. OF INDICES	644.160
MEAN RING WIDTH	.420	NUMBER OF ABSENT RINGS	3	PERCENT OF ABSENT RINGS	.03054
ANOVA PERIOD IS	1725-1975	NUMBER OF YEARS	251	MEAN INDEX VALUE	1.004
STANDARD ERROR	.072			SUM OF SQ. OF INDICES	299.808
R OF MEAN WITH STD ERROR	.671	(INTERCEPT)	.036	(SLOPE)	.0364
MEAN RING WIDTH	.369	NUMBER OF ABSENT RINGS	2	PERCENT OF ABSENT RINGS	.03085

correlation option first results in a list of each series used and the mean, standard deviation, mean sensitivity, and mean variation, as well as the averages of those statistics, for all series. The second optional output is a list of autocorrelations for each series (up to 15 orders) and the average per order for all series. The third output is a matrix of correlation coefficients and the average of each series with all others printed. Inspection of these results is usually made with a concern for whether one or more individual series have relatively low correlations with others or exceptionally high autocorrelations of any particular order.

In general, this is one final time in the analysis stream when aberrant specimens or even portions of specimens may be detected. The analyses described above can be repeated optionally for any specified subperiod throughout a chronology. The subperiod can also be lagged by an optional number of years. The default values in the program that are commonly used are 20-year subperiods with a 10-year lag. This permits a relatively refined perspective on the covariation in a chronology and may permit one to isolate problems that require re-analysis or additional collection. For example, it might be found that one specimen or tree has consistently low correlation with others in one subperiod or more. This should force re-examination of all statistics, curve-fitting options, measurements, and even field notes that pertain. Although this is time-consuming, and sometimes humiliating, it can lead to greater understanding and better chronologies for dendroclimatic research.

TECHNICAL CONSIDERATIONS

All of the programs described above are written in American National Standards Institute (A.N.S.I.) standard FORTRAN IV (A.N.S.I., 1966). Although A.N.S.I. FORTRAN V was standardised in 1977 (A.N.S.I., 1978), a survey of users indicated this was not yet widely utilised or available on many systems. I am in the process of developing a FORTRAN V version but this will not be completed or tested for some time.

Although the design of these programs was orientated toward modernisation and additions, it was also directed at continuity with the older programs so that users might easily adapt to a transition. Toward this end, the format of all data records such as those for ring widths and indices has remained constant. Additionally, the newer programs provide increased flexibility in terms of user control over the mechanics of data record storage and transmission. INDXA was almost solely orientated toward the use of Hollerith punched cards for input and output. The current programs emphasise the use of disc or tape files as alternatives. All programs now include extensive error-checking procedures. Problems such as reversed, missing, duplicate, or mixed data records are detected in the input stream while a number of other common processing errors are located. When errors are encountered each program attempts to specify the problems and print it on the output file before the program is stopped. The data handling capacity of the programs is greater than that of earlier versions. RWLIST will handle any number of specimens up to 2,000 years in length. INDEX and SUMAC can treat up to 80 specimens of 2,000 years length in a single execution. All programs have been extensively tested and are now operational on the Control Data Corporation Cyber series and IBM 370 machines. Their adaptation to other systems should be relatively straightforward.

COMMENT

M.K.Hughes

In order to use the climate-growth response in

Figure 1.9: ANOVA results from SUMAC.

RUN TITLE SHEEP RANGE, NEVADA P. LONGAEVA
DATE 06/24/80

ANALYSIS OF VARIANCE

SOURCE OF VARIATION		RAW SUM OF SQUARES	CORRECTED SUM OF SQUARES	DEGREES OF FREEDOM	MEAN SQUARE	VARIANCE COMPONENT	PERCENTAGE VARIANCE COMPONENT
GROUP MEANS	(G)	ONLY ONE GROUP IS BEING USED					
CORE CLASS MEANS	(C)	6579.078	-.010	1.000	-.010		
TREE MEANS IN GROUPS	(T X G)	6587.532	8.443	12.000	.704		
CORE MEANS IN GROUPS	(C X G)	ONLY ONE GROUP IS BEING USED					
CORE MEANS WITH TREES IN GROUPS	(C X T/G)	6588.318	.797	12.000	.066		
SUBSAMPLE	(C X T/G)	SUBSAMPLING NOT USED					
MEAN INDICES IN TOTAL CHRONOLOGY	(Y)	7794.999	1215.910	250.000	4.864	.178	53.901
CHRONOLOGIES OF GROUPS	(Y X G)	ONLY ONE GROUP IS BEING USED					
CHRONOLOGIES OF TREES IN GROUPS	(Y X T/G)	8505.385	701.943	3000.000	.234	.082	24.762
CHRONOLOGIES OF CORE CLASSES	(Y X C)	7912.934	17.946	250.000	.072	.000	.033
CHRONOLOGIES OF CORE CLASSES WITH GROUPS	(Y X C X G)	ONLY ONE GROUP IS BEING USED					
CHRONOLOGIES OF CORES WITH TREES IN GROUPS	(Y X C X T/G)	8735.255	211.138	3000.000	.070	.070	21.304
SUBSAMPLE	(Y X C X T/G)	SUBSAMPLING NOT USED					

TOTAL SUM = 6579.09 N OF ELEMENTS = 6526.
ERROR SQ. OF Y = .0091 ERROR OF Y = .0951

SIGNAL / NOISE RATIO = 15.2 / 1.0 PERCENT SIGNAL / PERCENT NOISE RATIO = 93.8 / 6.2

palaeoclimatology, we make certain assumptions:

a) the form of the climate-growth response is
   characteristic of a particular combination of site,
   habitat, and tree species;

b) influences other than climate introduce noise into
   the record found in each tree ring - this noise may
   be averaged out by increased replication, leaving
   the common, climatically induced, pattern enhanced;

c) age-related trends in growth rate may be identified
   and removed with little distortion of the climatic
   signal; and

d) both the climate-growth response and the statistical
   nature of non-climatic "noise" are stable in time.

These assumptions are now examined. The best evidence for
the existence of climate-growth responses characteristic
of particular site-habitat-species is the high degree of
cross-dating found within a site as compared to between

sites. However, the practical utility of this assumption
rests on the recognition of distinct sites and habitats.
This is not always possible without examination of the
tree-ring material. There may be value in defining
homogeneous sites as much in terms of tree-ring series
similarity as ecological conditions. The assumption that
non-climatic variability can be treated as Gaussian
"noise" is supported by both the improvement in cross-
dating and of climatic correlations seen as replication of
tree-ring series is increased. Which statistical model of
the signal/noise relationship should best be applied in
designing sampling at a stand will depend on an improved
understanding of the statistical nature of the noise.

It is difficult to test whether age-related trends
can be removed without distorting the climate signal.
Dendroclimatologists have often assumed that any trend in
ring series is of non-climatic and probably age-related

Figure 1.10: Analysis of percent variance accounted
for by climate (Y or signal, represented by solid
and dashed horizontal lines), percent variance
accounted for by non-climatic factors (error
variance or noise, represented by solid and dashed
curves), and signal-to-noise ratio (represented by
solid and dashed straight lines with positive
slopes) for western (W) and eastern (E) North
American tree-ring chronologies. The table at the
top of the figure gives average values for percent
variance due to climatic and non-climatic factors
and the average standard deviation of the core
series for all western arid-site chronologies and
for a sample of 20 eastern North American chronologies (from DeWitt & Ames, 1978).

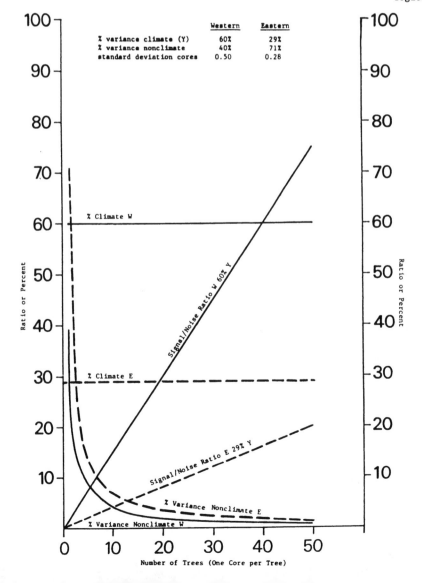

	Western	Eastern
% variance climate (Y)	60%	29%
% variance nonclimate	40%	71%
standard deviation cores	0.50	0.28

Ratio or Percent

% Climate W

Signal/Noise Ratio W 60% Y

% Climate E

Signal/Noise Ratio E 29% Y

% Variance Nonclimate E

% Variance Nonclimate W

Number of Trees (One Core per Tree)

origin and should be removed. Alternatively, it may be thought that age-related trend and low-frequency climate response cannot be separated. It may be better to risk discarding the latter in order to be sure of removing the former. The problem is particularly severe in tree species of mesic, closed-canopy forest where trends may vary within and between trees in a complex manner (Figure 1.11). There are circumstances, such as the growth of a tree through a pre-existing canopy, where a change in climate-growth response with time would be expected (Ogden, this volume). The problem is not insurmountable if the transition is recorded clearly in the wood. More difficult problems may arise from subtle changes in response related to age or gradual shifts in site conditions.

Each of the four assumptions discussed seems to apply, increasingly from (a) to (d), with greatest validity to the "classic" regions of dendroclimatology where a strong moisture or heat deficit plays a major role in controlling tree growth in open stands. Their validity is less in more mesic regions with closed-canopy forests. In such regions we need to place greater emphasis on population and stand level interactions in our models of the climate-growth response. Furthermore, it may be of value to consider these interactions explicitly in the design of methods of standardisation.

COMMENT

R.W.Aniol & B.Schmidt

The computer programs described by Graybill are useful instruments for the development of chronologies which retain maximal climatic information. The statistics of each sample and of the resulting chronologies are helpful for choosing suitable material. In Europe, oak chronologies of some length are made up from many fairly short ring-width series. To develop such a chronology those series have first to be synchronised. One procedure for this is the computation of "Gleichlaufigkeitswerte" (percentage of agreement), which can be done with or without standardising the tree-ring widths (Eckstein, 1969).

In some years, external influences on the growth of trees (Graybill calls these influences the macroclimatic signal C) are particularly strong. As a result many trees react in the same way to this influence and show the same sort of variation in their growth compared to the year before. In a chronology such years are called pointer years. A statistical definition for pointer years was given by Eckstein (1969). In a pointer year, more

Figure 1.11: Trend in mean ring width within trees. Each graph shows plots of 20-year mean ring-width plotted every ten years for each of several heights in a single tree. Growth curves are shown from the main hole (solid line) and from major boughs (dashed line) of five trees of _Quercus petraea_ at a site in North Wales.

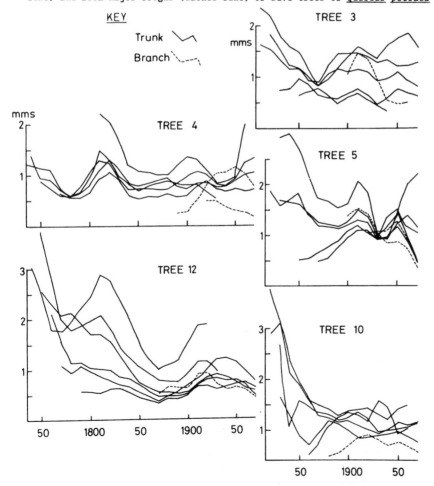

tree-ring series agree with the trend in the chronology than the greatest possible number of agreements that would be expected to occur by chance at a given level of significance. The biggest number of agreements that can occur by chance can be computed with a formula introduced by Graf-Henning (1960). In the Tree-Ring Laboratory of the University of Cologne, pointer years at a level of significance of 5%, 1%, and 0.1% are used (Schmidt & Aniol, 1981).

Synchronising a series of unknown age with a chronology is facilitated by using the percentage of agreement in the pointer years as additional information to the percentage of agreement in the whole overlap. In the synchronous position the percentage of agreement in the pointer years is greater than for the whole overlap. The agreement becomes increasingly better with a more rigorous definition in statistical terms of the pointer years. This sort of behaviour should be expected from all series which make up a given chronology. In this way the quality of all the separate series in a chronology can be tested.

Pointer years also help to compare the quality of two chronologies. It is necessary to make two comparisons, as the quality of one chronology is tested by assuming the other one to be correct. In Figure 1.12, a chronology from the Weser mountain range is compared with a Rhine chronology in both ways. Naturally, the percentage of agreement for the whole overlap is identical. The difference in the percentage of agreement of the pointer years is an indication that the Rhine chronology has a wider range than the Weser chronology. This can be concluded from the fact that the pointer years of the Weser chronology agree better with the Rhine chronology than vice versa. Computation of the percentage of agreement using pointer years is an additional source of verification when

developing chronologies. Furthermore, the very existence of pointer years is an interesting phenomenon which may yield information on the nature of the climate-growth response.

Figure 1.12: Comparison of two chronologies in synchronous position. A_1, A_2: percentage of agreement for the whole overlap; B_1, C_1, D_1: percentage of agreement for the pointer years of the Weser chronology compared with the Rhine chronology at the three levels of significance 95%, 99%, and 99.9%; B_2, C_2, D_2: percentage of agreement for the pointer years of the Rhine chronology compared with the Weser chronology at the three levels of significance 95%, 99%, and 99.9%.

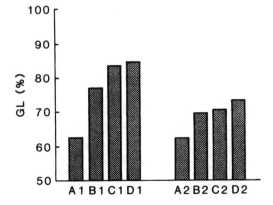

INTRODUCTION

The Editors

In the previous chapter, various methods of
enhancing the climate signal in tree-ring chronologies
have been mentioned. In order to reconstruct climate, it
is necessary to extract this signal - to separate climatic
factors from the many other environmental variables
limiting the plant processes which control growth. The
climate-growth response is complex, and reliable theoretical models have yet to be developed. Semi-empirical
techniques have, therefore, been developed in order to
extract the climate signal. Response functions are used to
describe associations between climate data and annual ring
measurements. Transfer functions are used to calibrate the
ring measurements with climate data in order to provide
regression equations for climate reconstruction. Both
techniques employ multivariate statistical methods such as
principal component or eigenvector analysis and canonical
correlation and regression.

The response of the growth of trees to climate and
other environmental factors is discussed by Fritts. He
highlights the need for semi-empirical techniques to
define and extract the macroclimatic signal from tree-ring
chronologies. The response function provides an empirical
method of describing the nature of the climatic factors
that influence tree growth. As Hughes and Milsom point
out, the response function does not measure the climate-
growth response but rather the effectiveness of a
particular statistical model at predicting the element of
tree-ring variation forced by external factors.

The response function has proved to be a valuable
tool for analysing the climate-growth relationship.
Guiot, Berger and Munaut review various methods of
calculating response functions. All are based on the
fundamental model described by Fritts et al. (1971) but
differ in the way in which they handle statistical
problems arising from the complex nature of the climate-
growth response. For example, the growth of the tree in a
particular year may depend not only on environmental
conditions during the growing season but also on
conditions in prior seasons. Climate data for prior
seasons, and the growth of the trees in prior years, have,
therefore, to be included as predictor variables in the
response function. Statistical questions arise concerning
the best way to include these variables in the response
function, and the interpretation of the role of prior
growth in the response function is problematic. This
matter is discussed further by Brett who gives specific
examples from his work on English Ulmus and Quercus.
Selective filtering techniques have been introduced by

Guiot and his colleagues in order to determine whether or
not trees respond in a different manner to climatic
fluctuations on different timescales. The possibility that
the trees response to climate changes as the trees age
has also been considered. In view of these and other
problems, Cropper questions the value of current methods
of calculating response functions and of interpreting
their results, and makes the constructive suggestion that
simulation techniques should be used to evaluate their
strengths and limitations.

While the response function is a means of predicting
tree performance from climate data, climate reconstruction
requires that climate data be predicted from tree-ring
data. The response function is used as a guide to the
climatic factors which influence tree growth and to the
direction and strength of the relationship; this information is then used in the design of the transfer function
model. Lofgren and Hunt describe the use of the transfer
function. Transfer functions are calibrated over a period
of time when both instrumental climate data and tree-ring
data are available and the resulting regression estimates
are then used to extend the climate record back in time
prior to the calibration interval. Again, because of the
complexity of the climate-growth response, multivariate
techniques are employed and various statistical problems
arise. Lofgren and Hunt note particular concerns, such as
the proper inclusion of the autoregressive and moving
average nature of the climate-growth relationship, the
need for statistical significance testing, and the use of
principal component (or eigenvector) analysis to maximise
the climate signal. Gray discusses the validity of the
assumption that the climate-growth response is stable with
time and considers the "no-analogue" problem. If a
particular climatic pattern was prominent prior to the
calibration interval, but not present during it, then that
pattern cannot be included in the transfer function model
and the resulting climatic reconstruction will not include
it unless it is a linear combination of other patterns.
Research described by Gray suggests that for European
temperature the no-analogue problem is unlikely to be
critical for the period since 1780. Non-linearity in the
climate-growth response can also cause no-analogue
problems.

The critical assessment of the reliability of any
dendroclimatic reconstruction is a crucial stage. Gordon
describes the various methods through which this can be
achieved. Data not used in the development of the transfer
function model are necessary. Instrumental climate data
not included in the calibration of the model are most
suitable but often limited in availability. This means
that withholding data for verification results in a loss
of data for calibration and, therefore, less reliable or
less stable regression estimates. Some compromise has to
be reached. Gordon suggests that subsample replication may
provide a solution to this problem. Another approach is to

use independent proxy climate data and historical, documentary information for verification. Gray and Pilcher note the irony of the fact that in Europe where climate data are relatively plentiful and thorough verification is possible, the tree-ring data base is very poorly developed. They stress the value of historical climate information in verification for the European region, and note that the problem of the lack of climate data for calibration could be partly alleviated by the use of transfer function models containing fewer variables.

The need for critical evaluation of the methodology and results of dendroclimatic reconstruction is stressed by many of the contributors to this chapter, both from dendrochronological and climatological backgrounds. Pittock describes what the climatologist requires of dendroclimatic reconstruction and questions the extent to which dendroclimatology can meet these requirements. He emphasises the need for better understanding of the biological processes underlying the reconstruction technique, and for continuous refinement of the techniques employed. He also notes that attempts should be made to extract information concerning more and different climate parameters and that longer records, of great value to climatologists, could be made available if dating standards were relaxed. Cropper points out that, when evaluating dendroclimatology's potential, the method by which the magnitude of the climate signal in tree rings is determined needs to be considered carefully.

Finally, to set the stage for the review of the global dendroclimatological data base, Kutzbach and Guetter report experimental results aimed at defining the density of the sampling network necessary to reconstruct mean-sea-level pressure from temperature and precipitation data. The methods employed were identical to those of dendroclimatic reconstruction; the results are, therefore, relevant to the design of dendroclimatological sampling networks.

It is unlikely that all the problems inherent in the statistical modelling of the climate-growth relationship will ever be solved, but reproducible results are a strong indication of reliability. Reproducibility can be assessed by analysing subsets of the data in the case of a single site, and by comparing the results of response functions for a large number of sites in a particular region. Likewise for the transfer function; subsample replication in space as well as time is important. The number of variables to be predicted needs to be balanced carefully with considerations of reproducibility and stability. The response function, or preferably the pooled results of a large sample of response functions, should be used as a guide to the design of the transfer function model: for example, how many and which climate variables to include, which seasons to select, and so on.

Because of the sophistication of the statistical techniques employed, critical evaluation of the results,

in terms of statistical, biological and physical significance, is of paramount importance. Verification of reconstructions is needed not only to place dendroclimatology on a firm scientific basis but also to assure potential users, mainly in the climatological community, of the validity and reliability of the reconstructions. The requirement for thorough verification must influence the entire reconstruction process from the design of the sampling network through to the design of the transfer function model. A well-verified but small set of proxy climate data is preferable to a large set of data of questionable quality. The documentation of verification results alongside the publication or dissemination of reconstructions is essential.

Most of the techniques discussed in this chapter were designed for ring-width data rather than density or isotope measurements. This reflects the relative progress in the use of these three types of tree-ring data to produce climate reconstructions. Although standard ring-width techniques have been applied to density measurements, it may be necessary to develop alternative techniques for parameters other than ring width. Techniques should also be developed which can handle more than one type of tree-ring measurement and, possibly, other proxy or historical climate data as additional predictors.

There has been a lively debate concerning the use and interpretation of the response function in recent years. While this is due in part to the statistical complexity of the techniques employed, it is largely a result of the expansion of dendroclimatology into many new areas with species, site types, and climate different from those for which the original techniques were developed. The extension of the basic dendroclimatological techniques to new regions has led to a necessary reassessment of the validity of the methodology and to many modifications and improvements. As dendroclimatology progresses in these new regions, it is likely that the debate, which is characteristic of the healthy diversity of a maturing discipline, will continue.

THE CLIMATE—GROWTH RESPONSE
H.C.Fritts
INTRODUCTION

Many tree-ring features can be correlated with climatic variations. These features include characteristics such as ring width, the proportion of earlywood and latewood, cell size, wood density, and chemical isotope ratios. How ring features become correlated with climatic factors and why certain dendrochronological procedures are used to enhance the tree-ring record of climate are now described.

There are basically two different types of climate response. One deals with chemical isotopes of carbon, hydrogen, and oxygen found in wood and involves temperature-dependent fractionation mechanisms affecting the isotopic ratios of water, oxygen, and carbon dioxide. These mechanisms, particularly those occurring within the plant, are still under scrutiny (Long, this volume; Wigley, this volume). The second type involves structural characteristics related to well-known environmental factors that limit plant processes such as photosynthesis, transpiration, root absorption, and transport. The ring characteristics most often studied by dendroclimatologists fall within this type, that is, ring width, earlywood width, latewood width, wood density, frost rings, reaction wood, and other related features of wood. The following sections explore this type of response and describe how climatic variations become recorded in a tree-ring chronology or, to state the problem in biological terms, how the characteristics of the ring can be influenced by variations in climate. More details and important references for the following discussion are found in Fritts (1976).

THE REALM OF CLIMATE FACTORS

It is conventional to view climate not as a static set of mean statistics describing the "normal" situation, but rather as a dynamic system of processes which not only control the means but also produce variations. Climate may be regarded as the aggregate of all meteorological phenomena that occur over a relatively long time period. Temperature, precipitation, pressure, wind, humidity, and sunshine can be considered factors of climate. It is common to speak of large-scale patterns of these factors as macroclimate, while the more local patterns surrounding a thermometer, person, tree, or leaf are considered microclimate. The magnitudes of variations in microclimatic factors can differ markedly from those of the macroclimate. However, the variations over time are more or less correlated in a statistical sense depending upon how well the microclimate is coupled to the macroclimate. For time periods over seconds and minutes, the coupling often is weak and the correlation low, but coupling and correlation generally increase for time periods from months to seasons, years, and decades.

Not all of the climate factors are important to a tree at any one instant. Only those that impinge upon or limit some process can affect growth (Mason & Langenheim, 1957). The well-known principle of limiting factors states that each process is governed by one factor at a time depending upon which one is available in least supply (Meyer et al., 1973). There are, however, a number of processes occurring in a plant organ such as a leaf, and each process can be limited by different conditions. Since there are many processes and parts of the plant, there conceivably can be more than one climatic factor limiting to an organism at a given time. Often other factors unrelated to climate, such as aging, essential mineral supply, competition, grazing, disease, and fire,

can be more limiting than climatic factors and impinge upon processes in the leaf. These factors will dilute any climate information recorded by the leaf and reduce its correlation with variations in macroclimate.

INTEGRATION WITHIN THE TREE

Thus far, we have considered the operational environment of a leaf (Mason & Langenheim, 1957), its microclimate, and the associated macroclimate. The operational environment of a leaf will influence the growth ring in the bole of the tree only by supplying foods, hormones, and other regulating substances which are transported out of the leaf through the stem and into the cambium. Other structures, such as the roots, contribute substances of different kinds. As these substances are translocated and distributed through the tree, they become mixed, and some of the more or less random variations among leaf and root microclimates and among different non-climatic factors are averaged out in the translocating process. It is this averaged stream of substances along with the temperature of the stem that becomes the cambium's operational environment.

The ring grows over a period of from one to several months within the year. Therefore, the ring character is determined by all substances or conditions that have limited cambial cell division, enlargement, and different-iation throughout the growing period. The growth-limiting conditions themselves are integrations of many different microclimates throughout the tree, and additional integrations occur as their effects are averaged over time. Consequently, ring characteristics in the bole of a tree are more likely to be correlated with variations in macroclimate than the environment of a single leaf. This occurs because of the averaging by the tree as a whole and also because of the stronger coupling between microclimates and macroclimate for long periods as compared to short periods of time.

Further integration occurs during the autumn and winter months as translocated foods and materials are stored away for future growth. Roots are produced, buds develop, and physiological changes occur such as winter hardening, which at a later date can prohibit or allow a factor to be limiting to growth. Factors operating in prior years also represent a level of integration which strengthens, alters, and further complicates the kind of response to climate (see Fritts, 1976, p.26). The multiple effects of the operational environments for a given year t can be translated into a ring-width response not only for year t but also for lags t+1 up to a theoretical lag limit of t+k. The climate of a growing season also affects the vigour of the tree as a whole, and more leaves, roots, and growing tissues, as well as wider rings are produced in favourable years. Less growth occurs in unfavourable years. Therefore, because the ring width in prior years is correlated with this prior vigour, and because of the lagging effects of climatic factors, the ring widths are autocorrelated.

However, I do not know of any direct physiological linkage or cause-and-effect relationship in which one ring width affects the next, except for the trend in width with time which is removed by standardisation. Autocorrelation in a standardised chronology is a statistic useful for time-series analysis; it is apparently not a physiological relationship. If autocorrelation is removed indiscriminately from tree-ring chronologies, meaningful climate information will be lost.

RESPONSE VARIATIONS IN ANATOMICAL FEATURES

Different ring features are formed at different times of the year and reflect various integrations of the environment. The earlywood width is a response to limiting environmental factors such as low temperature in spring as well as to those of the prior summer, autumn, and winter. Latewood width is more likely to reflect conditions in the spring and summer concurrent with the growing period, but information about other seasons is also present. The density of the last cells formed in the year (maximum density) is most likely to reflect the shortest time period restricted largely to summer and early autumn of the same year. However, the influence of stored substances and other preconditioning within the tree can be strong, so that even maximum density is correlated with limiting conditions at other time periods. Considerably more climate information is available if the features within the ring are studied along with ring width (Schweingruber et al., 1978a; Conkey, 1979b, this volume).

ENHANCING THE MACROCLIMATIC INFORMATION BY REPLICATE SAMPLINGS

All ring features vary around the circumference of a growth layer, and this variation is sampled by taking two or more cores from the stem. Variations between trees are sampled by obtaining materials from a number of trees. The law of large numbers (Wallis & Roberts, 1956) states that when large samples are obtained, the random differences among individuals (variations unrelated to macroclimatic factors in common with all trees) are more or less averaged out. If 10, 20, 30, or more trees are sampled from a given stand and habitat, the ring data can be averaged to obtain a more or less reliable estimate of the collective operational environments of the habitat. The average of many trees for each year is better correlated with the corresponding forest macroclimate than a single tree's record.

The quality and strength of the macroclimatic record (the climate signal) contained in a chronology partly depends upon how often microclimatic factors were limiting to growth-affecting processes in the sampled trees and how many trees and cores were included in the sample. There are analysis of variance techniques available (Fritts,

1976), and my students and I have used them to estimate the number of trees and cores that should be sampled to obtain a given level of confidence. These techniques are discussed by Cropper (this volume). However, if trees limited by contrasting operational environments, for example, trees from both wet and dry sites or north and south slopes, are sampled and the ring data averaged together, then the climate signal may be distorted or a portion lost. This is the rationale for sampling only one species in one particular type of habitat. Trees in other sites can be sampled, but the record should be treated as a different chronology because these trees are likely to have a different climate response.

TECHNIQUES TO MINIMISE NON-CLIMATIC RESPONSE

Even with careful selection, replication, and averaging, some non-climatic variation (noise) will remain in a chronology. Nevertheless, the importance of many non-climatic variations can be minimised. The ring widths of aging trees vary as a function of ring geometry, tree size, site productivity, and competition from neighbours. The literature includes a variety of standardisation procedures which are used to deal with this type of non-climatic variation (see Graybill, this volume). Evidence for fire or other factors damaging the tree can be noted, and the individual samples with marked effects of these non-climatic factors should be rejected for dendro-climatic analysis. The experienced dendroclimatologist uses a variety of selection procedures and field methods to obtain the strongest climate signal and to eliminate the most apparent non-climatic features in the growth record that would contribute to noise. There are other non-climatic factors that cannot be easily identified or are too small to be detected. Their effects occur more or less randomly and are simply dealt with by averaging the results from a large number of trees.

MEASURING THE CLIMATE RESPONSE

The complexity of the climate response is so great that we must use multivariate methods called response function analyses to study it. Temperature and precipitation for different months throughout the year are used as statistical predictors of ring width. Details of the procedure are considered elsewhere (Fritts et al., 1971; Fritts, 1974, 1976; Guiot et al., this volume). The size and sign of the partial regression coefficients (PRCs) are interpreted as a response of growth to the particular monthly climate variable. However, they are actually the result of the macroclimate's correlation with microclimates and the associated operational environments. I have plotted six response functions in a way designed to emphasise the kinds and degree of correlations that exist (Figures 2.1, 2.2, and 2.3).

The vertical axes in the figures correspond to the PRC values for temperature, and the horizontal axes, the

PRC values for precipitation. The position of the plotted
point represents the degree of correlation between growth
and the two variables for a particular month. All values
greater than the 95% confidence limits are designated with
vertical lines for temperature or horizontal lines for
precipitation drawn through the point. Points falling in
quadrant 1 (upper right) indicate high growth correlated
with warm, wet conditions; quadrant 2 (lower right), cool,
wet conditions; quadrant 3, cool, dry conditions; and
quadrant 4, warm, dry conditions.

The response functions in Figure 2.1 are for indices
of a drought-sensitive Pinus longaeva chronology at 3,000m
elevation in the White Mountains of California, U.S.A.,
and the average of six arid-site Pinus ponderosa chronologies in similar habitats at about 1,800m elevation in
Colorado. Figure 2.2 includes responses for a low-
elevation, old-growth temperate forest chronology of
Pseudotsuga menziesii from the coastal range in Oregon and

an eastern temperate forest chronology of Quercus alba
from Indiana. Figure 2.3 includes responses for one
mountain stand of Picea rubens at 945m elevation in Maine
sampled and analysed by L.E.Conkey (Laboratory of
Tree-Ring Research, University of Arizona). The upper plot
is for indices of maximum densities; the lower plot for
indices of ring width.

The wide scatter of points within the plots is
typical of response function results. Statistical uncertainties prevent an exact interpretation of individual
points, but the changing signs and magnitudes of the PRCs
from season to season are significant. With the exception
of Quercus alba where no negative precipitation PRCs are
evident, direct and inverse correlations for both climate
variables are significant in each response. For example,
the Pinus longaeva rings (Figure 2.1) appear to be wide
when the prior June and August are warm and prior July is
wet; October, November and May are wet and October,

Figure 2.1: The response functions for two ring-
width chronologies with coefficients plotted for
each month as a function of both temperature
(vertical axis) and precipitation (horizontal axis).
P stands for month of year prior to growth. Significant values for temperature are designated by a
vertical line and for precipitation by a horizontal
line drawn through the point. Upper plot is for
arid-site Pinus longaeva in the White Mountains of
California; lower plot is for the mean of six
chronologies of Pinus ponderosa along the front
range of the Rocky Mountains, Colorado (Fritts,
1976).

Figure 2.2: The response functions for two ring-
width chronologies with coefficients plotted for
each month as a function of both temperature
(vertical axis) and precipitation (horizontal axis).
P stands for month of year prior to growth. Significant values for temperature are designated by a
vertical line and for precipitation by a horizontal
line drawn through the point. Upper plot is for a
temperate forest stand of Pseudotsuga menziesii in
western Oregon and lower plot is for Quercus alba in
an open stand in northern Indiana (data unpublished
and from Fritts, 1976).

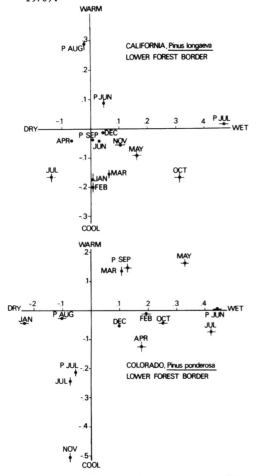

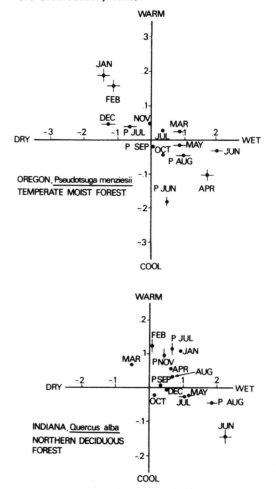

January, February, March, and May are cold; and July is
cool and dry. Two significant coefficients for temperature
indicate positive response, five are the reverse; four
significant coefficients for precipitation indicate
positive response, one is the reverse. There are no cases
(indeed, it seems to be a rare occurrence) where all
significant PRCs fall within one quadrant indicating a
similar response for all seasons. This confirms that a
variety of physiological processes are important. Since
statistical uncertainties are large in response function
work, interpretations of values for a specific response
should be tentative until confirmed with other types of
evidence or replicated response function results (Fritts,
1974).

CONCLUSION

I have attempted to point out why properly selected,
adequately replicated, and appropriately standardised

Figure 2.3: The response functions for two ring-
width chronologies with coefficients plotted for
each month as a function of both temperature
(vertical axis) and precipitation (horizontal axis).
P stands for month of year prior to growth. Significant values for temperature are designated by a
vertical line and for precipitation by a horizontal
line drawn through the point. Upper plot is for
indices of maximum density and lower plot is for
indices of ring-width, for Picea rubens on Elephant
Mountain, Maine (data courtesy L.E.Conkey,
Laboratory of Tree-Ring Research, University of
Arizona).

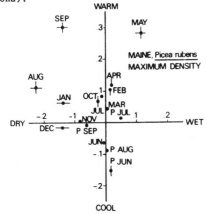

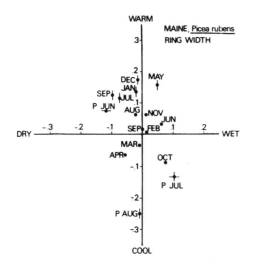

tree-ring chronologies will correlate with macroclimatic
factors that vary in kind from month to month and from one
season to the next. This causes the coefficients for
monthly temperature and precipitation to vary within and
among response function results. While part of this variation is due to statistical uncertainty, the remainder
can be attributed to differences in physiological responses. Therefore, the climate signal varies from one
chronology to the next, partly due to differences in
geographical locations and partly due to differences in
microclimates and varying physiological response. Because
of this variation in response, it is desirable to collect
a number of well-replicated chronologies representing
different ring features, different species, and
contrasting habitats throughout an area. The different
chronologies will probably contain different kinds of
information about variations in past climates and be
useful for a variety of problems.

COMMENT
M.K.Hughes & S.J.Milsom

In reconstructing past climate from tree rings our
basic tool is the complex of processes that records past
environmental conditions in annual rings of differing
sizes and properties. The better we understand this tool,
the more skilfully we can use it. Thus, the account given
above by Fritts, and more fully in Fritts (1976), is of
great importance to our work.

Measuring the climate response. Calculated response
functions do not measure the climate-growth response in a
simple sense. The climate-growth response links macroclimate at the sampled tree-ring site to tree performance
as recorded in the annual rings of a sample of trees.
Response function analysis produces more or less effective
predictive equations for tree performance. The nearest
available temperature and precipitation data are used as
the basis for predictors, the assumption being that they
give a close approximation to macroclimate as experienced
at the site. Hughes et al. (1978a) showed that the
percentage chronology variance reduced (%) changed from
34% climate and 31% prior growth to 45% climate and 28%
prior growth when rainfall data from 60km distant from the
tree site were replaced with data from 6km distant. This
increase was a result of improving the realism of the
response function input. In surveys of large regions, as
reported by Pittock (this volume), it is possible that
sites most stressful to trees will be distant from
meteorological stations. Consequently, there is a greater
likelihood of the percentage variance reduced by the
response function being an underestimate of the climate-
growth response.

Response functions do not measure the climate-growth
response but rather the effectiveness of a particular
statistical model at predicting that part of tree-ring

variation believed to be forced by factors external to the
stand. In the absence of a clear indication of the importance of other factors, the best available measure of
the importance of the climate-growth response in a
chronology comes from the analysis of variance techniques
described by Graybill (this volume). The greater the
climate-growth response, the higher the proportion of
variance attributable between years. Hughes et al.
(1978a) report the occurrence of opposite trends in mean
index and correlation among trees as well as a higher
percentage variance attributable between years in a
climatically variable, as opposed to an equable, 25-year
period.

Describing the climate response. Whilst response functions
do not necessarily give a good measure of the full extent
of the climate-growth response, they are a useful
descriptive tool in dendroclimatology. Chronologies
produced in different ways in the same stand may be
compared and, given that the same meteorological data are
input, guidance gained on effective sampling tactics. For
example, Milsom (1979), working at the same site as Hughes
et al. (1978a), compared chronologies from a group of 10
trees. The trees were sampled at a number of heights.
Chronologies taken at about 0.5m (14 cores) and 4.0m above
ground (20 cores) were compared. Chronology statistics
were broadly similar (Table 2.1), including percentage
variance attributable between years, but the effectiveness
of the response functions and the role of climate in them
differed. While percentage chronology variance reduced for
the 0.5m chronology was 33% climate and 38% prior growth,
that for the 4.0m chronology was 53% climate and 7% prior
growth. The form of the precipitation and temperature
elements of the two functions was similar (Figure 2.4).
The 0.5m chronology corresponds to the type most often
sampled in oaks in the United Kingdom, namely, slices from
stumps of felled trees or cores taken at and below breast
height. The response functions suggest the use of step-
ladders in sampling!

Table 2.1: Statistics for 0.5m and 4.0m chronologies.

	0.5m	4.0m
ANOVA period	1860-1910	1872-1931
Mean serial correlation	0.448	0.483
Mean sensitivity	0.291	0.236
Mean correlation between trees	0.524	0.506
% variance: years	42.06	39.58
% variance: years x trees	23.87	22.62
% variance: years x trees x cores	34.07	37.80
Error of Y	0.089	0.072

A further, more common, use of response functions is
to indicate months and variables for which the response is
particularly strong or particularly consistent. It is
assumed that these combinations of months and variables
will be those most effectively reconstructed (see
LaMarche & Pittock, this volume). For example, response
functions for five replicated oak chronologies in the
British Isles all show significantly positive temperature
elements in the October prior to growth (Table 2.2).
Similarly, they all show significantly positive temperature elements at the time of bud-opening and earlywood
formation in April and/or May. Four show significantly
negative temperature elements in December and all show
some significant elements for June or July rainfall in one
or other year. Whether these results indicate the
variables most suitable for reconstruction remains to be
tested.

RESPONSE FUNCTIONS
J.Guiot, A.L.Berger & A.V.Munaut
INTRODUCTION
The aim of the response function in dendroclimatology is to analyse the influence of climate on the annual
growth of trees. Because climate parameters are often
highly intercorrelated, classical multiple regression
cannot be applied directly. A simple trial shows the
failure of this method. Since Fritts et al. (1971), the
use of the powerful technique of principal components has
contributed to significant progress in dendroclimatology.
The principal components provide orthogonal variables on
which the regression can be performed. Although the choice
of this general method is not a problem, its application
is still controversial. The purpose of this discussion is
to compare different possibilities for using principal
components and to compare them with a new method based on
spectral analysis. This last method has been developed to
counteract one of the weaknesses of linear regression.
Linear multiple regression does not distinguish between
short-term climate variations (such as interannual
variability) and longer term variations. This new method
divides the frequency spectrum into bands and performs a
regression for each band.

In order to compare these different methods, a cedar
site in Morocco (site 301), where climate has been shown
to be a sufficiently limiting factor (Berger et al.,
1979), has been used. It is located at Tleta de Ketama in
the Rif and is presented in Munaut et al. (1978). The
tree-ring widths have been detrended through INDXA
(Graybill, this volume) and transformed into a master
chronology of indices extending from 1800 to 1975:
Index(t). The meteorological data are from the Tetouan
Airport (5°E, 36°N) and extend from 1941 to 1971.

SURVEY OF THE DIFFERENT METHODS
Fritts (1976) shows how the response of the tree to

climate can be estimated. Although multiple regression is a better technique than correlation analysis, it is not perfect. One of its assumptions requires that the predictors be uncorrelated and this is not often the case. For example, July mean temperature and August mean temperature in Tetouan have a correlation coefficient of 0.80 (significance level < 0.001). A consequence is the instability and variability of the standardised regression

coefficients in the response function. Furthermore, some of them, which, in fact, are partial correlation coefficients, become larger than one. In order to compare these methods, the following 24 predictors are used: monthly mean temperatures (T) and monthly total precipitation (P) from prior October, Oct(t-1), to current September, Sep(t).

The first method is stepwise multiple regression:

Figure 2.4: Response functions for ring-width chronologies at 0.5m and 4.0m above ground. The chronologies are for two levels in the same ten trees at Maentwrog, North Wales. The vertical lines through each element show 95% confidence limits.

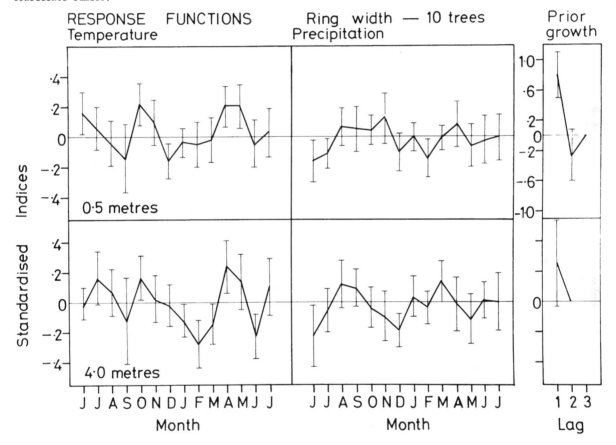

Table 2.2: Significant response function elements for British Isles oak chronologies.

SITE		TEMPERATURE			PRECIPITATION			PREVIOUS GROWTH
	MONTH	J J A S O N D J F M A M J J			J J A S O N D J F M A M J J			1 2 3
1. Raehills		P P P			N P P N			P
2. Rostrevor		P P P N N N P			P P P P P			P
3. Coed Crafnant		N P N N N P P N			N P P P P N P			P
4. Maentwrog		P N N N P P N			N N			
5. Peckforton		P P N N P			P N P P P P P			

P indicates a positive element significant at P < 0.05 and N indicates a negative element significant at P < 0.05.

Authors

1,2 Hughes et al. (1978b)
3   Luke, personal communication
4   Milsom (1979)
5   Leggett, personal communication

the predictors are introduced step by step. In a given step, a predictor is entered if its partial correlation with the dependent variable is significant at the level called "F to enter". In the following steps, it is removed if its partial correlation coefficient becomes non-significant at the level called "F to remove". The level F used here is defined as the probability of error when the hypothesis of null partial correlation is rejected. The results for Site 301 are given in Figure 2.5.

The second method uses principal components. The correlation matrix of the predictors and the eigenvectors of this matrix, the uncorrelated principal components, are computed. A multiple regression is performed on a limited number of the principal components and, finally, some calculations permit a return to the initial predictors. More details are given in Fritts (1976). The problem of this method is to know how many principal components to introduce. If all the principal components are used, the method reduces to a classical multiple regression (Figure 2.6a). Some components originate from errors or inaccuracies in observation or measurement and must be eliminated or else they will introduce instability in the estimation of the regression coefficients. The smallest principal components usually correspond to an eigenvalue (or variance) which is almost equal to zero.

Some of the components must be eliminated using criteria related to a fixed level. As this level is often arbitrarily chosen (and sometimes selected on a "trial and error" basis), we have tried to build a new procedure to define it objectively. It is based on the fact that, when the predictors are independent, the product of the eigenvalues is 1.0. As a consequence, it has been decided to select the principal components of which the cumulative eigenvalues product (PVP, sometimes abbreviated as CEP) is greater than one (Berger et al., 1979). The results are shown in Figure 2.6b. In this selection, some principal components have a non-significant weight on the dependent variable and hence they can be rejected. Fritts (1976) uses a stepwise procedure to enter the useful principal components. Because they are uncorrelated, it uses the

principal components which have a significant correlation with the dependent variable at a level PROB. This is done in Figure 2.6c with PROB = 0.50 (median). The level PROB used here is defined as the probability of rejecting the true hypothesis of null correlation when it is false. Different trials have shown that this value is the best level when the number N of observations is small (Guiot et al., 1979); N = 30 here. In fact, 0.50 is recommended when N is almost equal to the number of predictors and 0.80 or 0.90 when N is much greater than this number.

The third method (the "rejection" method) consists of first selecting all the variables in the regression and then deleting the least significant in a stepwise fashion. This method combined with the second one is even more powerful and will be illustrated in a later section.

The "stepwise" and "rejection" methods still have a fault: they lead only to a subset of variables selected from among the input variables. They do not really give a response function, because they are not able to describe any relationship between the removed variables and tree rings. The second method, which uses some of the principal components of all the climate variables, is very good. If the levels PROB and PVP are well defined, this leads to properly estimated and stable regression coefficients.

Figure 2.6: Response function of Index(t) on 24 climate parameters (monthly temperatures and precipitation from Oct(t-1) to Sep(t) with regression after extracting principal components: (a) all components are introduced, $R^2 = 0.99$, $F = 13.7^{**}$, 24 components; (b) all the components with PVP > 1.0 are introduced, $R^2 = 0.81$, $F = 2.6^{**}$, 18 components; (c) all the components with PVP > 1 and PROB > 0.50 are introduced, $R^2 = 0.78$, $F = 7.3^{**}$, 9 components.

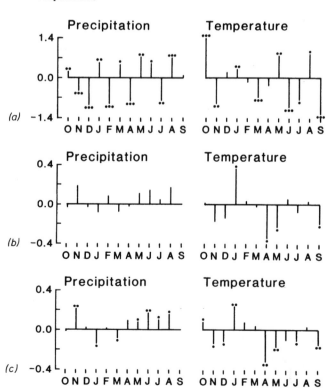

Figure 2.5: Regression of indices (Index(t)) on 24 climate parameters (monthly temperature and precipitation from Oct(t-1) to Sep(t)) through stepwise multiple regression with F to enter = 0.10 and F to remove = 0.15. The vertical axis gives standardised coefficients. P.G. = prior growth. Dots indicate significance at the 0.05^{*}, 0.01^{**}, or 0.001^{***} level. $R^2 = 0.77$, $F = 20.7^{***}$.

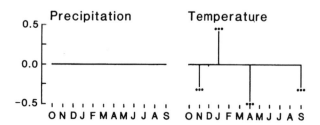

These coefficients are consistent with the stepwise method, whereas the classical multiple regression sometimes
gives opposite coefficients: for example, P[Jul(t)] and
T[Jul(t)]). In addition to the problem of method
selection, the a priori selection of input variables needs
to be discussed. When the serial correlation of the
tree-ring indices from Site 301 is investigated (serial-r
= 0.45), the effect of persistence is significant at the
level 0.001. It is thus necessary to take into account the
growth of previous years. The different ways of
considering persistence and the input climate variables
are the subject of the following paragraphs.

Choice of the global set of input predictors. The
relatively high value of the serial-r of the cedar
chronology suggests that climate may influence tree-ring
width over one year or more. In our case, additional tests
(Guiot et al., 1979) have shown that persistence is
limited to one year. Thus Index(t-1) has been introduced
as a predictor. This can be performed in different ways.

The first is to introduce Index(t-1) as an input
predictor on the same level as the climate variables and
then to compute the principal components of these 25
predictors (Figure 2.7a). A comparison with Figure 2.6c
does not show large differences for the climate parameters, but Index(t-1) has a significant weight in the
response function.

The second method is to compute principal components
of the climate parameters and to put Index(t-1) on the
same level as these orthogonal functions. This is the
technique described in Fritts (1976). This method (Figure
2.7b) gives its true importance (that is, the coefficient
almost equal to the serial correlation) to Index(t-1), but
the weights (regression coefficients) of the climate
parameters are not really different from the weights given
in Figure 2.7a.

The third method (Figure 2.7c) was reported by
Berger et al. (1979). The persistence of the whole
chronology is taken out before starting to compute the

Figure 2.7a: Response function of Index(t) on 24
climate parameters and Index(t-1). The principal
components are computed on the 25 predictors. R² =
0.77, F = 9.0***, 8 components.

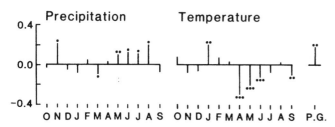

Figure 2.7c: Response function of Index*(t) on the
24 climate parameters. Index*(t) is computed through
a simple regression of Index(t) on Index(t-1). The
period used to compute Index*(t) is 1800-1975, while
the response function is computed on the period
1941-1971. R² = 0.74, F = 7.3***, 8 components.
R²_TOTAL = 0.79.

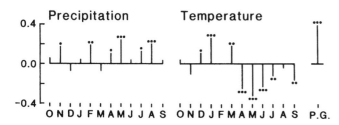

Figure 2.7b: Response function of Index(t) on 24
climate parameters and Index(t-1). The principal
components are computed on the 24 climate parameters
only: Index(t-1) is taken on the same level as the
components. This method follows Fritts (1976). R² =
0.76, F = 8.5***, 8 components + Index(t-1).

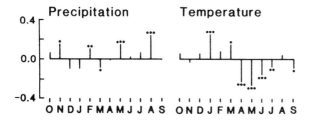

Figure 2.8a: As Figure 2.6c but with 32 climate
parameters. R² = 0.77, F = 7.4***, 9 components.

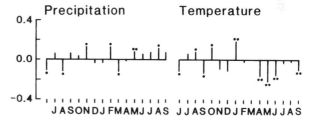

Figure 2.8b: As Figure 2.7a but with 32 climate
parameters. R² = 0.74, F = 11.1***, 6 components +
Index(t-1).

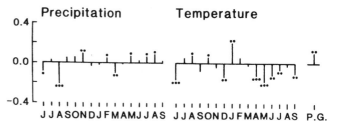

Figure 2.8c: As Figure 2.7c but with 32 climate
parameters. R² = 0.74, F = 6.4***, 9 components.

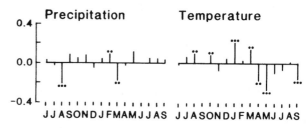

response function. This leads to a chronology directly
related to the climate. Here too, the difference between
Figures 2.7a and 2.7b is not significant. The coefficient
of Index(t-1) is also similar to the serial-r and,
furthermore, the coefficient of determination, R^2_{TOTAL}, is
the highest. This method thus seems to be preferable
statistically. The separation of climate and persistence
is also preferable biologically (but see Fritts, this
volume).

A fourth possibility on the selection of input
variables is to introduce the climate of the prior growth
seasons: for example, Jun(t-1), Jul(t-1), Aug(t-1), and
Sep(t-1). This leads to a consideration of 32 climate
parameters. Using this set of input variables, we have
applied the three methods (Figures 2.8a, 2.8b, 2.8c) which
are related, respectively, to Figures 2.6c, 2.7a, 2.7c.
Among the 24 climate parameters which are common to all
these figures, no large differences can be detected.
Moreover, when the parameters related to year t-1 are
introduced, the explained variance of the indices is not
increased. Thus, the importance of these additional
variables does not seem real. Nevertheless, some weights,
P[Aug(t-1)], T[Jun(t-1)], T[Aug(t-1)]... , are significant; it can be seen, for example, that the weight of
P[Aug(t-1)] is opposite to the weight of P[Aug(t)]. It is
the same for P[Jun]. This may be related to a biological
phenomenon, but it can also be explained by the interactions between Index(t-1) and the climate parameters of
the previous year. It is often difficult to decide which
variables have to be selected and kept to build the final
response function. It is for this reason that another
method is needed.

Selection of the most significant variables.   The
regression coefficients of Figure 2.7c are not all
significantly different from zero at the 0.10 level.
Moreover, the stepwise procedure described earlier allows
the determination of the most significant predictors. For
the indices

$$\text{Index*}(t) = \text{Index}(t) - 0.45 \text{ Index}(t-1) - 0.55 \quad (2.1)$$

Figure 2.9a: Evolution of R^2 in the reject
procedure.

1 → 12: P[Oct(t-1)] → P[Sep(t)]

13→ 24: T[Oct(t-1)] → T[Sep(t)]

See text for explanation.

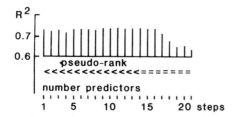

the most significant predictors are, respectively,
P[Aug(t)], T[Jan(t)], T[Apr(t)], T[May(t)], at the levels
0.10 (F to enter) and 0.20 (F to remove). But, between the
step where the 24 climate parameters are used and the step
where only four climate parameters are selected, it is
interesting to know the order of importance of the
predictors. In order to do that the rejection method is
useful in combination with the regression on principal
components. At the first step, the least significant
predictor is deleted, except for the four pre-selected
variables, and the regression after extracting principal
components is computed with the 23 remaining predictors.
Then this procedure is iterated until the minimal
selection of these four parameters is achieved. The
evolution of R^2 is given in Figure 2.9. The number of
principal components with PVP greater than 1.0 is a
picture of the rank of the matrix of the predictors,
sometimes called the pseudo-rank (Lawson & Hanson, 1974).
Since where the pseudo-rank is equal to the number of
predictors, the predictors are sufficiently independent,
step 14 seems to be a very characteristic step. Step 19,
where all the predictors are significant at the 0.10
level, is also interesting. The weights characterising
these steps are represented in Figure 2.9b. After step 14,
R^2 decreases. This illustrates that all the information
given by the 24 predictor variables can be effectively
summarised by 11 principal components. Another proof of
the importance of PVP is the compatibility of the weights
in Figure 2.9b with the weights in Figures 2.7c, 2.6b, or
2.6c, but not always with the weights in Figure 2.7a.

After the method has been chosen, the climate

Figure 2.9b: Standardised coefficients in characteristic cases of Figure 2.9a Step 14 where the
number of components with PVP > 1.0 is equal to
number of variables, R^2 = 0.75, F = 7.6**, 8
components, and Step 19 where all the predictors
have a significant weight on the dependent variables
significant at level 0.99, R^2 = 0.65, F = 16.1***, 3
components.

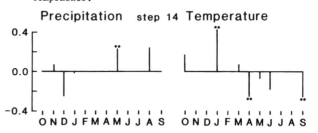

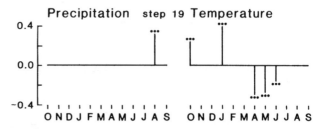

parameters selected, and persistence taken into account, the response function which is obtained can be considered to be reliable and stable, but is it complete? Another program must be designed in order to analyse the residuals of this function.

A NEW RESPONSE FUNCTION PROGRAM

The aim of this new program is to separate the different influences acting on tree growth. First, it tests if the tree-ring chronologies and the climate parameters have the same trend and if they are homogeneous. These tests of trend (stability of the mean) and homogeneity (stability of the dispersion) are described in Sneyers (1979). The results related to our example are given in Table 2.3, where it can be seen that some

Table 2.3: Probability of error α_i when the H_o : no trend (no dispersion) is rejected for monthly precipitation, monthly temperatures (Tetouan), and indices for site 301 (period: 1941-1970). A global test allows the rejection of the hypothesis of trend (dispersion) of climate. $X \sim \chi^2$ with 2k d.f. where k = no. of terms in the sum X.

	PRECIPITATION		TEMPERATURE	
	TREND	DISPERSION	TREND	DISPERSION
J	0.46	0.36	0.06	0.30
F	0.20	0.24	0.84	0.20
M	0.62	0.24	0.76	0.28
A	0.86	0.01 (−)	0.24	0.68
M	0.86	0.20	0.92	0.84
J	0.02 (+)	0.36	0.16	0.62
J	0.002 (−)	0.06	0.68	0.99
A	0.10	0.04 (−)	0.92	0.24
S	0.54	0.76	0.92	0.76
O	0.48	0.76	0.84	0.54
N	0.20	0.42	0.62	0.10
D	0.68	0.92	0.36	0.76
$X=2\Sigma \ln\alpha_i$	37.8 (1)	37.2 (2)	17.6 (3)	19.6 (4)

Probability

$(\chi^2 > X)$	< 0.05	< 0.05	>> 0.10	>> 0.10
Climate	> 0.10 (1)+(3)	> 0.10 (2)+(4)		
Indices	~0.25	~0.75		

parameters have a significant trend and some are significantly inhomogeneous at the 0.05 level. A "global" analysis shows an instability of the mean and of the dispersion for the precipitation data only. In fact, the instability of the dispersion is mostly present for the dry months (P[Apr(t)], P[Aug(t)]) and, as the accuracy for light precipitation is very poor, it is impossible to draw any conclusion. For the trend, it is also difficult to draw any conclusions because sometimes it is positive (P[Jun(t)]) and sometimes it is negative (P[Jul(t)]). For the temperatures, the agreement in stationarity and homogeneity between the temperature data and tree indices is shown by the lack of significant trend and significant change in dispersion.

The persistence tests for precipitation and temperature data do not provide significant results (Table 2.4), but those for the tree-ring chronology do. It is, therefore, necessary to take this into account and this is the second step of the procedure (Figure 2.10). Here, the elimination of this persistence in tree rings is done using the period 1941 to 1971 over which the climate variables exist and not the whole chronology for 1800 to 1975. It is performed through a simple regression

$$Index(t) = 0.63 + 0.38 \; Index(t-1) + Index^*(t) \quad (2.2)$$

As the residuals of this relation, Index*(t-1), are not significantly autocorrelated at the 0.05 level, they may be related to the climate.

The predictors will be the 24 climate parameters used earlier: monthly total precipitation and monthly mean temperatures from previous October to current September. Twenty-four weights and residuals, RES(t), are so obtained. As these residuals may be correlated with other previous parameters like precipitation and temperature of Jun(t-1), Jul(t-1), Aug(t-1), Sep(t-1), the partial correlation coefficients between these eight parameters and RES(t) must be compared. In this case, none of these correlation coefficients is significant at the 0.05 or 0.10 levels. As only some are significant at the 0.20

Table 2.4: Serial correlation for the 24 climatic parameters (Tetouan); global test (as in Table 2.3) for climate and test for indices chronology of site 301.

	J	F	M	A	M	J	J	A	S	O	N	D	
P	0.17	0.11	−0.15	0.30	0.24	0.02	−0.18	−0.20	−0.04	−0.28	0.12	0.09	Serial-R
	0.26	0.44	0.52	0.06	0.14	0.76	0.44	0.36	0.98	0.18	0.40	0.52	α_i
	0.03	0.05	0.19	0.06	0.23	0.36	0.05	0.33	−0.02	0.11	0.22	0.21	Serial-R
T	0.74	0.92	0.22	0.62	0.14	0.04	0.64	0.04	0.96	0.44	0.18		α_i

$X = -2\Sigma \ln\alpha_i$

P	27.0	(~0.25)
T	31.0	(~0.10)
C	58.0	(~0.10)
I	9.2	≈ 0.01

T = Temperature, P = Precipitation, C = Climate, I = Indices chronology, α = probability of error when hypothesis of no persistence is rejected.

level, it must be concluded that the influence of the
climate parameters is limited to one year. Figure 2.10
shows that a wrong interpretation could exist if the 32
parameters had been used first. Indeed, some of the
supplementary parameters have a significant weight while
the procedure described above (Step 3, Figure 2.10) leads
to the opposite conclusion. It is thus important to
proceed step by step and the new response function program
outlined here seems to be ideal for doing that.

SPECTRAL MULTIVARIATE ANALYSIS

The methods used until now are linear and thus they
can be responsible for losing a part of the available
information. Nature is more complex and feedback mechanisms are generally present. For example, an unfavourable
year in a favourable period does not have the same effect
as when it occurs in an unfavourable period. It is thus
important to make a distinction between short-term

climatic changes (interannual variability) and long-term
climatic variations.

Digital filtering provides us with adequate techniques for analysing these effects. A high-pass filter is
used to eliminate the low frequencies and a classical
least squares regression is computed between the remaining
high frequencies of the dependent variable and the high
frequencies of the predictors. A low-pass filter
eliminates the high frequencies, and, because of the
problem of autocorrelation of the residuals, a method of
regression based on generalised least square and principal
components is applied to the resulting series. The two
filters used are complementary, so the sum of the high
frequencies and low frequencies gives back the initial
series. If particular bands of frequencies must be
explored, it is possible to split the series in several
frequency bands. More details about this method are
provided by Guiot (1980a).

This technique has been applied to the analysis of
the indices of Site 301 using the 24 climate variables.
The filters used have a cut-off period of four years. The
characteristics of these filters and the response
functions are given in Figure 2.11. The interpretation of
the response function in the high frequencies is the same
as for the preceding response functions. It describes
relationships between the interannual variability of
tree-growth and climate variables. In the high-frequency
response function, precipitation has a positive influence
throughout the year. The low-frequency response function
reveals a different effect, namely, that the persistence
of wet and warm conditions in late winter and spring for
several years will be associated with lower tree-ring
indices. Previous techniques could not reveal such
relationships.

To summarise, the high-frequency relations define
which climate in comparison with the mean of the period is
beneficial for the trees and the low-frequency relations
help to define a favourable period for the trees. For
example, the period 1962 to 1970, with wet autumns, warm
and dry early winters, cold and dry springs, and cold and
wet summers, is definitely a period favourable to the
growth of the trees. This conclusion was not at all
evident from the results obtained through preceding
methods and it is particularly true for the role of
precipitation whose regression coefficients were often
non-significant. Nevertheless, this method is very
efficient with a reduced number of predictors as shown in
Guiot et al. (1981). This new program thus allows the
clarification of the climatic response of the tree.
Moreover, it is also very efficient in reconstructing past
climate, as demonstrated in Guiot et al. (1979). The only
problem is to have sufficiently long instrumental climate
series for the reliable analysis of the long-period
component.

Figure 2.10: Response function of indices on precipitation and temperature in Tetouan computed in
three steps. Step 1: regression of Index(t) on
Index(t-1) and computation of residuals, Index*(t) =
Index(t) - 0.38Index(t-1) - 0.63, R^2 = 0.16, F =
5.1**, DF (degrees of freedom) = 28. (a) Step 2:
regression of Index*(t) on 24 climate parameters.
Precipitation and temperature from Oct(t-1) to
Sep(t) and computation of residuals RES(t), R^2 =
0.72, F = 7.1***, 7 components, DF = 19; (b) Step 3:
regression of RES(t) on precipitation and temperature from Jun(t-1) to Sep(t-1), R^2 = 0.13, F = 0.5,
4 components, DF = 14; (c) response function
computed with fusion of Steps 2 and 3 (same method
as in Figure 2.8c), R^2 = 0.71, F = 8.0**, 6
components, DF = 20.

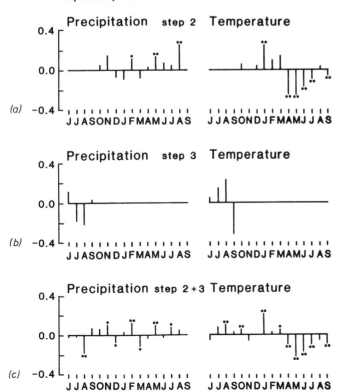

REDUCING THE NUMBER OF PREDICTORS

In all these methods, the high number of predictors
is a problem. To alleviate this difficulty, the months
which have a similar influence on the growth can be
grouped in seasons. This can be done on the basis of the
preceding response functions. This seasonal grouping
reduces the number of predictors, eight in our case, in
such a way that it is now possible to include climate
parameters related to the year prior to growth, and so to
compute the response function without introducing
Index(t-1). We have used a water budget parameter, defined
by the potential evapotranspiration estimated by
Thornthwaite's (1948) method. The response function
obtained explains 63% of the growth variance. For further
details, see Guiot et al. (1981).

CONCLUSION

In conclusion, in order to analyse the climate-
growth response, the best method seems to be regression
after extracting principal components, but it is equally
important to proceed step by step. Trend and persistence
in the tree-ring chronology can first be analysed and both
must be eliminated before the climate response is studied.

Figure 2.11a: Frequency responses of the high-pass
and low-pass filters used in the spectral multivariate regression.

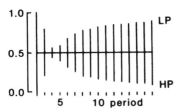

Figure 2.11b: Response function of the high-
frequency component of Index(t) on 24 climate
parameters (monthly temperature and precipitation
from Oct(t-1) to Sep(t)), $R^2 = 0.87$, $F = 6.0^{**}$, 14
components.

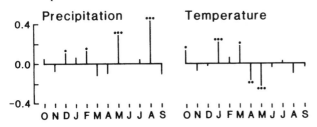

Figure 2.11c: Response function of the low frequencies through the generalised least squares,
$R^2_{TOTAL} = 0.97$. $R^2 = 0.98$, $F = 18.7^{***}$, 15 components.

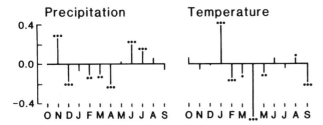

As a result of the statistical interference between
variables, we think that it is dangerous to introduce
directly more than one year into the response function.
If the residuals so obtained are thought to be related to
previous years' climate parameters, it seems statistically
more appropriate to compute a regression between these
parameters and the residuals. It is interesting to know if
some predictors can represent the largest part of
tree-growth variance. To answer that question, a
combination of stepwise multiple regression, regression
after extracting principal components, and rejection of
the least significant predictors is useful.

An important limitation of all these methods is that
they do not allow for differentiation between climate
parameters that induce a fast (high-frequency) response of
tree growth and those that may have a long lasting
(low-frequency) effect. This is why the multivariate
spectral analysis has been introduced. This new technique
is able to display the influence of some important
parameters not shown by other methods.

COMMENT

D.W.Brett

Although the general method for deriving response
functions (Fritts, 1976) has been shown to be applicable
to trees in the British Isles, some features of the
published functions for elm (Brett, 1978) and oak (Hughes
et al., 1978b) indicate that some modification of the
method may lead to more realistic results. In particular,
response functions based on the 14-month period covering
prior June to current July and including three prior
growth indices in the regression, Index(t-1), Index(t-2),
Index(t-3), show the sort of interaction between the
independent variables which leads to inflated, usually
negative, coefficients for weather during the prior
growing season and inflated coefficients for the prior
growth indices. This problem has been discussed by Fritts
(1976, p.382) in connection with response functions for
<u>Pinus sylvestris</u> in Sweden and Guiot et al. suggest ways
to overcome it. The weather variables relating to the
previous season may be omitted from the analysis
altogether, but there are likely to be some persistent
influences of the previous season that are not manifest in
the width of the ring formed that year, so the prior
growth season cannot be ignored. I support the suggestion
that only Index(t-1) be used for prior growth; in my own
trials, Index(t-2) or Index(t-3) only enter the stepwise
regression when the period of analysis includes the
weather for the prior growing season.

The response functions referred to here have been
computed using four modifications of the GENSTAT program
outlined in Brett (1978). In the original version (A),
which follows Fritts (1976), one or more vectors for prior
growth, Index(t-1), Index(t-2), and so on, are offered to
the stepwise regression with the principal components of

the weather variables; the second version (B) includes
Index(t-1), Index(t-2), and so on, in the principal
components analysis along with the weather variables; the
third version (C) follows Berger et al. (1979) and uses
the residuals from a preliminary regression of Index(t) on
Index(t-1) in the stepwise regression as dependent
variable instead of Index(t). A fourth version uses the
reciprocal digital filters given by Fritts (1976, p.270).
At present, this version (D) otherwise follows C and only
the high frequencies have been used for response
functions, that is the variance in the indices and weather
with wavelengths less than approximately eight years. Any
of the GENSTAT options may be used for the stepwise
regressions: for example, adding the best predictors,
fitting all the predictors and dropping the worst,
minimising by adding or dropping. The elm chronology used
in based on 10 trees from London parks (Brett, 1978), the
oak chronology is a new one and is based on 10 trees from
Nashdom Abbey, near Burnham in Buckinghamshire (about 24km
west of London); precipitation and temperature data are
for Kew, Surrey.

Essentially similar results are obtained for the
period Oct(t-1) to Sep(t) with programs A, B, and C:
examples for elm are shown in Figures 2.12 and 2.13.
Larger differences will result if three prior growth

indices are used with B since there will always be weights
for all three in the response functions; when only
Index(t) is used, the results from A and B are alike. The
12-month period referred to is the one favoured by Guiot
et al. and seems to be generally suitable for British
trees which are unlikely to be forming wood in October but
may be in September, at least in some years. I measured a
considerable increase in diameter of an elm through
September 1976 after prolonged drought was broken at the
end of August. Much of this expansion must have been due
to wood formation after the initial rehydration. We still
need much more information about the growth of the trees
we use for dendroclimatology.

In fact, we are faced with the problem of choosing
Sep(t) or Sep(t-1) for the analysis if we wish to avoid
including both, yet there might be a post-growth effect
associated with the prior September weather. In a full
16-month analysis, Jun(t-1) to Sep(t), significant
positive weights result for Sep(t) precipitation and
temperature and a significant negative weight for Sep(t-1)
temperature using programs A and B. Using the two stage
analysis, with the residuals from the Oct(t-1) to Sep(t)
analysis being used for the regression with Jun(t-1) to
Sep(t-1) weather (program C), no significant weights are
obtained for the prior season weather; the weights for
Sep(t) precipitation and temperature, however, remain as
in the full analysis. These results refer to elm. The oak
chronology gives a significant result for the prior
season's weather in the second stage of analysis using the
C program: a negative weight for Jul(t-1) and positive for
Aug(t-1) (Figure 2.14).

Guiot et al. recommended that only those principal
components down to the cumulative eigenvalue product > 1.0
be used in the regression. In the normal stepwise regression routine, the independent variables (the scores of the
principal components of the weather variables in the
response function programs) will be selected in order of
the rank of their correlation with the dependent variable
(ring-width index) and will enter the regression so long

Figure 2.12: The response functions shown in Figures
2.12 to 2.17 are for London elm or Burnham oak,
precipitation and temperature data are for Kew,
Surrey. P.G. is prior growth (Index(t-1)). Dots
indicate significance at the 0.05˙, 0.01˙˙, or
0.001˙˙˙ level. The period of analysis is 1902-1950
except for Figure 2.16 (1906-1950) and Figure 2.17
(1907-1950). This figure shows the response function
for elm, calculated using program A. 24 climate
components and 3 previous growth vectors available
to the regression. Step 11 shown at F = 9.23˙˙,
DF = 35, R^2 = 0.74 (0.69 climate). P.G. is
Index(t-1).

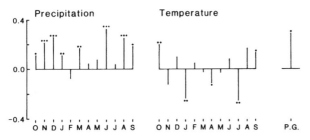

Figure 2.13: Elm, program C. P.G. is autoregression
coefficient. 24 climate components available to
regression in this stage. Step 9 shown at
F = 10.3˙˙, DF = 34, R^2 = 0.71 (climate).

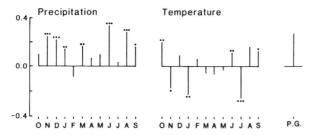

Figure 2.14: Elm, principal components restricted to
PVP > 1.0. This response function was computed with
program B (I(t-1) in principal component analysis)
but program A gives similar results with restricted
components. The major differences between this
result and Figures 2.12 and 2.13 are effectively
caused by the elimination of the 22nd component from
the regression. Step 8 shown at F = 11.2˙˙, DF = 38,
R^2 = 0.70 (including P.G.).

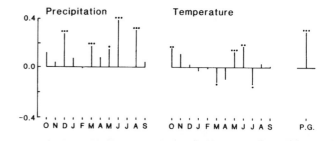

as this reduces the residual mean squares. The response function is computed for each step and any final limit for F for the entry of a variable (change in sum of squares/residual mean squares) can, therefore, be selected. Examination of the correlations between the dependent and independent variables shows, for Kew weather and London trees at least, that, with the possible exception of the first two and the last two principal components, there exists no relationship between the rank of the components and their rank as predictors of ring width. For the principal components analysis of Kew precipitation and temperature Oct(t-1) to Sep(t), 1902 to 1950, restricting the components to a cumulative eigenvalue product > 1.0 gives 19 out of the 24, taking those with roots greater than 0.5 gives 16, and those with not less than 1% of the variance gives 20. In the analysis of elm with program A, for example, the vector of eigenvalue 22 enters the regression at the eighth step at F = 5.03 (well above the 5% significance level): thus, a latent root accounting for only 0.8% of the variance of the weather accounts for almost 5% of the chronology variance. The practical effect of restricting the entry of components of the regression in this case is to reduce the weights in the response function for Nov(t-1) and Sep(t) to a negligible value whilst enhancing the weights for May(t) and Jun(t) temperature among other smaller changes. An example of a response function derived from a restricted number of components is shown in Figure 2.15. I believe there is a strong case against restricting the

components on the basis of their statistical properties and for selection on the basis of their value as predictors in the regression.

The method of spectral, or selective filtering, analysis outlined by Guiot et al. will provide useful information about tree growth and weather through the separation of the high- and low-frequency responses. Using program D the response function for the high frequencies of the elm chronology shows some clear differences from the all-frequency response function. In particular, it has lost the weights for the late summer precipitation (positive influence of rainfall in August and September) while the positive weights for May and June are enhanced, and the strong negative weight for July temperature and Nov(t-1) temperature disappear while a significant weight for April temperature appears (Figure 2.16). The high-frequency response function for oak (Figure 2.17) shows a large negative weight for March rainfall and significant negative weights for February and May temperature, none of which show in the all-frequency response function. Interannual temperature variations thus appear much more important than is suggested by the response functions derived from unfiltered data (Figure 2.14). It also appears that high March rainfall will be followed by lower than average diameter growth, though it is not easy to see a physiological connection between the two events at this stage. The low-frequency components of the elm and oak chronologies are very similar, consisting of four low waves with troughs around 1921, 1933, and 1944, so both are clearly expressions of a common regional influence. It will be interesting to analyse these frequencies in the manner suggested above by Guiot et al.

COMMENT

J.P.Cropper

The basic principle of all dendrochronological work is that a group of trees respond in a similar fashion to common external environmental factors. These factors constitute a signal that is common to the trees and subsequently allows cross-dating. In dendroclimatology, the trees are selected in such a way that the signal is

Figure 2.15: Oak, using program C. P.G. is auto-regression coefficient. Step 9 of the regression for Oct(t-1) to Sep(t) shown at F = 3.87, DF = 34, R^2 = 0.41 (climate).

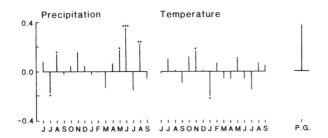

Figure 2.16: Elm, high-frequency response function program D. P.G. is for autoregression. The high frequencies account for 67% of the variance of the residuals. Step 12 shown at F = 9.89**, DF = 38, R^2 = 0.70 (including P.G.).

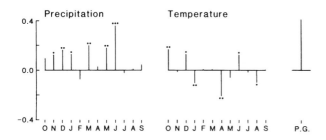

Figure 2.17: Oak, high-frequency response function as Figure 2.16. High frequencies account for 69% of the variance of the residuals. Step 9 shown at F = 6.2*, DF = 26, R^2 = 0.55, 0.44 (climate), with respect to original chronology variance.

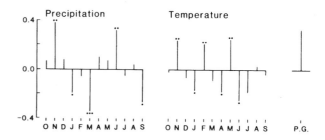

predominantly a climate variable. To distinguish which portions of the climate parameter of interest are important to tree growth, the technique called response function analysis has been developed. There are several methods now available to obtain a measure of climate-growth response ranging from simple correlations (Brett, 1978), multiple linear regressions (Fritts, 1976), regressions involving principal components (or eigen-vectors) of climate (Fritts et al., 1971), and even regression after spectral filtering as described above by Guiot et al. The development of new methods would indicate a degree of uncertainty or dissatisfaction with those already in existence. How should these regression and correlation techniques be evaluated?

One of the methods available to answer such a question is that of simulation. Simulations have been little used in tree-ring analysis up to now (Stevens, 1975), but proposals for their increased usage are being made in order to investigate various problems. In the use of simulations, the input data are generated with known properties and the techniques being investigated can be

interpreted in the light of this prior knowledge.

Two possible ways of using simulations to evaluate a response function technique will be discussed here. The first method involves generating a synthetic chronology of known climatic associations, a pseudo-chronology. In this way we know what the response function should look like and can evaluate the output of the technique being tested with this in mind. It is not required that strict agree-ment occur between the response as it was constructed and the response as it was computed. The technique would be accepted as satisfactory if the margin for error is small (Ulam, 1966). The margin for error can be reliably evaluated by repeating the simulation over many trials, each involving a newly generated synthetic data set. The second method is simply a matter of generating a synthetic chronology with no climate association, an independent-chronology, in order to assess how many "significant" responses are chance phenomena.

Using a pseudo-chronology. The simplest possible response in biological terms would be that of a group of trees responding to, say, a single month's temperature. For an

Figure 2.18: Simulation results using a pseudo-chronology developed from 80 years of February temperature. (a) Response function plot produced using a multivariate regression procedure (Fritts et al., 1971) with 95% confidence limits on each

variable, regression step 18, F level = 4.3, total varianc explained = 0.92; (b) plot of simple correlation coefficients formed betwen pseudo-chronology and each month of the climate variables. 95% confidence intervals are indicated.

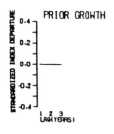

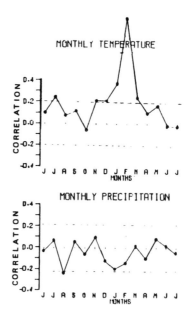

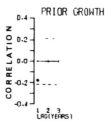

(a)   (b)

80-year climate record from Durango, Colorado, U.S.A., the record of February temperature was converted into an index series, the pseudo-chronology, by dividing each value by the mean. Two analyses were subsequently performed: (a) the multivariate regression technique described by Fritts et al. (1971) allowing the first 20 principal components of climate as possible predictors; and (b) a simple correlation analysis between the pseudo-chronology and each of the monthly climate variables.

Figure 2.18a is a plot of the response function produced by technique (a). February temperature is indeed the most significant element. However, it differs from the results expected by construction as there are 17 other significant elements indicated. Figure 2.18b contains the simple correlation analysis with the 2-tailed 95% confidence limits (78 degrees of freedom) placed about the zero line (Lentner, 1975, p.353). This plot can be inter- preted, due to the nature of the pseudo-chronology, as a plot of the correlations between February temperature and the other climate variables. The patterns of weights in Figures 2.18a and 2.18b are very similar. In Figure 2.18b, however, there are only seven significant parameters, five of which are just barely significant. Any adjustment in

the confidence limits for the loss of degrees of freedom due to autocorrelation within the time series would probably make them insignificant.

Using an independent-chronology. A computer generated "independent-chronology" (that is, one generated from random numbers without regard for the climate data to be used) would not be expected to have more than chance associations with a climate data set. Figure 2.19a shows a rerun of the Owl Canyon response function against regional climate data for the 32-year period 1931 to 1962, similar to that shown as Figure 4 in Fritts et al. (1971). Figure 2.19b shows the result with the same climate data but using the synthetic independent-chronology (TEST59) instead of that for Owl Canyon. The analyses were run in exactly the same way (as in Fritts et al., 1971) with both of these plots coming from the same step in the multiple linear regression process.

In Figure 2.19a, 88.3% of the chronology variance was accounted for exclusively by climate and 15 climate variables were deemed significant at the 5% level. In Figure 2.19b, slightly less variance was accounted for (68.0% climate and 12.7% prior growth, 80.7% total), yet

Figure 2.19: Real and independent chronology resp- onse function plots. (a) Response function plot of Owl Canyon chronology against regional climate data for the period 1931 to 1962 with 95% confidence limits indicated, regression step = 19, F level =

1.22, total variance explained = 0.88 (climate = 0.88); (b) response function plot for the independent-chronology TEST59 against the same climate data as in (a) with 95% confidence limits on each variable, regression step = 13, F level = 2.03, total variance explained = 0.81 (climate = 0.68).

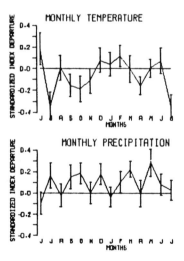

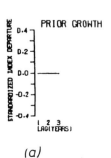

(a)

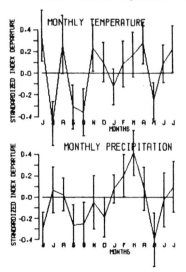

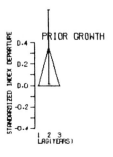

(b)

16 variables were significant. For the sake of complete-
ness, Figure 2.20 contains the plots of simple correlations between both the Owl Canyon and independent-
chronology and the climate data used in the response
function. For Owl Canyon, four elements of the plot are
significant at the 5% level (29 degrees of freedom), but
if adjustment is made for loss of degrees of freedom due
to the first order autocorrelation (Mitchell et al., 1966;
Fritts, 1976) then only one element remains significant.
In Figure 2.20b, there are no significant correlations
indicated between the monthly climate data and the
independent-chronology, as expected.

Conclusions. The two samples presented indicate that the
interpretation of a response function has to be performed
carefully. This care is not only because of our limited
degree of understanding of the climate-growth response,
but also because there is the potential for the response
function technique results to be misleading by attributing
significance to non-significant variables.

It is hoped that this discussion will bring to the
attention of other workers some of the potential of using
simulations to assist in evaluating response function
techniques. It must be stressed that the examples presented here are the result of only two simulation runs. A

great many simulations are required before an adequate
comparison of response function techniques can be made.
By developing a set of pseudo-chronologies of known
responses, such as multivariate responses or those with
different sized noise components, it should be possible in
the future to assess better some of the limitations of the
response function technique. With these limitations
identified, the interpretation of climate-growth relationships from response functions may be accomplished with
more certainty.

TRANSFER FUNCTIONS
G.R.Lofgren & J.H.Hunt
INTRODUCTION
Transfer functions are statistically derived
equations which relate two sets of variables for which a
causal relationship can be described but for which an
analytical expression cannot be derived. In the development of transfer functions for dendroclimatic
research, one set of variables consists of tree-ring
chronologies and the other of climate records. Principal
components of one or both sets may replace the actual
data.

Transfer functions are determined by matching annual
tree-growth information and the corresponding yearly

Figure 2.20: Plots of simple correlations using same data as in Figure 2.19. (a) Simple correlations using Owl
Canyon chronology against regional climate data; (b) simple correlations using independent-chronology against
regional climate data. The 95% confidence intervals are indicated.

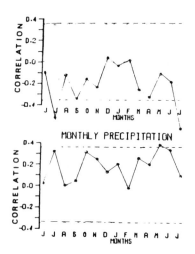

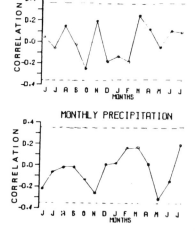

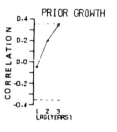

(a)   (b)

climate record during a particular interval of time called the calibration interval. This is usually during the 20th century. The calibrated function can then be used to transfer annual anomaly records or patterns of tree growth into yearly anomaly records or spatial patterns of climate for periods of time when climatic information is not available. Fritts et al. (1971) have presented a transfer function developed for dendroclimatic reconstructions of sea level pressure. Blasing (1978) has described the canonical relationships associated with this transfer function. Webb & Clark (1977) have reviewed and compared transfer function techniques used for estimating past atmospheric and oceanic conditions from palaeontological data.

Transfer functions used for dendroclimatic reconstructions are calibrated using several time series of tree-ring chronologies distributed over space with several time series of a climate variable also distributed over space. Reconstructed time series of the climate variable are obtained from time series of the dendrochronological data. In contrast, transfer functions for palaeontological reconstructions are calibrated using a spatial array of palaeontological indicators with a spatial array of climate means. Climatic variability is reconstructed at temporal intervals using dated stratigraphic layers of the palaeontological indicators.

This paper will consider:

a) some basic assumptions;

b) the conceptual development of transfer functions;

c) regression techniques used to calibrate transfer functions;

d) evaluation of calibrated transfer functions; and

e) canonical correlation analysis used to calibrate transfer functions.

Underlying our discussion is the importance of forming appropriate transfer functions and of the critical use of statistics to evaluate calibration results.

It is necessary to assume that three conditions are met before the transfer function can be derived and used for the reconstruction of climate:

a) that the Uniformitarian Principle applies, that is "the physical and biological processes which link today's environment with today's variations in tree growth have been been in operation in the past" (Fritts, 1976);

b) that climatic conditions which produced anomalies in tree-growth patterns in the past have their analogues during the calibration period; and

c) that the systematic relationship between climate as the limiting factor and the biological responder can be approximated by a linear mathematical expression.

The predictand set need not, however, be a linear function of climate (for example, the logarithm of the precipitation record could be used). Thus, the linearity of the transfer function is not as severe a restriction as it might first appear. It need hardly be said that these assumptions are not completely met by the climate-growth relationship, but in reality form the limitations of the system. Hutson (1977), for example, discusses the limitations associated with the "no-analogue" condition.

Continuity of climate is also assumed in the calibration and reconstruction of climate in areas outside the domain of the climatic indicator. In the middle latitudes, the continuity results from wave-like structures associated with migratory cyclones. Kutzbach & Guetter (this volume) use the spatial continuity in two climate variables (temperature and precipitation) to reconstruct a third climate variable (sea-level pressure). We have calibrated 20th century sea-level pressure anomalies with temperature and precipitation anomalies from western North America. Figure 2.21 shows that over 70% of the pressure variance is explained by this calibration for a large area off the west coast of North America. The continuity of weather systems thus makes it possible to use temperature and precipitation over North America to reconstruct pressure over the North Pacific.

TRANSFER FUNCTION CONCEPT

Tree growth during a particular year is the result of many contributing factors (Fritts, 1976). Normally, climate prior to and during the growing season limits growth for that year. A functional relationship is conveniently stated as

$$G(t) = f[C(t)] \qquad (2.3)$$

where $G(t)$ is growth for year t, and $C(t)$ is the climate factor for year t.

The climate in prior years may also affect physiological processes which limit growth. Fritts (1974) states, "The width of the ring represents the net effect of the climatic factors on processes that influence growth, including the influence of climatic factors in prior years that modify the tree's capacity to respond in later years to factors of climate." The general functional relationship, expressed by this statement, allows an expansion of Equation (2.3)

$$G(t) = f[C(t), C(t-1), C(t-2),\ldots C(t-n)] \qquad (2.4)$$

In principle, this functional relationship may be inverted to solve for the climate of year t

$$C(t) = g[G(t), C(t-1), C(t-2),\ldots C(t-n)] \qquad (2.5)$$

This relation no longer expresses a biological causative relationship as in (2.3) and (2.4) but the moving average nature of the tree's ability to integrate climate (Stockton, 1975). Replacing part or all of the prior climate terms in (2.5) with an inverse relationship derived from (2.3) introduces the autoregressive property of the tree's sensing of climate

$$C(t) = h[G(t), C(t-1), G(t-1),... C(t-n), G(t-k)] \quad (2.6)$$

or

$$C(t) = s[G(t), G(t-1), G(t-2),... G(t-k)] \quad (2.7)$$

This conceptual development suggests many different transfer functions, one or more of which may be applicable to a particular situation. The user is faced with the responsibility of selecting specific predictor sets in the development of the specific transfer function. These functional relationships are approximated using regression theory since their analytical form has not been determined.

CALIBRATION OF TRANSFER FUNCTIONS

A linear relationship between the dendroclimatic indicator x (predictor set) and the climate variable y (predictand set) may be expressed as

$$y = \phi(x) + \epsilon \quad (2.8)$$

where ϕ is assumed to be a linear function and ϵ, an error component which includes non-modelled physiological and climatic interactions.

We will first consider the multiple linear regression (MLR) of one predictand with K predictors. For this case, Equation (2.8) may be expressed as

$$y(i) = \sum_{k=0}^{K} x(i,k)\beta(k) + \epsilon(i) \quad (2.9)$$

for each observation i from 1 to N (years). When departure or normalised data are used, $\beta(0)$ is 0. When actual observational data are used, $x(i,0)$ is 1 and $\beta(0)$ is a constant, the y-intercept. Equation (2.9) may be expressed in matrix form as $Y = XB + \epsilon$.

The transfer function is calibrated when the constants $\beta(k)$, or estimates denoted by $\hat{\beta}(k)$, are determined. A set of N observations is chosen for which the dendroclimatic indicators and the climate variable are both recorded. Estimates $\hat{B}$ of B are determined from these data by requiring that the residual or observed error matrix

$$e = Y - X\hat{B} = Y - \hat{Y} \quad (2.10)$$

has the least square property for the N calibration observations. With minimal mathematical restrictions, $\hat{B}$ is given by (Johnston, 1972)

$$\hat{B} = (X^TX)^{-1}X^TY \quad (2.11)$$

where X^T denotes the transpose of matrix X and $(X^TX)^{-1}$ denotes the inverse of matrix X^TX. Once $\hat{B}$ is obtained the equation $\hat{Y} = X\hat{B}$ may be applied to predictor sets outside the calibration interval to reconstruct estimates of past climate.

Assumptions of normality and independence of the true error terms, $\epsilon(i)$'s, (Neter & Wasserman, 1974; Stockton, 1975) must be made for significance testing of the $\beta(k)$'s. Plots of the distribution of the observed errors can be used to ascertain the validity of these assumptions (Draper & Smith, 1966; Howe & Webb, 1977).

EVALUATION OF THE CALIBRATION

The regression estimates ($\hat{Y} = X\hat{B}$), although guaranteed to be least square, need to be evaluated in order to determine what fraction of the variance in the predictand is accounted for by the regression and to compare with other regressions using different predictors.

The measure of how much of the variance in the predictand is accounted for by the predictors is called the explained variance or calibrated variance. The explained variance is defined as the ratio of the sum of squares due to regression, SSREG, to the total sum of squares, SST:

$$R^2 = SSREG/SST \quad (2.12)$$

where

$$SSREG = \sum_{i=1}^{N} [\hat{y}(i) - \bar{\hat{y}}]^2 \quad (2.13)$$

and

$$SST = \sum_{i=1}^{N} [y(i) - \bar{y}]^2. \quad (2.14)$$

The means of the measured and estimated predictand during the calibration period are denoted by $\bar{y}$ and $\bar{\hat{y}}$, respectively. The error or residual sum of squares (SSE) is

$$SSE = \sum_{i=1}^{N} [y(i) - \hat{y}(i)]^2. \quad (2.15)$$

Basic properties of regression (Seber, 1977) guarantee that

$$SST = SSREG + SSE \quad (2.16)$$

Equation (2.16) allows two equations to be written which give expressions for the calibrated variance which are equal only during the calibration period. The first expresses R^2 in terms of the residuals

$$R^2 = 1 - SSE/SST \quad (2.17)$$

The second expresses R^2 as the square of the Pearson correlation coefficient between Y and $\hat{Y}$ (Seber, 1977)

$$R^2 = \sum_{i=1}^{N} [y(i)-\bar{y}][\hat{y}(i)-\bar{\hat{y}}]^2 / \sum_{i=1}^{N} [y(i)-\bar{y}]^2 \sum_{i=1}^{N} [\hat{y}(i)-\bar{\hat{y}}]^2 \quad (2.18)$$

The value of R^2 obtained from a particular regression

indicates only how well the data sets are matched during the calibration period. This R^2 value may not indicate that the regression estimates can adequately model relationships between the data sets over all possible observations or over an independent set of observations (Gordon, this volume). Selecting a predictor set which maximises R^2 does not necessarily minimise the independent period residual sum of squares.

In order to compare results with other regressions using different predictors, an adjusted R^2, $R^2(adj)$, is defined using mean squares. A mean square is a sum of squares divided by the appropriate degrees of freedom. We define the total mean square, MST, as

$$MST = SST/(N-1) \qquad (2.19)$$

and the mean square error, MSE, as

$$MSE = SSE/[N - (K+1)] \qquad (2.20)$$

The mean square error degrees of freedom is $N-(K+1)$ if the regression includes the intercept term $\beta(0)$, or $N-K$ if it does not. The adjusted R^2 is then (Neter & Wasserman, 1974)

$$R^2(adj) = 1 - MSE/MST \qquad (2.21)$$

$R^2(adj)$ is a "more conservative" estimate of the explained variance (Nie et al., 1975).

ARTIFICIAL PREDICTABILITY

Chance correlations might cause the calibration period explained variance to overestimate the usefulness of the regression when the number of predictors, K, is not small compared to the number of observations, N (Lorenz, 1977). A strong relationship inferred when a weak one or when none at all exists is aptly called "artificial predictability" by Davis (1976). Artificial predictability may allow the calculated R^2 to exceed that which would be obtained had the entire population been included in the regression. If one could use the entire population in the regression, B would be estimated exactly, that is, $\hat{B}(pop)$ = B, and a population value for the explained variance, $R^2(pop)$, could be obtained. In general, the expected value of the explained variance for a finite size sample has been shown (Kendall & Stuart, 1979) to be

$$E(R^2) \geqslant R^2(pop) \qquad (2.22)$$

indicating that any particular value of the explained variance is likely to overestimate the population explained variance. R^2 is called a biased estimator of $R^2(pop)$. The bias can be clearly illustrated for the case where 15 predictors are used in a regression over 31 observations. If no true population relations exist between data sets, that is, $R^2(pop) = 0$, the expected

value of R^2 (Morrison, 1976) is

$$E(R^2) = K/(N-1) \qquad (2.23)$$

An R^2 of 0.5 in this case is meaningless.

Kutzbach & Guetter (1980, this volume) suggest reporting the adjusted R^2 as an approximate unbiased estimate of the population explained variance. Artificial predictability, when it occurs, should become apparent during verification of the reconstructions (Lorenz, 1977; Gordon, this volume).

STABILITY

It is important to consider the stability of estimates obtained by the regression technique. The regression estimates and the sample explained variance, R^2, are derived for a particular sample of N observations. Webb & Clark (1977) state that if the calibration "data are typical... additional observations would not greatly change the estimates". This is a crucial point as large variations weaken the validity of a particular transfer function. One can test for stability by carrying out calibrations with other predictor sets or by varying the length of the calibration period.

PREDICTOR DATA SETS

The reconstructions resulting from a regression analysis using only one tree-ring chronology can often be improved by using additional chronologies if each new chronology has additional information about the predictand. The predictor set in Equation (2.9) can take the form of K chronologies. Alternately, principal components of the predictor data can be determined and $K < K'$ of the total K' principal components used in the calibration. The use of principal components reduces the number of predictors, excludes small scale variations that are presumably due to noise, and results in orthogonal variables.

Since the literature is not consistent in its definition of principal components (Stockton, 1975), we shall define principal components in terms of the eigenvector analysis. Principal components result from the transformation of a given set of variables into a new set of variables which are orthogonal or uncorrelated with each other. The matrix product of the data set in normalised or departure units and the time invariant eigenvectors form the matrix of eigenvector amplitudes (Sellers, 1968; Julian, 1970; Blasing, 1978) defined here and by others (Tatsuoka, 1971; Nie et al., 1975; Morrison, 1976) as the principal components.

The lagging relationships suggested in Equations (2.6) and (2.7) can be incorporated in the predictor set by including an additional L predictors lagged from the first K predictors. Since this causes the number of predictors to increase by L, the risks associated with artificial predictability are correspondingly increased.

PREDICTAND DATA SETS

Equation (2.9) is the application of Equation (2.8) for a single predictand which is normally an anomaly time series of a climate parameter. Spatial climatic variability can be determined when the predictand data set of Equation (2.8) takes the form of a matrix of N observations of J climate records. The linearity of φ in Equation (2.8) may be expressed in terms of the regression coefficients β(k,j) as

$$y(i,j) = \sum_{k=1}^{K} x(i,k)\beta(k,j) + \varepsilon(i,j) \qquad (2.24)$$

for each j from 1 to J.

The climate data may alternatively be regional averages for a climatologically homogeneous region, the logarithm transformation of total precipitation, or an index such as the Palmer Drought Severity Index (Stockton, 1975; Cook, this volume). They need not be measured climate data provided the predictand data set is related to climate as is the predictor data set. Physical or biological data such as sea-surface temperatures (Douglas, 1976) or fish-catch data (Clark et al., 1975) have been successfully calibrated against tree rings.

The predictand data set may consist of principal components taken from the original climate data set. The formation and selection of the J most important eigenvector amplitudes eliminates, it is hoped, more of the error noise in the data than climate information. With J < J´ of the amplitudes entered in the regression, less than the total variance of the original data set is contained in the predictand data set. Therefore, the regression can be evaluated by calculating the amplitude

explained variance or by calculating the variance of the original data explained after converting the estimated amplitudes back into real world variables, that is, the original units.

When the predictand data set takes the form of a spatial distribution of climate time series, other ways of evaluating the calibration become possible. The estimated spatial distribution can be correlated with the actual spatial distribution for each year in the calibration interval. Alternatively, an equation analogous to Equation (2.17) can be applied to the estimated and actual spatial distributions for each year. Over the spatial distribution, Equation (2.17) is not equal to Equation (2.18) and offers an additional evaluation of the success of the calibration.

SELECTING PREDICTOR VARIABLES

We have suggested that improved calibration may result when several chronologies representing a spatial distribution are included in the predictor data set. Risks of artificial predictability are enhanced when the number of chronologies is increased. When the predictor set is composed of eigenvector amplitudes, the use of lagging amplitudes to improve the calibration results invokes the same dilemma. Selectivity of predictor variables is desirable in both of these situations. A number of a priori schemes for determining significant eigenvector amplitudes have been presented (Tatsuoka, 1971; Johnston, 1972; Preisendorfer & Barnett, 1977; Blasing, 1978) but there is little agreement on their relative merits.

Regression analysis may be used to select predictor

Figure 2.21: The spatial distribution of the explained variance (%) of winter sea-level pressure was calibrated with 46 temperature and 52 precipitation stations from the western United States. The calibration interval was from 1899 through 1963. Five principal components of precipitation were calibrated with eleven principal components of sea-level pressure. The stepwise canonical correlation technique was used to calibrate this transfer function. This calibration shows that much of the winter pressure variance over the eastern North Pacific Ocean can be significantly related to the variance in temperature and precipitation over the western United States.

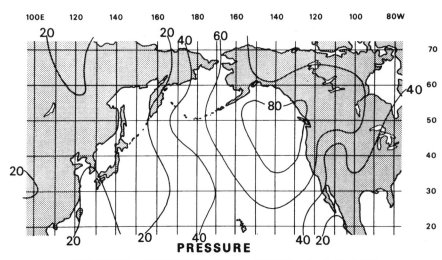

CALIBRATED VARIANCE FOR MODEL 11P 5T12R

variables rather than using a priori criteria. One scheme
is to form all possible combinations of the K predictors.
This involves 2^K regressions. Alternatively, stepwise
regression schemes require the formation of many fewer
combinations. The goal of stepwise regression is to select
the most important predictor variables needed to estimate
the behaviour of the predictand variable. However, the
stepwise selection procedure can be greatly influenced by
intercorrelations among the predictors. The amount of
variance in the predictand attributed to any predictor
variable is dependent upon the order in which the variable
is entered, that is, upon which variables are already in
the equation. Thus, the final model may not be unique but
may be dependent upon the stepwise procedure used.

Three alternative stepwise procedures have been
suggested in Nie et al. (1975).

Forward inclusion. The predictor variable explaining the
most predictand variance is entered first provided the
variance explained is significant. At each succeeding step
the variable explaining the most additional variance in
the predictand is entered and retained if the additional
explained variance is significant. The process ceases when
no additional variable is accepted.

Backward elimination. All predictor variables are entered.
The variable contributing the least additional explained
variance, given that all other variables are entered, is
deleted provided its explained variance is insignificant.
The variable which then contributes the least explained
variance, given the set of variables still entered, is the
next candidate for deletion. This process continues until
no more variables can be deleted.

Stepwise solution. This is "an improved version of the
forward-selection procedure" (Draper & Smith, 1966). It
proceeds as in the forward inclusion process. However,
after each new variable is accepted, the others already
entered are retested for possible deletion as in the
backward elimination process. This continues until no
further variable is accepted or deleted.

These stepwise procedures select appropriate
predictors for each predictand independently. The transfer
function developed includes only those predictors which
have been found to be significant for each predictand.

CALIBRATING SEVERAL VARIABLES

Canonical correlation analysis (Hotelling, 1936) was
introduced to analyse the relationship between two sets of
variables. Canonical correlation analysis transforms the X
and Y matrices into new matrices X^c and Y^c, whose columns
are called canonical variates. A given column of X^c is uncorrelated with every other column of X^c and all columns
but the corresponding column of Y^c. Corresponding columns

of X^c and Y^c form pairs of canonical variates which are
linear combinations of the original variables. These
linear combinations are essentially equivalent to
principal components whose selection criteria has been
altered from the eigenvector analysis to maximise the
relationship between the two sets of variables (Nie et
al., 1975).

The original variables, even though they are eigenvector amplitudes, are necessarily uncorrelated only when
the calibration interval is identical to the eigenvector
analysis period. Also, inclusion of lagged amplitudes as
predictors will introduce correlated variables into the
data set. Canonical variates constructed from eigenvector
amplitudes reintroduce orthogonality over the calibration
interval.

Fritts et al. (1971) first used canonical correlation analysis with tree-ring data to calibrate with
sea-level pressure. The mathematical procedure for
obtaining the canonical variates in each data set, and the
canonical correlations between the corresponding pairs of
canonical variates is outlined by Glahn (1968), Clark
(1975), and Blasing (1978). It is described in more detail
by Anderson (1958) and Johnston (1972). Each predictand
canonical variate is regressed separately against its
corresponding predictor canonical variate. Each estimated
predictand canonical variate value $\hat{y}^c(i,p)$ is given by

$$\hat{y}^c(i,p) = x^c(i,p)\lambda(p) \qquad (2.25)$$

where $\lambda(p)$ is the correlation of the "p"th canonical pair.
The canonical-pair regression estimates are given in
matrix form by

$$\hat{Y}^c = X^c \Lambda \qquad (2.26)$$

where Λ is the diagonal matrix of canonical correlations.
Glahn (1968) suggests that only significant canonical
correlations may be used in the regression by setting the
insignificant $\lambda(p)$'s to zero in Equation (2.26). The estimated canonical predictands $\hat{Y}^c$ may then be transformed
back into the predictand variables $\hat{Y}$.

Fritts developed a stepwise canonical-pair regression procedure which examines the contribution of each
canonical variate pair to the predictand variance in
contrast to selecting variate pairs by significant
canonical correlations. Each canonical variate pair is
entered into regression in order of decreasing canonical
correlation by including the appropriate $\lambda(p)$ in Λ, the
diagonal matrix of Equation (2.25) (Fritts et al., 1979).
Only those canonical variate pairs which contribute
significantly to the explained variance are retained.
When a pair is accepted, pairs with larger canonical
correlations previously rejected are then re-entered and
retested one-by-one. This stepwise process continues until
all canonical variate pairs have been entered and tested.

This procedure eliminates those canonical variate pairs which may be highly correlated but explain too little variance in the predictand to be significant.

The significance test for retaining a pair of canonical variates is an F-test on the increased predictand explained variance due to that pair. The test uses the sum of squares due to regression

$$SSREG = \sum_{j=1}^{J} [\sum_{i=1}^{N} [\hat{y}(i,j) - \bar{\hat{y}}(j)]^2] \qquad (2.27)$$

where $\bar{\hat{y}}(j)$ is the mean of the "j"th predictand estimate, and the residual or error sum of squares

$$SSE = \sum_{j=1}^{J} [\sum_{i=1}^{N} [y(i,j) - \hat{y}(i,j)]^2] \qquad (2.28)$$

For testing each step, one needs the increase in the sum of squares due to regression, SSINC, over the previous step that passed. The F-ratio for the significance test, using mean squares, is

$$F = [SSINC/K]/[SSE/(NJ-MK)] \qquad (2.29)$$

where M is the total of canonical variate pairs included at the step being tested, (NJ-MK) is the residual degrees of freedom, and N, J, and K have the same meanings as before.

Autocorrelation in a time series can be an important problem in regression analysis. Autocorrelation in the errors, $\varepsilon(i,j)$, Equation (2.24), violates the basic assumption of independence of the error terms (Johnston, 1972). Calculation of the autocorrelations of the observed errors

$$e(i,j) = y(i,j) - \hat{y}(i,j) \qquad (2.30)$$

serves to test the independence assumption. If r is the average of the first order autocorrelation of the J residuals and is positive, the residual degrees of freedom are adjusted by multiplying by $(1-r)/(1+r)$ in analogy with the adjustment for effective sample size suggested by Mitchell et al. (1966).

Canonical-pair regression including all canonical variates results mathematically in the same estimates as does multiple linear regression wherein all steps of the regression are included. Thus, canonical-pair regression can provide different estimates only when some of the $\lambda(p)$'s have been set to zero. Estimates obtained from the canonical-pair stepwise regression do in fact hold up better on independent data sets than those estimates produced by including all canonical variate pairs.

SUMMARY

The underlying consideration of this discussion has been the importance of the appropriate use of data and critical use of statistical methods. A summary of some of our concerns is presented here.

a) Proper consideration of the autoregressive and moving average nature of the tree-growth-climate relationships is important in the formation of an optimum transfer function.

b) Normality and independence of the error terms, which are needed to justify significance testing, can be checked by plotting the time series of residuals or by calculating the serial correlation of residuals. Allowance for autocorrelation can be incorporated into the statistical methods.

c) Artificial predictability may indicate relationships in the calibration data which do not occur in the population. The adjusted R^2 gives a better indication of the true predictive ability of the transfer function than does R^2.

d) Lack of stability between transfer function calibrations when calibration intervals have been varied only slightly may further indicate artificial predictability.

e) Eigenvector analysis may be used to form predictor and predictand data sets with reduced noise. Additionally, elimination of insignificant variance is obtained through stepwise multiple linear regression.

f) Canonical correlation analysis maximises the relationship between the two data sets and re-introduces orthogonality into the predictors. The stepwise canonical-pair regression technique eliminates insignificant variance.

Proper use of transfer function methods allows a transfer of the climate information integrated by the tree into estimates of past climatic variability. Application of these methods may enhance knowledge of past climatic variability and may help to enlighten expectations of future conditions.

COMMENT

B.M.Gray

In order to use transfer functions to reconstruct past climate from measurements of tree-ring growth, the Uniformitarian Principle, as described by Lofgren & Hunt above, is assumed to apply. Some evidence that this is so for the European region over the last two centuries exists, and is described below.

Spatial patterns of climate. Principal component analysis can be used to reduce both the number of predictor and predictand variables in climate reconstruction. The component patterns or modes of spatial variation of the data can be examined to see whether they have changed over the period for which observations exist. Clearly, the frequency of occurrence of the patterns (the component amplitudes) changes in time, and it is these changes that are derived in the reconstructions of climate. If the patterns themselves have altered, because of changing boundary conditions such as widespread glaciation, the reconstructed frequencies would be invalid.

To test the stability of temperature patterns, a set of European temperature records (1761 to 1969) was analysed in 30-year periods. The 17 stations used lie between 47 to 63°N and 4°W to 25°E, and are well distributed in this area. Each period contained 360 monthly mean observations. If more than one-third of the observations were missing for any one station, that station was omitted, but in every case at least 13 stations were present. The first six principal components accounted for over 90% of the variance of the data in every period. By correlating the spatial loading of each component over all the periods, a quantitative estimate of the similarity of any pattern in time can be made. If the patterns are identical, the correlation coefficient will be 1.0 or -1.0. Where the correlation coefficient is greater than 0.5, the similarities in pattern can be detected easily by eye. Values over 0.5, significant at above the 2% level, were obtained for all seven periods for the first three component patterns (Table 2.5).

Periods 4, 5, and 6 covering 1851 to 1940 show hardly any change at all in the six patterns considered. Lower correlations for patterns 5 and 6 in the earlier periods may be partly attributed to poor quality observations. In general, the results support the hypothesis that the major patterns of temperature variation have not changed over the last 200 years. The critical factor in observing such stability appears to be the high ratio of the number of time observations to the number of stations used.

Local tree-climate response. The reproducibility of response functions in time bears directly on the Uniformitarian Principle. Three oak chronologies were tested: two site chronologies from Maentwrog in North Wales (Leggett et al., 1978) and Fontainebleau near Paris (Pilcher, unpublished data), and an unindexed area chronology from Hereford and Cumberland (Giertz-Sieberlist, unpublished data). Three different tests were used to compare the response functions, F1 and F2, from two periods for each series. None of the tests is completely satisfactory, but the combined results give an indication of the stability of the response function in time. The tests were:

a) Correlation coefficients were calculated for the temperature and rainfall elements in F1 and F2. A significant correlation indicates the similarity of the responses.

b) Two kinds of matching elements are considered, those where the error bars do not cross the axis and those where the error bars do: that is, significant and non-significant elements. Comparing F1 and F2 is considered equivalent to tossing two loaded coins and calculating the chance number of matched heads or tails that arise in N throws.

c) Each element of F1 is tested against the corresponding element of F2. If the 95% error bars overlap, a match is designated. The significance is tested using a Poisson distribution.

The Maentwrog response functions for 1907 to 1940 and 1941 to 1974 gave the following results:

a) Temperature elements showed a correlation of 0.55, significant at the 10% level.

b) The samples were similar at the 3% significance level.

c) A significant difference was found at the 9% level.

The Hereford-Cumberland series was divided into the periods 1755 to 1839 and 1855 to 1949.

a) Temperature results were correlated at better than the 10% significance level (r = 0.5).

b) F1 and F2 were similar at the 2% level of significance.

c) No significant difference was found between the samples.

Response functions for Fontainebleau were calculated for 1800 to 1839 and 1870 to 1939.

a) No significant correlations were found.

b) F1 and F2 were similar at the 3% level of significance.

c) No difference at all was found between F1 and F2, when the multiple regressions were halted at steps 14 and 16 respectively. If the regressions were halted at step 9, when the greatest number of significant months occurred, 4 differences were found; a result which was significant at the 5% level.

Table 2.5: Similarity of European temperature patterns over seven time periods from 1761 to 1970.

COMPONENT	PERIODS						
	1:2	2:3	3:4	4:5	5:6	6:7	1:7
1	0.97	0.99	1.00	1.00	1.00	1.00	0.98
2	0.79	1.00	0.98	1.00	0.96	0.97	0.99
3	-0.60	1.00	-0.97	0.99	0.95	0.86	0.94
4	0.52	0.90	0.80	0.83	0.90	0.44	0.42
5	-0.40	-0.31	-0.20	0.71	0.91	0.69	0.70
6	0.02	0.51	0.30	-0.83	0.94	0.45	-0.02
Number of stations	13	15	17	17	17	16	13

Although the agreements between response functions calculated for non-overlapping periods is good in general, the
results from Fontainbleau were disappointing. A possible
cause of the discrepancy between the two functions from
this site is the age of the trees in the second period.
If the physiological processes in an older tree differ
from those in a younger tree, this might account for the
differences in the response functions. A younger chronology for 1870 to 1969 was obtained from the same site and
the response function calculated. No age effects were
detected in the response function for climate, although
the previous growth factor increased from 6% to 25%. The
two chronologies of different ages were similar at the
0.1% level, indicating good site reproducibility.

VERIFICATION OF DENDROCLIMATIC RECONSTRUCTIONS

G.A.Gordon

INTRODUCTION

Current dendroclimatic methods of reconstructing
past variations in climate involve the use of several
forms of multivariate regression techniques (Fritts et
al., 1971; Webb & Clark, 1977). The climate parameter to
be reconstructed, the predictand, is algebraically expressed as some function of one or more dendrochronological
parameters, the predictors. Such a function can assume
many forms, each being a different combination of
predictors. A model is one particular combination that
experience, theory, or both, suggest will result in a
successful reconstruction of climate.

Each predictor in a model is associated with a
coefficient which must be evaluated. Regression techniques
optimally evaluate the coefficients using a set of
calibration data, which is a body of data incorporating
matched observations of the predictors and predictand.
The coefficients are unique to the calibration data and
are independent of any observations not included in the
calibration data. The model with its empirically derived
coefficients can be used to estimate values of the predictand for any of those independent data. The reliability
of a calibrated model is some measure of how dependable or
accurate the independent predictand estimates are. To be
successful a model must be shown to be reliable when
applied to independent data. Reliability should be
estimated prior to the climatic interpretation of the
predictand estimates.

It is the purpose of this discussion to illustrate
this need for proper assessment of model reliability and
to suggest approaches to assessment which are open to the
dendroclimatic researcher. These remarks are intended to
be generally applicable to any situation where regression
is used to reconstruct climate. They are illustrated from
the author's experience with the reconstruction of spatial
climatic anomalies over North America (Fritts et al.,
1979).

THE NEED FOR MODEL ASSESSMENT

The use of the regression techniques involves
assumptions which may critically affect reliability
(Lofgren & Hunt, this volume). One very important
assumption is that analogues of the predictor-predictand
relationship are present in the calibration data. If the
coefficients which are uniquely determined by the
calibration data are to be meaningful, then analogues must
be present in sufficient numbers and variety. The
coefficients should be stable and should not vary greatly
with changes in the calibration data. A successful model
does not represent a simple mimicry of the calibration
data, but instead represents a universal property of the
particular climate-growth response.

Multivariate regression models which accurately
estimate the predictand for the calibration data are often
found to be unreliable when they are applied to independent data. This phenomenon was recognised and
appropriately referred to in early statistical literature
as "the shrinkage of the coefficient of multiple
correlation" (Larson, 1931; Wherry 1931; Anderson et al.,
1972; Stone, 1974). Because the predictive power of a
regression model can at times fail dismally when applied
to independent data, it becomes imperative that some
attempt be made to assess the deterioration in accuracy of
independent predictand estimates. The assessment can then
be used either to improve the model or to provide the
proper perspective with which to view the resultant
climate reconstruction. Any procedure that is used to make
such an assessment of reliability is defined here as a
verification procedure. The one question that a
verification procedure must address is: How well does the
model, calibrated with the available body of data, perform
when applied to independent data?

Statistical theory does not provide a means to
answer that question (Mosteller & Tukey, 1968). The
process of optimisation (least squares) used to evaluate
the coefficients virtually ensures that the model will be
more accurate for the calibration data than for any other
body of data that may arise in practice. Thus, any
evaluation of performance that is made with calibration
data will probably result in overconfidence in the
predictive power of the model. The model must accurately
be applied to independent data in order to verify its
reliability.

Some insight into the likely independent performance
of the model can be made by simply discounting or adjusting the fit of the optimised regression model (Wherry,
1931; Lorenz, 1956, 1977; Kutzbach & Guetter, this
volume). Essentially this adjustment amounts to recog-
nising the effects of optimisation on the independent
performance of the model and accordingly to reducing the
model's explained variance on the calibration data. The
adjusted level of explained variance should reflect the
expected independent performance of the model. However,

this type of adjustment is directed primarily at accounting for the loss of degrees of freedom in constructing the regression model. In some situations this adjustment is insignificant. A small adjustment, therefore, does not necessarily indicate a reliable model, but may indicate instead one where the number of degrees of freedom is not a limiting factor. The fact that the reliability of a model cannot be safely verified by using the calibration data or statistics derived from them is unavoidable. Some form of independent data must be used (Anderson et al., 1972).

INDEPENDENT VERIFICATION

Independent verification may be approached in three ways. The predictand estimates can be compared with proxy data other than tree rings, with historical accounts of events occurring in independent periods, and with independent predictand observations. The first approach involves inference from the proxy data to climate. Similarly, the second relies heavily on the interpretation of records of specific events. The subjectivity inherent in these first two methods compromises their value in evaluating the reliability of a model.

Verification with independent predictand observations remains a more objective approach to verification. Furthermore, this last approach allows the use of comparative statistics and the direct measurement of errors on independent data. Correlation coefficients, non-parametric sign tests, contingency analysis, and reduction of error statistic (Fritts et al., 1979; Gordon, 1980) quantify the reliability of a model in a way that can be directly compared to the results of other investigators. Statistical verification, therefore, is both an effective and adaptable means of assessing the reliability of the model.

The remainder of this discussion will deal only with statistical verification. This should not be taken to mean that proxy and historical verification are not of value. On the contrary, the proxy and historical approaches constitute the only means to gain information on reliability at very long timescales. Their use, however, awaits the further development of appropriate techniques (see, for example, Lawson, 1974; Moodie & Catchpole, 1975; Ingram et al., 1981; Smith et al, 1981).

In examining the problem of statistical verification, the conclusion is quickly reached that the greatest difficulty to be overcome is the scarcity of independent predictand records that are suitable for verification. The success of the regression technique of reconstruction depends upon having a sufficient number of calibration observations. In complex modelling situations too few calibration data would certainly lead to high levels of instability in the model coefficients and, hence, an unreliable model. There is naturally a temptation, therefore, to use all available predictand

data as calibration material. The need to use all available calibration data is particularly acute in most palaeoclimatic applications of regression, where the size of calibration data sets is often small. Independent data in sufficient quantity usually cannot be obtained by shortening the calibration period for fear of relying on too few observations with which to calibrate the model.

THE USE OF EARLY INDEPENDENT CLIMATE RECORDS

Some palaeoclimatic modelling situations do provide a ready source of independent data. For example, the length of the calibration record in reconstructing spatial variations of climate is dictated by the shortest record in the spatial network of predictand stations. This inevitably leaves some stations with predictand observations occurring prior to the period of calibration. These records provide an obvious source of independent data for verification, and they should be thoroughly exploited. While the use of these early records is usually straightforward (Fritts et al., 1979; Gordon, 1980), they have four attributes that may affect the conclusions drawn from verification: limited spatial distribution; unknown quality; limited length; and limited selection.

By definition, these early data are not available at all stations, which precludes verification of the model uniformly over the predictand network. In North America, for example, the longest climatic records, and hence most independent verification data, are concentrated in eastern North America. Current reconstructions by Fritts et al. (1979), wherein the predictor sites are located exclusively in western North America, are expected to be poorest in the east. Therefore, verification results obtained from early eastern records may be negatively biased.

The quality of early data can also influence verification results. These data are often of unknown quality and contain inhomogeneities with the calibration data, such as are caused by changes in station location, observation times, or instrument exposure (Kennedy & Gordon, 1980). Resources permitting, these early data should be subjected to the same close scrutiny applied to the calibration data. Unrecognised inhomogeneities are a source of error that can influence the apparent reliability of a reconstruction. The length of the early records may present another difficulty. The majority of the early observational records in North America are so short that their statistical evaluation becomes somewhat ambiguous. Because of dependencies on sample size, the interpretation of many statistics derived from fewer than 30 observations can be held suspect. It is best to maintain a conservative position when examining such results.

Finally, a thorough verification of a model would involve examining its performance in selected situations that represent the deviations from the calibration data that are expected when applying the model to independent data. It would be desirable, for example, to examine estimated predictands corresponding to no-analogue situations, that is, variations of the predictors that do not appear in the calibration period (Hutson, 1977). With limited amounts of early independent data, little control can be exercised over the situations in which the model is tested. Just as important, however, a random selection of the verification data over many situations is also prohibited. This last fact introduces a possible source of positive bias into the verification results in that the few observations available are likely to be immediately adjacent in time to the calibration data and therefore, similar. The model is more likely to perform well in similar instances.

These difficulties will not completely compromise the information to be gained about model performance from early records. Verification with these types of data should be pursued wherever possible. Situations may often arise, where there are simply no early independent data suitable for verification. In such cases, there are alternatives open; alternatives that still allow a rigorous and thorough evaluation of the model. These are the various subsample replication techniques (Mosteller & Tukey, 1968; Stone, 1974; McCarthy, 1976).

SUBSAMPLE REPLICATION

Subsample replication might be more familiar to some as a modified "jack-knife" technique (Mosteller & Tukey, 1977). In its most simple form, subsample replication involves dividing the available body of calibration data in half. The model being examined is then calibrated on the first half and its performance is tested on the second half. The roles of the two halves are then reversed with the model being calibrated on the second half and tested on the first. This results in two estimates of the reliability of the model calibrated on a sample half as big as the available calibration data. Of course there are a great many different ways that the data may be halved, and the selection of a proper subset of these possibilities has been discussed in some detail by McCarthy (1976). These reliability estimates may give a satisfactory indication of the reliability of that particular model. If the model is found to be reliable, it can be recalibrated with the entire calibration data set and then used to reconstruct climate. This may be attractive if the body of calibration data is so large that halving it does not reduce the calibration data to an inadequate size. Palaeoclimatic researchers are rarely so fortunate.

It would be preferable in any event to evaluate the performance of the model calibrated on all of the data because that model should be the most reliable and will be used to make the final reconstruction. Dividing the calibration data into a number of equal parts will better approximate the final reconstruction situation. If 10 divisions are made, for example, any nine parts can be combined for calibrating the model and the results used to estimate the observations in the tenth that remains. After 10 replications, the body of independent predictand estimates and observations is as large as the original calibration data set. The model can then be verified on this large body of independent data. This procedure could even be carried to its limit, where the observations in the calibration period are left out one at a time and calibrations are made on the remainders. This might be called a "leave-out-one" procedure. Such a procedure which uses the available data to the maximum extent would come closest to verifying the final model calibrated on all the data.

There are two apparent limitations to subsample replication procedures. First, the amount of computation necessary to carry out suitable replications can increase enormously the amount of computation necessary to select a reliable model. Second, the model that is used to make the final reconstructions is never actually verified. On further examination, the first limitation is not seen to be a serious obstacle, but the second does require consideration when reviewing verification results.

The amount of computation necessary to carry out a subsample replication procedure may at first appear overwhelming. The speed and efficiency of present-day computers, however, have created a situation where 10 calibrations may be had for little more trouble or expense than one. This is true even for the large data sets encountered in palaeoclimatic research. With the proper modifications to the optimising algorithms, even the limiting "leave-out-one" procedure may become a viable alternative (Mosteller & Tukey, 1968). The implementation of subsample replication procedures need be neither inconvenient nor come at a high cost.

The second limitation poses a more serious problem. The model that is used to make the final reconstructions is calibrated on the entire body of available data. The model that is verified, however, is calibrated on smaller pieces of those data. Because of the optimisation procedure, the model coefficients are unique to the calibration data and will be altered by changes in that data set. Therefore, the verification results will not apply to the final reconstruction that is produced, but will apply to the form of the model only. The subsample replication statistics only indicate which combination of predictors provides the most reliable estimates of climate. Thus, the form of the model is verified; the final regression equation, calibrated using all the data, remains unverified. The reconstructions produced by the final calibration can only be evaluated through the use of a body of early records. Where none exists, the indication of reliability obtained by subsample replication must suffice.

SUMMARY

Dendroclimatic reconstructions obtained from regression models do contain some level of error. If the extremely valuable information that the reconstructions contain is to be used intelligently, some assessment of the pervasiveness and magnitude of the inaccuracies must be made. The blind acceptance of a reconstruction without some indication of its limitations will surely lead to some erroneous conclusions about past climate. That misfortune becomes less likely with the continued investigation and establishment of verification procedures.

Any verification procedure must involve some body of independent data; this represents the greatest obstacle to verification. Limited amounts of early independent predictand data will usually be available and should be utilised. Subsample replication techniques offer a viable alternative to verification with early independent data. The value of any statistical verification procedure relies upon the use of appropriate statistics and their careful interpretation. Simple correlation-type statistics will not provide sufficient information on the reconstruction models. More sensitive measures such as the reduction of error statistic (Lorenz, 1956, 1977; Fritts et al., 1979; Kutzbach & Guetter, this volume) should be used. More research into the behaviour and interpretation of verification statistics is much needed.

Any attempt at verification will be of value. Historical, proxy, and statistical verification will all reveal useful information. Statistical verification will be particularly valuable, especially when a combination of early records and subsample replication techniques is used. These methods will provide a necessary perspective on the reliability of the climatic signal contained in dendroclimatic reconstructions.

COMMENT

B.M.Gray & J.R.Pilcher

Statistical verification. Subsample replication appears to be the optimum method at this time by which limited data can be used to both calibrate and verify the form of a transfer function model. This method reduces the problem that a formula good for one sample may be poor for the population (Lorenz, 1956). The final reconstruction should certainly use all the available data for maximum statistical stability, and for this model the parameters themselves are not verifiable. However, this need not be a total disadvantage. The parameters of many models, such as ARMA (autogressive moving average), show slight changes as the data base is altered. By comparing the parameters in a subsample replication, it might be possible to estimate their scatter. The parameters for the final model can then be examined to see if they lie within reasonable limits. If the subsamples do not overlap, it might be possible to detect some structure in the parameter values, which could indicate omissions in the model.

Limiting the number of variables in the model should improve the stability of the parameters. The commonest way to do this is to use only a few of the principal component amplitudes of the predictor, rather than the full set of raw data. This method could be useful in subsample replication where the number of observations limits the number of degrees of freedom available. It should be noted, however, that Conkey (1979a) found higher calibration and verification in models using untreated data than in models using principal component analysis.

Historical verification. The problems of verification are very different in different parts of the world. At one extreme, there are areas of northwestern Europe with instrumental records of climate back to the late 18th century and detailed historic records before that. At the other extreme, there are areas where even to obtain enough climate data to calibrate a dendroclimatic model would prove a major difficulty. It is ironic that Europe, with the best instrumental climate record and the best chance of verification, has, at present, a very poorly developed dendroclimatic programme.

Where really long climate records are available, it is possible to carry out a verification on as large a block of independent data as was used for the calibration. However, even in Europe, the reliable long records do not form a uniform and satisfactory grid for calibration of a close grid of tree-ring stations. The long records do, however, provide very good spot verification over long periods. This spot verification can be extended further using the indices derived from historical data such as crop yields, and wine vintages. In spite of the misgivings some climatologists have about these indices and other historical or documentary information, these data are going to be very important, particularly if a decision is taken to include chronologies from building timbers in any European dendroclimatic study. In much of Europe the living trees are young, and long chronologies can only be constructed using such timbers. To use such chronologies adds a considerable measure of uncertainty to a dendroclimatic study as it is possible that the trees used for calibration are not a good representation verification that would extend beyond the living trees in the chronology into the historic building period.

The methodology of historical reconstruction has been clarified greatly in the last few years (Wigley et al., 1981a). The criteria for an ideal weather source are that it should be written by a first-hand observer, within a short time of the event noted. In practice, the criteria may be relaxed slightly to include observations documented later and even hearsay if it corroborates other evidence. If these criteria are not observed, many possible sources of error can be identified, including errors in transcription or translation and self-interested exaggeration. For instance, Bell & Ogilvie (1978) have cited transcription errors where one event has generated seven differently dated references in later chronicles; and it is likely that officials may have exaggerated the severity of the weather to excuse the non-performance of their duties.

There is a large amount of information which meets the necessary criteria for the period 1600 to 1750 in England, Flanders, Germany, and the Paris area. Cross references by independent authors occur in these regions for this period, but are rare for other regions and periods. Scattered data are available from Valencia, Venice, Rome, Leipzig, and other places in Europe, and probably additional data exist in the mass of unpublished material.

CLIMATIC RECONSTRUCTIONS FROM TREE RINGS
A.B.Pittock

THE NEED

Climatologists need climate reconstructions from tree rings in order to extend their time series of climate data backwards in time and hence improve the statistical and physical understanding of climate, its variations, and extremes. Longer data sets are needed to test numerical models of climate, and to study spatial patterns of variation on longer timescales. Historians and other palaeoscientists also need dendroclimatic data to aid the interpretation of historical, archaeological, and other past events.

Ideally, each of these purposes requires specific palaeoclimatic data. This may be maxima or minima of temperature, precipitation, atmospheric pressure, atmospheric circulation parameters, or riverflows. Data may be required on an annual mean basis or for specific seasons or even for transient extreme events such as frosts or other critical temperature levels. Data may be needed for one or more given sites or, as in the case of numerical models of climate, covering the whole globe. Moreover, the ideal of climatologists would be for reconstructed data which represent at least 70 to 80% of the variance in the instrumental climate series, and which have basically similar statistical properties to the instrumental data. For example, they should have similar variability about the mean, frequency behaviour (including autocorrelations), and spatial properties (for example, correlation between different sites).

REALITIES AND UNCERTAINTIES

The reality is that tree rings from particular trees at particular sites can give us only limited types of information (for example, concerning precipitation or temperature) in particular seasons of the year, and they can give only a fraction of the total variance in whatever variable is considered. Typically, only 30 to 40% of the total variance in a tree-ring series is climate-related, and only a similar percentage of the variance in indep-

endent climate data is accounted for by a reconstruction. The types of climate variables which can be obtained from tree rings, and also the spatial coverage of any chronology network, is limited by the ecological ranges of sensitive and otherwise suitable trees. The lack of suitable trees, especially of long-lived species in many areas and of old individual specimens in many long-settled areas, is a major factor even in temperate latitudes. Gaps in coverage occur over large areas of climatic importance, such as eastern and central Argentina and mainland Australia (Pittock, 1980a,b; LaMarche, this volume).

We must ask ourselves difficult questions concerning the validity of dendroclimatic reconstructions, particularly about the frequency response of the reconstructed data, the post hoc selection of the results from a multiplicity of reconstruction models, the identification of biological growth curves, and biological persistence or tendencies toward quasi-periodic behaviour (notably biennial, but also that due to lobate growth). Are dendroclimatologists justified in applying complicated statistical models and transforms, particularly when they do not yet understand the complex climate-soil-hydrological-biological interactive system? What is the precise effect in the frequency domain of using a canonical regression transform on tree-ring and climate data? Does the climate-growth response of an individual tree remain constant over the life of the tree? Given that a tree necessarily grows in a microscale environment which is changing as it grows through the forest canopy and as its roots spread wider and deeper into different soil and subsoil environments, the assumption of a constant climate-growth response is very questionable. How is this accounted for in statistical sampling when, for any given site, the ring-width chronology is normally based on trees of increasing age?

How well are biological growth curves separated from real climatic variations? Even when simple exponential growth curves are used, does this procedure in fact change the frequency behaviour of the derived time series? How serious is this, and how well do dendroclimatologists stay on guard against this affecting their (or other's) conclusions from their work? This, it seems to me, is a problem with even a 200-year time series, but how much more serious is it for the very long time series which it is possible to derive from bristlecone pine, Irish bog oak, the riverine oak in Germany, and eventually the New Zealand bog kauri? How should the data from individual trees be joined together into these long series without artificially supressing real changes in means and variances? And how can real climatic fluctuations be separated from other fluctuations such as in the hydrology or nature of the soil due to forest clearing and soil erosion of human origin? These questions and many others have been discussed in this volume and elsewhere; some have yet to be answered satisfactorily.

How well do dendroclimatologists understand even in the broadest terms the "black box" of the climate-growth response or of the growth-climate calibration? It is reasonable to assume that the better the trees at a given site respond to climate variables the greater will be the proportion of their individual variations in growth rate which is common between the trees at that site. This can be demonstrated by correlating these two variables for 33 tree-ring sites in Argentina and Chile (LaMarche et al., 1979a,b). This is shown in Figure 2.22. The average percentage variance in the individual cores at each site which is related to climate has been calculated as the product of the percentage variance in the average chronology which is explained by climate (as found by the response function analysis) and the percentage variance of all the cores which goes into the average chronology (that is, the percentage variance in common between cores). There is a positive correlation, as expected, but the percentage variance related to climate falls a long way short of the percentage variance in common (compare the equivalence line, "a", with the line of best fit, "b"), and there is a considerable amount of scatter. As should be apparent from the manner in which the ordinate has been calculated, the assumed relationship is, in fact, axiomatic. There are a number of reasons why the shortfall

and scatter should occur:

a) considerable variance in common is due to common influences on the trees at a given site other than climate;

b) insufficient replication at the site has left too much noise in the estimated variance in common;

c) the climate data used are from too far away or are of too poor quality;

d) not all the relevant climate variables are included in the response function analysis (for example, cloudiness, wind speed, or data for further months prior to the growth season); and/or

e) non-linear combinations of, or dependencies on, climatic factors should be used.

The existence of large percentages of variance in common between widely separated sites under common climate influences suggests that (a) is not important other than at a few "poor" sites, while high levels of replication now routinely used at most sites minimise (b). (c) is undoubtedly important for some tree-ring sites in Argentina and Chile. However, my belief is that the shortfall in estimated response to climate is in most cases largely due to the use of inadequate or inappropriate combinations of climate data in the response function analysis. In other words, there is more climate information in the tree rings than is modelled by the response function analysis. In order to improve results dendroclimatologists need to experiment with other combinations of climate data for calibration, guided by better biological and physical insights into how trees respond to climate. Further research into biophysical models of tree growth is needed.

One other more general point should be made about reconstructions: a time series of proxy climate data from one tree or one site, or even instrumental data from one location, may have little or highly qualified general significance. The palaeoclimatic literature contains many examples of general climate inferences made about large regions, even whole continents, latitude zones, hemispheres, or the globe, based on highly localised data. A primary question each worker should ask is whether, or in what way, the data is representative of the region or of some wider area.

Figure 2.22: Scattergram of average percentage variance in individual cores which is related to climate according to the site chronology response function analysis (see text) and percentage variance in common between cores. Data are for 33 sites in Argentina and Chile including three different species of tree, viz, <u>Araucaria</u> <u>araucana</u>, <u>Austro-cedrus</u> <u>chilensis</u>, and <u>Pilgerodendron</u> <u>uviferum</u>. This diagram was compiled with the assistance of R.L.Holmes and modified after discussions with G.R.Lofgren and J.P.Cropper (all of the Laboratory of Tree-Ring Research, University of Arizona).

THE DENDROCLIMATOLOGISTS' RESPONSE

The grudging response of the modern quantitative meteorologist to reconstructed climate series which typically represent only a quarter or a third of the variance in a fraction of the variables which she or he wants must be that it is better to have some data than none at all. Perhaps dendroclimatologists have a right to say the same thing with more enthusiasm, but only as long as they realise that they are in fact falling far short of the need and that their results are far from giving the whole story. They need to be both modest about their results and motivated to try to do better.

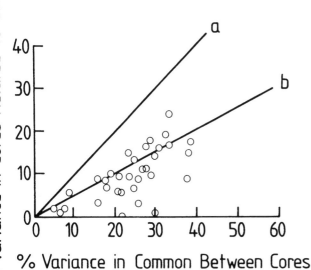

Let us therefore agree and state categorically that the present methodology is not final. It is based on inadequate understanding of a complex physical-biological system and on complex statistical techniques, the precise effects of which are not fully understood. The more dendroclimatologists understand of these biophysical systems and of the statistical models they use, the better they will be able to design their sampling techniques and transforms to get better, more reliable, and more credible results. It follows that there must be constant emphasis on testing and verifying assumptions, methods, and reconstructions on independent data. Workers should not delude themselves by thinking that selecting the reconstructions from one statistical model (out of many) which tests out best on independent data is necessarily the best in general or that it will necessarily stand up as well against another set of independent data. Post hoc selection of results is an insidious, dangerous, and widespread practice (see, for example, Pittock, 1978).

The question of representativeness is answered either by obtaining more proxy data from different localities, or by examining the spatial patterns present in the modern climatological record. The latter show that different or opposing anomalies or trends can and do occur at different sites, often surprisingly near each other. For example, principal component analysis of 30 years of

annual precipitation data for Argentina and Chile (Pittock, 1980a) reveals components 1, 2, and 3 as illustrated in Figure 2.23. These components account for 20.3, 13.1, and 8.1%, respectively, of the total variance in precipitation over the whole network (and much higher percentage variance at stations near the centres of action of each component), and the annual amplitudes of these components correlate significantly with various indices of the atmospheric circulation. Dendroclimatic chronologies at present exist in Argentina and Chile essentially along the Andes between latitudes 33 and 43°S. We see therefore that these chronologies may give us information about components 2 and 3, but not about component 1, and that in order to get information about component 1 we need palaeoclimatic data from eastern Argentina and possibly southern Brazil. Component 3 shows that opposing trends may well occur as close as 10 degrees of latitude apart along the Andes. Different or opposing trends may even occur in the same climatic variable (for, example, precipitation) at the same site in different seasons, as is the case with component 3 in Figure 2.23). Dendroclimatic data must, therefore, be interpreted in the context of what is known about modern patterns of climatic variability and taking into account data from the most climatically significant localities where possible. This will require collaborative and interdisciplinary research

Figure 2.23: Maps of the first three principal components, or eigenvectors, of the correlation matrix of annual precipitation data from a network of 87 stations in Argentina and Chile for the 30 years 1931-60. Stations are indicated by triangles. Components 1, 2, and 3 (a, b, and c, respectively) account for 20.3%, 13.1%, and 8.1% of the total network variance, respectively. From Pittock (1980a).

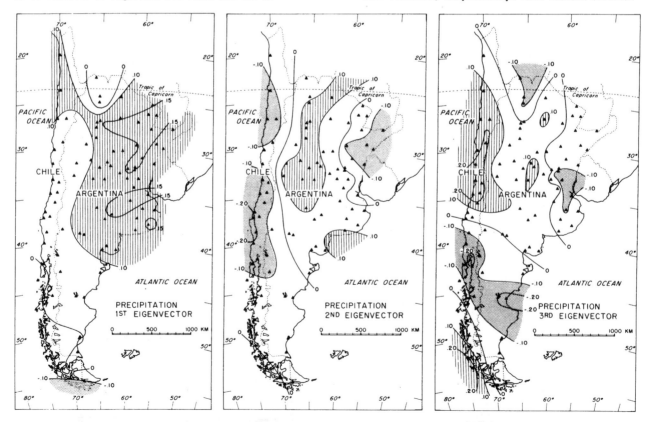

in order to obtain the maximum value from the
dendroclimatic research.

As dendroclimatologists try to meet the need for
data from a wide area (ultimately the whole globe) and for
all seasons and climate variables, they must obviously
search for widely dispersed sites and trees giving a
variety of response functions. In this they might well pay
special attention to analyses such as Kutzbach & Guetter
(1980, this volume) on palaeoclimatic data network design,
and analyses of spatial patterns in instrumental climate
data (Pittock, 1980a,b) which will suggest priority areas
in which to look for data. Parenthetically, care must be
taken to identify differing frequency responses in data
from different biological sources, whether they be merely
different species of trees or even different proxy
indicators.

Realistically, we should not expect that dendroclimatology alone will provide all the palaeoclimate data
which climatologists and others want. In order to get even
a little nearer the ideal, dendroclimatologists will need
to encourage and participate in the integration of
dendroclimatic data into networks of other proxy data such
as those derived from glacial, pollen, and perhaps even
ocean core analyses. This, however, will pose great
problems, particularly in the time domain. It may be that,
given the less accurate dating and slower response of
other palaeoclimate indicators, the range of usefulness of
dendroclimatic data can be extended by selectively
relaxing dating standards where necessary to a degree
comparable with that obtainable in ice cores or varves.
This may enable less "well-behaved" species to be used,
for example, in the tropics, Africa, or mainland
Australia.

Finally, I believe much greater effort should be put
into the extraction of greater and more varied information
from dated tree-ring samples, particularly to obtain data
for different climate variables on intra-annual, as well
as interannual, timescales. Isotope ratios and densitometric methods are the obvious candidates, but there may
well be others as yet unknown.

Spending two years at the Laboratory of Tree-Ring
Research at the University of Arizona has deepened my
respect for anyone who can extract climate information
from such a complex system as a tree. Therefore, I want to
pay my respects to those who have made dendroclimatology
the science I believe it is today. Nevertheless, it is
still in its youth. I believe it is robust enough to be
asked, and to ask itself, tough questions and to grow and
mature as a result. As a climatologist, I remain impatient
for the answers, but am still sufficient of an outsider to
ask rude questions which may threaten some well entrenched
beliefs and assumptions.

COMMENT

J.P.Cropper

There are problems in the analysis and interpretation of any scientific data, whether from a Saturn
space probe or a chemical experiment performed in the
laboratory. Dendroclimatology is not exempt from these
problems and it does us good to bear the limitations of
our science in mind when interpreting our results.

The method used by Pittock to assess the quality of
chronologies as climate predictors (Figure 2.22) is valid,
but only if the estimates of percentage common variance
and climatically related variance can be considered
reliable. If the reliability of these estimates is
questionable or variable (see Hughes & Milsom, this
volume), then the analysis has little interpretive value.
With the present method of producing a final chronology by
forming the arithmetic mean, the reliability of the final
chronology as being representative of that data set
increases with sample size (the Central Limit Theorem), as
does that of the analysis of variance. I have elsewhere
(Cropper, this volume) indicated the possibility of
obtaining seemingly significant climate calibrations from
response function analysis using random numbers in place
of tree-ring data. Thus the quality of the final chronology is important in obtaining meaningful results for the
amount of climate calibration (using the response function
method).

If the chronologies in a set were all developed from
a large sample size (say, over 20 trees), then the
estimates from the analysis of variance, mean chronology,
and response function could all probably be viewed as
being reasonably meaningful. In such a case, the type of
analysis performed by Pittock would be valid. It is rare,
however, to find a data set in which every chronology is
developed from a large sample size. This is particularly
true for the South American data used in the example; one
chronology was developed from as few as three trees
(LaMarche et al., 1979a; chronology ID KIL799) while
another contained data from 29 trees (LaMarche et al.,
1979a; chronology ID RUC799). The reliability of the
estimates of the points in Figure 2.22 is highly variable.

If the sample size could be taken into account in
some way, then the usefulness of the analysis would
increase. In some of the papers presented in this volume,
the concept of the signal-to-noise ratio has been
discussed, both implicitly and explicitly. Here, this idea
will be used as a means of evaluating, or scaling, the
amount of information in the final chronology. The method
employed will distinguish between, say, a chronology from
two trees with 30% common variance and one from 20 trees
with 30% common variance. The ordinate for each point is
calculated from the equation

% common signal = $100S/(S+N) = 100(S/N)/(S/N+1)$ (2.31)

where S/N is the signal-to-noise ratio for the final chronology. The signal-to-noise ratio is calculated using the equation

$$S/N = N \times \%Y/(100-\%Y)$$ (2.32)

where %Y is the percent variance common to all trees in the sample, (100-%Y) is the variance attributed to differences within and between trees, and n is the number of trees included in the final chronology. The above equation assumes a single core sample per tree. This assumption both simplifies the calculations and results in a minimal estimate of S/N if multiple samples per tree are incorporated into the final chronology.

The vertical axis can be interpreted as the potential information in the chronology, or the potential percent variance related to climate. The horizontal axis is derived from response function analysis and represents the amount of chronology variance accounted for by a linear combination of principal components of temperature and precipitation (Fritts, 1974; Guiot et al., this

volume). Presently, there is no method available for assessing whether the climate calibration is against the chronology signal or noise (or both) without performing a reconstruction and verification. Figure 2.24, generated from the same data used by Pittock, is a first step in trying to assess which portions of the chronology variance have been calibrated against climate. The figure can be interpreted as follows:

Quadrant A: Chronologies in this quadrant have less calibrated variance than they contain as noise. Thus, there is a possibility that the calibration is totally against the noise component in the final chronology.

Quadrant B: In this quadrant once again all of the climatic calibration could be against the chronology noise.

Quadrant C: To be plotted in this quadrant, the climate data must have calibrated a mixture of both signal and noise. This is because more variance has been calibrated than can be attributed to either signal or noise separately.

Quadrant D: In this quadrant, there must be a calibr-

Figure 2.24: Common signal in final chronology plotted against the variance in the final chronology explained by climate. Quadrants A and B: potentially calibration against chronology noise; Quadrant C:

definitely calibration against signal and noise; Quadrant D: definitely calibration against signal with possibility of some noise; Point E: the ideal chronology plot.

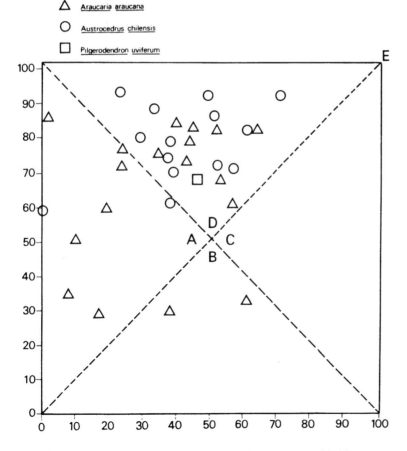

△ Araucaria araucana

○ Austrocedrus chilensis

☐ Pilgerodendron uviferum

ation against signal and need not be any noise involved. The climatic calibration accounts for more variance than could be noise alone, but is within the range that could be accounted for by signal alone.

The closer a chronology is plotted to the top right-hand corner of the figure (point E), the better the quality of the chronology and its climate calibration. Also, there is greater likelihood that a reconstruction made from such a chronology would verify against independent data.

There are various reasons why a chronology may be considered poor for climatic analysis, that is, be plotted in quadrant A or B. The chronology may not contain any climate signal, the climate data may not be reliable (Trenberth & Paolino, 1980; Kennedy & Gordon, 1980), the chronology sample size may be insufficient to allow good clarification of the common signal, the correct (limiting) climate parameters may not have been used in the response function analysis, or the groupings of months used in the response function may not have been biologically correct.

THE DESIGN OF PALAEOENVIRONMENTAL DATA NETWORKS FOR ESTIMATING LARGE-SCALE PATTERNS OF CLIMATE

J.E.Kutzbach & P.J.Guetter

Based on an article first published in Quaternary Research (14, 169-187, 1980)

INTRODUCTION

The purpose of this analysis is to determine guidelines for the spatial density and location of climate variables that are appropriate for estimating continental-to hemispheric-scale pattern of the time-averaged atmospheric circulation. The estimation of spatial patterns of sea-level pressure from observations of temperature and precipitation at particular sites is dealt with, and the density and location of sites is varied in order to evaluate alternative sampling strategies. To the extent that instrumental records of temperature and precipitation simulate botanical records, the guidelines for sampling strategy will have applications for studies of palaeoclimate.

The accuracy of mapping of the patterns of sea-level pressure must depend upon the density and the location of samples of temperature and precipitation. Site density controls spatial resolution, and spatial resolution controls the scale of circulation features that can be resolved. The location of samples determines the area within which accurate estimates of circulation can be made. However, in some situations it may be possible to estimate the circulation outside the region of samples because the stationary-wave patterns of the atmospheric circulation are of continental scale.

By evaluating the accuracy of estimation of the pattern of sea-level pressure from a wide variety of climate data networks, a number of sampling strategy guidelines will be determined. The application of the

guidelines to palaeoclimatic research is based on two fundamental assumptions:

a) instrumental records of monthly or seasonal-average temperature and precipitation (or some other climate variable) simulate the climate information that may be recovered from tree-ring (or other proxy or written documentary) data; and

b) the statistics of the spatial structure of past climates are not radically different from the statistics of spatial structure of climate determined from the instrumental records.

The first assumption is well-justified by the progress in palaeoclimatic reconstruction from tree-ring records (for example, Fritts, 1976), pollen records (Webb & Bryson, 1972), and foraminiferal records (Imbrie & Kipp, 1971). The second assumption should be justified for the climates of recent centuries or millenia. For one particular sampling strategy, it will be shown that the guidelines developed in this analysis agree closely with the results of Fritts (1976) for the reconstruction of patterns of sea-level pressure from a sample network of tree-ring records.

REGION OF STUDY

Sampling areas for sea-level pressure. The pattern of sea-level pressure will be estimated for five areas within the latitude band 30 to 60°N (Figure 2.25). The five areas span progressively broader longitudinal sectors:

a) Western North America, WNA, 130 to 110°W;

b) Western plus Central North America, WNA + CNA, 130 to 90°W;

c) North America, NA, 130 to 70°W;

d) Eastern Pacific plus North America plus Western Atlantic, EP + NA + WA, 150 to 50°W;

e) Pacific plus North America plus Atlantic, P + NA + A, 160°E to the Greenwich Meridian.

The sampling strategy analysis will evaluate the accuracy of estimation of the pattern of sea-level pressure for these five areas as a function of the density and location of temperature and precipitation records. Observations of monthly pressure for 78 Januarys (1899 to 1976) are sampled on a 5 degree latitude and longitude diamond grid (Figure 2.25). There are 18 observations (grid points) for WNA and 144 observations for P + NA + A.

Site network for temperature and precipitation data. Climate stations have been identified for six grids: Western North America (WNA), Western plus Central North America (WNA + CNA), North America (NA), North America plus Europe (NA + E), North America plus Japan (NA + J), and North America plus Europe plus Japan (NA + E + J). These six grids span progressively broader longitudinal sectors; gaps occur in the network over the oceans (Figure 2.25). These grids can be considered as approximations to palaeoenvironmental data networks. Observations of temperature and precipitation for 78 Januarys (1899 to 1976) are available for each site.

There are three levels of site density: low, medium, and high, in the approximate proportions 1:2:4. All sites in the low density network are included in the network of medium density and all of its sites are included in the network of high density (Figure 2.25). This procedure is similar to the "nesting strategy" described by Barnett & Hasselman (1979). If the site density is expressed as sites per million square kilometres (sites/Mkm^2), then low-, medium-, and high-density correspond to about 0.62 sites/Mkm^2, 1.25 sites/Mkm^2, and 2.5 sites/Mkm^2, respectively. For comparison, the global-average site density for upper-air meteorological observations is about 4 sites/Mkm^2 and for surface observations about 15 sites/Mkm^2. Site density for tree-ring records in western North America is greater than 10 sites/Mkm^2 (Fritts, 1976); the site density would be much lower if averaged for the entire continent because the eastern sites have not yet been incorporated in a joint network with those from the west. The highest site density of this study, 2.5 sites/Mkm^2, implies a characteristic distance between sites of about 600km so that the smallest resolvable

feature would be about 1,200km. The methodology developed here could be used for evaluating the effectiveness of higher or lower site densities.

DATA ANALYSIS

The analysis of the temperature, precipitation, and sea-level pressure data involved several steps: determination of the period of record and estimation of missing data; representation of the data in terms of principal components; determination of regression equations for estimating the pattern of sea-level pressure from the temperature and precipitation records for various site networks; and description of the results in terms of explained variance, F-test criteria, and reduction of error. These steps are described below.

Period of record. The sampling strategies were evaluated for two periods, 1899 to 1950 and 1925 to 1976. The results were then verified with data from the periods 1951 to 1976 and 1899 to 1924, respectively. The 52-year records are the "dependent" data sets (because the coefficients of the regression equations depend on the data); the 26-year records are the "independent" data. In a few cases where the station data were incomplete, data

Figure 2.25: Six climate grids (site locations) and five sea-level pressure areas for sampling strategy experiments. The grids consist of low density (●), medium density (● + ⊗), and high density (● + ⊠ + ○) site networks. The sea-level pressure variable is specified at latitude-longitude intersections (+). The grid of sea-level pressure is regular but the crosses (+) are not shown at some places in North America because of overlap with the station network symbols. See Table 2.6 for summary.

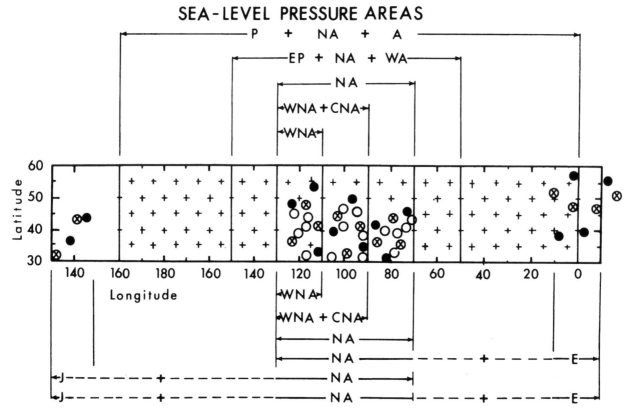

SEA-LEVEL PRESSURE AREAS

CLIMATE SITE LOCATIONS

from a station a few kilometres away were used to fill
gaps. Where there were no data available from a nearby
station, the station mean was substituted. This occurred
for five of the 46 stations for one or at most two
Januarys.

**Principal component analysis of sea-level pressure
data.** Principal component analysis (or eigenvector
analysis) was used to reduce the number of sea-level
pressure observations. In each of the five areas
(Figure 2.25, Table 2.6a), the observations of sea-level
pressure were expressed as departures from the 52-year
January mean of each dependent data set, at each point on
the 5 degree x 5 degree diamond grid. This procedure
yielded a total of 10 sets of pressure departures (five
areas x two dependent data sets). Each of the sets of
pressure departures were defined in terms of a weighted
sum of the principal components, where the component
scores are the weights

$$P(i,1) = \sum_{j=1}^{Mp} Y(i,j)e(j,1), \quad i = 1,2,\ldots,52 \qquad (2.33)$$

In this equation, P(i,1) is the "i"th sea-level pressure
departure (the "i"th January observation) for the "1"th
grid point as determined from the representation of the
data with Mp principal components; e(j,1) is the portion
of the "j"th principal component corresponding to P(1)
(the pressure departure at the "1"th grid point); and
Y(i,j) is the component score of the "j"th principal
component of sea-level pressure for the "i"th observation.
The component scores (eigenvector coefficients) give the
time series of the magnitude and sign of the principal
component patterns, and it is the component scores of the
principal components of sea-level pressure that will be
estimated via regression analysis from the climate
information at sites.

In Equation (2.33), Mp denotes the number of
principal components of sea-level pressure that are
retained. If all components are retained, then P(i,1) is
an exact fit of the observations. Using the criterion that
enough components should be retained to account for 90% of
the variance of the sea-level pressure, Mp varied from
three to eight (Table 2.6a): three principal components

explained 90% of the variance of sea-level pressure on the
18-point grid of the WNA area; eight principal components
were required to explain 90% of the variance on the
144-point grid of the largest area, P + NA + A.

**Principal component analysis of temperature and precip-
itation data.** The temperature and precipitation data at
each site were expressed in standardised or normalised
form

$$t(i) = (T(i) - \bar{T})/\sigma_T \qquad (2.34)$$

where T(i) is the "i"th January observation of a variable,
temperature or precipitation at a station, $\bar{T}$ is the
52-year January mean of the variable, and σ_T is the
standard deviation. Principal component analysis was done
on these standardised temperature and precipitation
variables for each of the temperature and precipitation
grids (six) and for each site density (three site
densities for three of the grids and two site densities
for the remaining three grids).

For each of the 15 data sets representing the
different site densities and grids, the equation for the
temperature and precipitation variables is

$$t(i,1) = \sum_{j=1}^{Mt} X(i,j)e*(j,1), \quad i=1,2,\ldots 52 \qquad (2.35)$$

where t(i,1) is the "i"th standardised temperature or
precipation (the "i"th January observation) from the "1"th
climate site as determined from the representation of the
data with Mt principal components, e*(j,1) is the portion
of the "j"th principal component corresponding to t(1)
(standardised temperature or precipitation at the "1"th
site), and X(i,j) is the component score of the "j"th
principal component of temperature and precipitation for
the "i"th observation. Note that the maximum value of 1 is
twice the number of climate stations because both
temperature and precipitation are variables at each
station (site).

The climate data at sites were expressed in terms of
the principal components that explained more than the
average variance per component. For example, if the total
variance (100%) of the climate data for a particular site

Table 2.6a: Summary of sea-level pressure areas and principal components (see text for abbreviations).

SEA-LEVEL PRESSURE AREA	LONGITUDINAL RANGE, WEST TO EAST	NUMBER OF PRINCIPAL COMPONENTS REQUIRED TO EXPLAIN APPROXIMATELY 90% OF THE VARIANCE
WNA	130°W – 110°W	3
WNA + CNA	130°W – 90°W	4
NA	130°W – 70°W	5
EP + NA + WA	150°W – 50°W	6
P + NA + A	160°E – 00	8

location and observation period was explained by 20 principal components, then only those principal components that explained 5% or more of the total variance were used as predictors in the regression analysis. Depending upon site density and site location, the number of principal components (Mt) varied from 2 to 14 (Table 2.6b). This restriction of predictor variables was intended to reduce the possibility of "overfitting" the regression equation (Davis, 1976).

Regression analysis. The component scores for each principal component of sea-level pressure were estimated from linear combinations of component scores for the set of principal components of the temperature and precipitation data at sites. The equations for the estimated component scores for the principal components of sea-level pressure are

$$\hat{Y}(i,j) = \sum_{k=1}^{Mt} b(j,k)X(i,k), \quad j=1,2,\ldots Mp \qquad (2.36)$$

where $\hat{Y}(i,j)$ is the regression estimate of the "i"th observation ("i"th January) of the "j"th principal component score of sea-level pressure; $X(i,k)$ is the "i"th observation of the "k"th principal component score of temperature and precipitation; Mt and Mp are the total number of predictors and predictands, respectively; $b(j,k)$ is the coefficient found by linear regression. Because the X´s are orthogonal within each dependent data set, each b may be found by bivariate regression of a Y(j) on an X(k)

$$b(j,k) = S(j,k)/S(k) \qquad (2.37)$$

where $S(j,k)$ and $S(k)$ are the covariance between Y(j) and X(k), and the variance of X(k), respectively

$$S(j,k) = (1/51)\sum_{i=1}^{52} Y(i,j)X(i,k) \qquad (2.38)$$

and

$$S(k) = (1/51)\sum_{i=1}^{52} [X(i,k)]^2 \qquad (2.39)$$

The variance of the principal component score for sea-level pressure, $\hat{Y}(j)$, due to the temperature and precipitation predictor, X(k), is

$$V(j,k) = [S(j,k)]^2/[S(j)S(k)] \qquad (2.40)$$

where $S(j)$ is the variance of the Y(j)´s

$$S(j) = (1/51)\sum_{i=1}^{52} [Y(i,j)]^2 \qquad (2.41)$$

and the total variance of $\hat{Y}(j)$ is

$$V(j) = \sum_{k=1}^{Mt} V(j,k) \qquad (2.42)$$

Explained variance and reduction of error. One measure of the accuracy of the regression model is the fraction of the variance of the Y´s explained by the $\hat{Y}$´s, denoted as explained variance, EV, where

$$EV = \sum_{j=1}^{Mp} V(j)/\sum_{j=1}^{Mp} S(j) \qquad (2.43)$$

An F-test on the significance of the fractional explained variance was made for each predictor-predictand combination. The hypothesis that there was no statistical

Table 2.6b: Summary of climate grids, site density, and principal components.

GRID	APPROXIMATE LONGITUDINAL RANGE	SITE DENSITY	NUMBER OF STATIONS	NUMBER OF PRINCIPAL COMPONENTS USED AS PREDICTORS, DEPENDENT DATA FOR:	
				1899-1950	1925-1976
WNA	130°W - 110°W	low	3	2	2
		medium	6	4	4
		high	11	6	6
WNA + CNA	130°W - 90°W	low	6	5	4
		medium	12	7	7
		high	23	9	9
NA	130°W - 70°W	low	9	5	5
		medium	18	8	9
		high	34	10	10
NA + E	130°W - 70°W; 10°W - 20°E	low	13	7	8
		medium	26	12	14
NA + J	130°W - 70°W; 130°E - 150°E	low	11	7	7
		medium	22	11	12
NA + E + J	130°W - 70°W; 10°W - 20°E; 130°E - 150°E	low	15	9	10
		medium	30	13	14

relation between the predictors and predictands could be rejected with 99% confidence in all cases.

The fractional explained variance statistic, EV, is evaluated from the dependent data sets. In contrast, the reduction of error statistic, RE, is based upon independent data; that is, it is a measure of how closely the model-predicted principal component scores of sea-level pressure, $Y(i,j)$, derived from independent temperature and precipitation data sets, agree with the principal component scores of the corresponding independent sea-level pressure data, $Y(i,j)$. The expression for reduction of error (RE) is

$$RE = 1.0 - \sum_{j=1}^{Mp} \sum_{i=1}^{N'} [Y(i,j)-\hat{Y}(i,j)]^2 / \sum_{j=1}^{Mp} \sum_{i=1}^{N'} [Y(i,j)]^2 \quad (2.44)$$

where N' is the number of observations in the independent data set (26). If the agreement between prediction and observation is perfect, $RE = 1$; if $\hat{Y}(i,j)$ were always predicted to be equal to the mean of the dependent data set (in this case, $\hat{Y}(i,j) = 0$), then $RE = 0$; negative values for RE can result if the predictions are poorer than the "climatology" prediction, that is, $\hat{Y} = 0$. A positive value of RE is an indication of some skill in the sense that the predictions are better than the "climatology" prediction that the climate does not change.

Usually RE will be less than EV and the discrepency tends to become greater when the ratio of the number of predictors (M) to the number of observations in the dependent data set (N) increases, or as EV decreases (Lorenz, 1956; Davis, 1976). According to Lorenz (1956), two approximate relations can be used to estimate the expected values of EV and RE, which will be referred to as adjusted explained variance (EV') and adjusted reduction of error (RE')

$$EV = EV' + [M/(N-1)][1-EV'] \quad (2.45)$$

and

$$RE' = EV' - [M/(N+1)][1-EV'] \quad (2.46)$$

EV' is the expected value of EV in a (hypothetical) large population from which the dependent data sample was taken. The second term in Equation (2.45), $[M/(N-1)][1-EV']$, may be considered to be spurious explained variance which would decrease if the dependent data sample could be enlarged. It arises because correlation between the predictors and predictands can occur by chance. Equation (2.45) may be solved for EV'

$$EV' = [EV(N-1)-M]/[N-M-1] \quad (2.47)$$

The expression for EV' in Equation (2.47) may be substituted into Equation (2.46) to give the adjusted

reduction of error (RE') as a function of EV, M, and N. In the absence of RE statistics, the value of RE' is an indication of whether a regression model will yield good predictions on independent data. A numerical example: a regression model using 20 predictors (M = 20) gives a fractional explained variance (EV) of 0.60 when 50 observations are used (N = 50). By Equation (2.47), the adjusted explained variance (EV') in a much larger sample would be 0.32. By Equation (2.46), the adjusted reduction of error (RE') is 0.05. In this example, one is alerted to the fact that the explained variance of the regression equation for the dependent data set (EV = 0.60) may be inflated and that a more conservative estimate of the expected value of EV would be 0.32. If applied to independent data the model might be expected to show only slightly more skill (RE' = 0.05) than would be obtained by predicting the mean value.

DISCUSSION OF RESULTS

The results contrast the accuracy of various regression models for estimating patterns of sea-level pressure using temperature and precipitation data for six different grids, three different site densities, and the two dependent data sets. Because three grids had three site densities and three grids had two site densities, a total of 15 different site networks were used to estimate sea-level pressure over five different areas, giving a total of 75 regression analyses for each of two dependent data sets. These 150 regression equations were then tested on the corresponding independent data sets. The tables summarise these 300 computations.

The adjusted explained variance (EV') of the component scores of sea-level pressure, as computed by Equation (2.47), is summarised in Tables 2.7a and 2.7b (dependent data). The explained variance (EV), as computed by Equation (2.43), is not tabulated but values of EV for individual cases are noted for comparison with other published results. Reduction of error (RE) is summarised in Tables 2.8a and 2.8b (independent data). The adjusted reduction of error (RE') is computed from Equation (2.46) using the values of EV' found in Tables 2.7a and 2.7b.

Certain general features of the results, first for dependent data and second for independent data, are enumerated.

Dependent data. For each of the sea-level pressure areas, the adjusted explained variance (EV') tends to increase with site density of the predictor information. For example, for the average of the 1899-1950 and 1925-1976 data sets and for the WNA + CNA grid, adjusted explained variance of sea-level pressure for WNA + CNA increases from 0.51 to 0.60 to 0.68 as the site density increases from low to medium to high (Figure 2.26). The number of predictors (Mt) increases from four to seven to eight or nine but the use of EV' rather than EV will tend to correct for increased explained variance due to "over-fitting". Averaged over both dependent data sets and all sea-level pressure areas, the adjusted explained variance of sea-level pressure for low-, medium-, and high-density sites is 0.46, 0.56, and 0.57, respectively. There is a tendency for a smaller increase of EV´ in going from medium to high site density than in going from low to medium site density (Figure 2.26), an indication of "diminishing returns". In isolated instances, EV´ decreases with increasing site density (P + NA + A sea-level pressure area, WNA + CNA grid, 1899 to 1950).

For most grids and site densities, adjusted explained variance decreases as the area of sea-level pressure expands, and vice versa. For example, for the WNA + CNA grid and for medium site density, using the average of the 1899-1950 and 1925-1976 data sets, the adjusted explained variance of sea-level pressure in the area WNA + CNA is 0.60; for the larger sea-level pressure areas, the adjusted explained variance is: NA, 0.51; EP + NA + WA, 0.53; P + NA + A, 0.36. For the smaller sea-level pressure area, WNA, EV´ is 0.69 (Figure 2.27). The only exception to the general trend is the slightly smaller value of EV´ for area NA than for area EP + NA + WA. This sea-level pressure area WNA + CNA (a 40 degree longitude sector), EV´ is 0.38, 0.60, and 0.64 for grids in WNA (a 20° longitude sector), in WNA + CNA, and in NA (a 60° longitude sector), respectively (Figure 2.27). For still broader longitude sectors, EV´ is approximately constant. Only for the largest sea-level pressure area

(P + NA + A) does EV´ increase for each progressive increase in the longitudinal extent of the grids (Figure 2.27). For the smallest sea-level pressure area (WNA), EV´ is approximately constant for all grids that are broader in extent than WNA + CNA (Figure 2.27). To maximise the adjusted explained variance of sea-level pressure for a particular area while using a fixed number of predictor sites, a good spatial sampling strategy is to distribute those sites as uniformly as possible over the entire sea-level pressure area. For example, for the area P + NA + A, a low density of predictor sites spread over NA + E (a total of 13 sites) produces a higher adjusted explained variance (average EV´ = 0.46) than a medium site density over WNA + CNA (12 sites, average EV´ = 0.36) or a high density over WNA (11 sites, average EV´ = 0.30), Figure 2.28. However, it appears that high site densities in restricted grids can in many instances produce explained variance results that are equivalent to results obtained using low site densities over more extensive longitudinal sectors (Figure 2.28): compare EV´ for area NA based upon high site density in WNA (0.45) and low site density in NA + E (0.49).

The explained variance (EV) of sea-level pressure over the area EP + NA + WA that was obtained from the high site density of temperature and precipitation data in WNA was 0.55 (average of 0.49 and 0.60). Fritts & Lofgren (1978) obtained an explained variance of 0.53 for sea-level pressure for winter for the region 100 E to 80 W (Pacific + North America) using as predictors a high-

Table 2.7a: Adjusted explained variance (EV´) for January sea-level pressure.

DEPENDENT DATA: 1899-1950

PREDICTORS: JANUARY TEMPERATURE & PRECIPITATION				SEA-LEVEL PRESSURE AREA (LONGITUDE BAND) (NO. OF PRINCIPAL COMPONENTS)					
Grid	Site Density	No. of Stations	No. of Principal Components	WNA (130°W-110°W) (3)	WNA+CNA (130°W-90°W) (4)	NA (130°W-70°W) (5)	EP+NA+WA (150°W-50°W) (6)	P+NA+A (160°E-00) (8)	Average over all areas
WNA	low	3	2	0.31	0.28	0.26	0.34	0.21	0.28
	medium	6	4	0.49	0.42	0.37	0.44	0.26	0.40
	high	11	6	0.69	0.63	0.56	0.55	0.35	0.56
WNA + CNA	low	6	5	0.59	0.56	0.49	0.52	0.38	0.51
	medium	12	7	0.68	0.56	0.56	0.39	0.56	
	high	23	9	0.72	0.67	0.59	0.59	0.37	0.59
NA	low	9	5	0.57	0.56	0.49	0.50	0.35	0.49
	medium	18	8	0.67	0.63	0.56	0.55	0.38	0.56
	high	34	10	0.73	0.66	0.59	0.59	0.42	0.60
NA + E	low	13	7	0.55	0.54	0.49	0.51	0.46	0.51
	medium	26	12	0.66	0.62	0.57	0.58	0.52	0.59
NA + J	low	11	7	0.58	0.55	0.48	0.50	0.40	0.50
	medium	22	11	0.69	0.66	0.59	0.58	0.45	0.59
NA + W + J	low	15	9	0.56	0.53	0.49	0.53	0.50	0.52
	medium	30	13	0.65	0.62	0.57	0.58	0.56	0.60
Average over all grids				0.61	0.57	0.51	0.53	0.40	0.52 (grand average)

density tree-ring network in western North America. Because these western North America trees respond significantly to winter moisture conditions, it should be expected that the "simulated" tree-ring records (January temperature and precipitation) would produce a similar level of explained variance of sea-level pressure. The close agreement (0.55 as against 0.53) supports the notion that sampling strategy guidelines can be derived from instrumental records and usefully applied to palaeoenvironmental records. A more detailed comparison between these results and those of Fritts & Lofgren (1978) would require a more precise equivalence of predictor grids and sea-level pressure area; in addition, one could use climate variables that would simulate more accurately the tree-ring growth response to climate in this region. Fritts & Lofgren (1978) computed the fractional explained variance with reference to total variance of the pressure field and not the 90% level used here. Our result, EV = 0.55, should, therefore, be reduced by at most 10%, that is, to about 0.50. The reduction of error (RE) is 0.21 for this experiment; that is, there is a considerable drop in the accuracy of a climate reconstruction when the regression model is applied to independent data.

Independent data. The results of estimating patterns of sea-level pressure from independent data (that is, temperature and precipitation data not used to develop the regression coefficients) provide some indication of how well a pattern of past climate could be specified from palaeoenvironmental information at sites, where the regression model had been developed from modern-day records. Using Equation (2.44) for reduction of error (RE), the principal component scores for the principal components of sea-level pressure, as estimated by regression, were compared to the principal component scores of the principal components of the sea-level pressure from the independent data set. (The principal component patterns that were derived from the dependent data were used to describe the independent data.)

In both independent data sets, Tables 2.8a and 2.8b, the average reduction of error (RE) is 0.33. This represents a large drop from the average explained variance of the dependent data set (EV = 0.59) and also a substantial drop from the average adjusted explained variance (EV´ = 0.52). The average adjusted reduction of error (RE´) is 0.44. For these data sets the adjusted reduction of error gives a fairly accurate estimate of the behaviour of the regression models on independent data (Figure 2.29), although it overestimates skill somewhat. The discrepency between RE´ and RE could result from several factors: the expressions derived by Lorenz are approximate; the expressions assume that the N dependent data observations are statistically independent, and that need not be so; the regression models may be incorrect; and the climate fluctuations of the dependent and independent data intervals may not be from the same population.

Since the drop in RE (compared to EV and EV´) may

Table 2.7b: Adjusted explained variance (EV´) for January sea level pressure.

DEPENDENT DATA: 1925-1976

PREDICTORS: JANUARY TEMPERATURE & PRECIPITATION				SEA-LEVEL PRESSURE AREA (LONGITUDE BAND) (NO. OF PRINCIPAL COMPONENTS)					
Grid	Site Density	No. of Stations	No. of Principal Components	WNA (130°W-110°W) (3)	WNA+CNA (130°W-90°W) (4)	NA (130°W-70°W) (5)	EP+NA+WNA (150°W-50°W) (6)	P+NA+A (160°E-00) (8)	Average over all areas
WNA	low	3	2	0.33	0.28	0.23	0.29	0.20	0.27
	medium	6	4	0.42	0.34	0.27	0.37	0.25	0.33
	high	11	6	0.54	0.43	0.34	0.42	0.26	0.40
WNA + CNA	low	6	4	0.54	0.47	0.38	0.41	0.29	0.42
	medium	12	7	0.71	0.58	0.46	0.50	0.34	0.52
	high	23	9	0.81	0.68	0.56	0.60	0.39	0.61
NA	low	9	5	0.58	0.50	0.40	0.45	0.29	0.44
	medium	18	9	0.76	0.65	0.53	0.58	0.38	0.58
	high	34	10	0.78	0.71	0.64	0.65	0.42	0.64
NA + E	low	13	8	0.64	0.57	0.50	0.55	0.47	0.55
	medium	26	14	0.78	0.72	0.66	0.66	0.57	0.68
NA + J	low	11	7	0.63	0.55	0.48	0.51	0.40	0.51
	medium	22	12	0.78	0.69	0.58	0.61	0.41	0.61
NA + E + J	low	15	10	0.68	0.60	0.51	0.56	0.50	0.57
	medium	30	14	0.78	0.70	0.59	0.63	0.59	0.66
Average over all grids				0.65	0.56	0.48	0.52	0.38	0.52 (grand average)

apply to all procedures for palaeoclimatic reconstruction, the RE values (or RE´ values) rather than the EV´ values should be used to estimate the accuracy of reconstructions. Some of the general relationships between reduction of error (RE) and grid and site density are summarised as follows.

The reduction of error usually increases with increasing site density for a given data and sea-level pressure area. For example, for area WNA + CNA for the average of both independent data sets (1899-1924 and 1951-1976), and using the WNA + CNA grid, the reduction of error increases from 0.36 to 0.46 to 0.53 as the number of sites increases from six to 12 to 23 (Figure 2.29). Exceptions to this rule are more numerous for reduction of error than they are for adjusted explained variance. Averaged over both independent data sets and all site locations and sea-level pressure areas, the reduction of error for low-, medium-, and high-density sites is 0.29, 0.35, and 0.37, respectively. The relative magnitudes of the EV, EV´, RE´, and RE statistics are illustrated for area WNA + CNA (Figure 2.29).

For a given grid and site density, reduction of error usually decreases as the sea-level pressure area expands, and _vice versa_. For example, for the WNA + CNA grid and medium site density, the average RE for area WNA + CNA is 0.45; for the smaller area (WNA) the average RE is 0.52 (Figure 2.30); for the larger areas (NA, EP + NA + WA, and P + NA + A) the average RE is 0.37, 0.30, and 0.15, respectively (Figure 2.30). There are exceptions. In the case of sea-level pressure area NA, using the NA grid and 1951-1976 data, the values of RE for

low-, medium-, and high-density sites are 0.28, 0.29, and 0.34, respectively; the corresponding values of RE for sea-level pressure area EP + NA + WA are 0.32, 0.37, and 0.39. This particular exception was also apparent in the dependent data set. Averaged over all site densities and both independent data sets for the NA grid, the reduction of error is 0.33 for the NA area; it increases for smaller areas (WNA + CNA, RE = 0.37; WNA, RE = 0.46) and decreases for larger areas (EP + NA + WA, RE = 0.29; P + NA + A, RE = 0.18).

Holding site density fixed, there is an advantage (in terms of RE statistics) in keeping the longitudinal extent of the grid approximately twice the longitudinal extent of the sea-level pressure area. Note also (Figure 2.30) the large drop in RE (compared to EV´) for area P + NA + A and the WNA grid (EV´ = 0.25; RE = 0.04). To maximise reduction of error for a particular area while using a fixed number of predictor sites, a good spatial sampling strategy consists in distributing the sites as uniformly as possible. Over the sea-level pressure area P + NA + A, the low-density site network in NA + E (a total of 13 sites) has a higher reduction of error (0.24 using 1899-1924 data, and 0.33 using 1951-1979 data) than the medium-density site network in NA (a total of 12 sites and RE of 0.10 and 0.21) or the high-density site network in WNA (a total of 11 sites and RE of 0.02 and 0.06). A considerable advantage in reduction of error statistics is

Figure 2.26: Adjusted explained variance (EV´) of January sea-level pressure for three areas, WNA, WNA + CNA, and NA, as a function of site density (sites per million km ; L = low, M = medium, H = high) in the corresponding three grids (site locations). Predictors at each site are January temperature and precipitation. The plotted values are averages of both dependent data sets.

Figure 2.27: Adjusted explained variance (EV´) of January sea-level pressure for five areas (WNA, WNA + CNA, NA, EP + NA + WA, P + NA + A) as a function of five grids (site locations, WNA, WNA + CNA, NA, NA + E, J + NA + E); medium site density for each location. Grid Δλ is expressed in degrees of longitudinal extent. Predictors at each site are January temperature and precipitation. The plotted values are averages of both dependent data sets. Circles indicate situations where the longitudinal extent of the grid corresponds approximately with the longitudinal extent of the sea-level pressure area. The EV´ information for the area EP + NA + WA is indicated with dashed lines because this area does not follow the general trend for EV´ to decrease as sea-level pressure area increases (for a given grid).

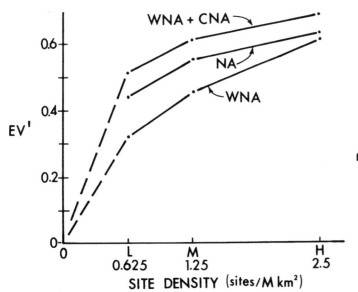

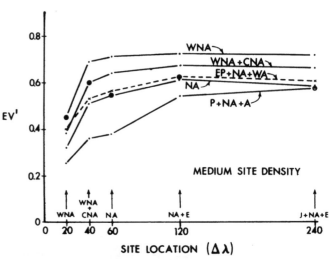

therefore achieved in this example by the use of the
low-density but longitudinally dispersed site network.

Summary. The general nature of the results of these
sampling studies can be explained in terms of the large
space-scale of the time-averaged waves of the atmospheric
circulation. The longitudinally-dispersed sites, even with
low site density, permit the amplitude and phase of the
largest-scale waves to be estimated with relatively high
accuracy. Sites that are longitudinally restricted, even
with high site density, permit relatively larger errors in
the estimation of the amplitude and phase of "downstream"
or "upstream" wave features. While this reasoning is
generally valid, there are exceptions which have been
noted. As a final example, the inclusion of climate sites
for Japan (NA + E + J) seems not to have led to apprec-
iably greater RE in sea-level pressure area P + NA + A
than that achieved with the North America and European
site network (NA + E). More experiments are needed to
determine the reason for this result.

Figure 2.28: Adjusted explained variance (EV′) of
January sea-level pressure for five areas as a
function of grid (site location) and site density:
WNA, high site density; WNA + CNA, medium site
density: NA + E, low site density. Predictors at
each site are January temperature and precipitation.
The plotted values are averages of both dependent
data sets.

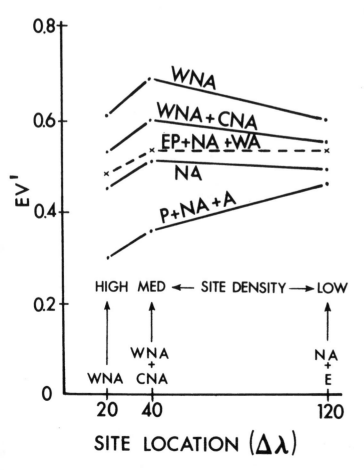

CONCLUSIONS

We have developed quantitative guidelines for the
spatial density and location of climate variables that are
appropriate for estimating the continental- to
hemispheric- scale pattern of atmospheric circulation.
Although the guidelines were developed from meteorological
records, they should be applicable to palaeoenvironmental
records. Further refinements of the procedure are
desirable. For example, palaeoenvironmental data may not
possess "signal-to-noise" ratios equal to those of actual
climate data. It would be possible to simulate various
noise levels and to allow for differences in
"signal-to-noise" ratios among sites. The use of January
precipitation at individual stations served to introduce
some "noise" due to well-known sampling problems for
precipitation; winter precipitation, regionally averaged,
would have been a less "noisy" data set. The January
precipitation data did not contribute much to the rather
good regression results reported here; that is, most of
the predictability came from the temperature observations.

Another place for refinement is in the development
of the regression model. Although it was desirable for
this study to set aside a substantial part of the data for
independent testing, future work might aim at using the
entire data set for model development (thereby developing
a more "robust" model) and relying on the EV′ or RE′
statistic for guidance on application to independent data.
Another suggestion would be to develop the regression
model in terms of a Fourier-series representation in order
to capture more explicitly the space-scale patterns of
atmospheric circulation.

The sampling strategy guidelines are not optimised
for a particular application but they are probably
illustrative of general relationships. Five results are
broadly verified (assuming uniform information content of
the climate data at each site).

First, given a fixed number of sites, a good
sampling strategy is to distribute the sites more or less
uniformly over the area for which estimates of atmospheric
circulation are desired, rather than placing them in a
restricted portion of the area. The resulting RE′s for
these two extremes of sampling strategy can differ by a
factor of two or more. In some cases, however, high-
density sites in restricted longitudinal sectors can
produce results that are of equivalent accuracy to those
produced by low-density sites that are spread over broader
longitudinal sectors.

Second, over the range of site densities tested
(roughly 0.5 to 2.5 sites/Mkm^2), the reduction of error
increases with site density. The increase is smaller for
medium- to high-density than from low- to medium-density
but further increase in RE for still higher site densities
might be possible.

Third, a good sampling strategy is to include sites
beyond the limits of the region for which estimates of the

Table 2.8a: Reduction of error (RE) for January sea-level pressure.

INDEPENDENT DATA: 1951-1976

PREDICTORS: JANUARY TEMPERATURE & PRECIPITATION				SEA-LEVEL PRESSURE AREA (LONGITUDE BAND) (NO. OF PRINCIPAL COMPONENTS)					
Grid	Site Density	No. of Stations	No. of Principal Components	WNA (130°W-110°W) (3)	WNA+CNA (130°W-90°W) (4)	NA (130°W-70°W) (5)	EP+NA+WA (150°W-50°W) (6)	P+NA+A (160°E-00) (8)	Average over all areas
WNA	low	3	2	0.20	0.15	0.11	0.17	0.11	0.15
	medium	6	4	0.37	0.29	0.19	0.15	0.08	0.21
	high	11	6	0.39	0.31	0.21	0.19	0.06	0.23
WNA + CNA	low	6	5	0.50	0.41	0.31	0.32	0.21	0.35
	medium	12	7	0.54	0.49	0.39	0.32	0.21	0.39
	high	23	9	0.63	0.52	0.41	0.41	0.31	0.45
NA	low	9	5	0.46	0.35	0.28	0.32	0.18	0.32
	medium	18	8	0.53	0.40	0.29	0.37	0.21	0.36
	high	34	10	0.53	0.43	0.34	0.39	0.19	0.38
NA + E	low	13	7	0.43	0.38	0.32	0.31	0.32	0.35
	medium	26	12	0.51	0.41	0.35	0.41	0.41	0.42
NA + J	low	11	7	0.42	0.34	0.26	0.26	0.14	0.28
	medium	22	11	0.52	0.40	0.33	0.38	0.17	0.36
NA + E + J	low	15	9	0.41	0.37	0.32	0.28	0.32	0.34
	medium	30	13	0.50	0.38	0.32	0.38	0.41	0.40
Average over all grids				0.46	0.37	0.30	0.31	0.22	0.33 (grand average)

Table 2.8b: Reduction of error (RE) for January sea-level pressure.

INDEPENDENT DATA: 1899-1924

PREDICTORS: JANUARY TEMPERATURE & PRECIPITATION				SEA-LEVEL PRESSURE AREA (LONGITUDE BAND) (NO. OF PRINCIPAL COMPONENTS)					
Grid	Site Density	No. of Stations	No. of Principal Components	WNA (130°W-110°W) (3)	WNA+CNA (130°W-90°W) (4)	NA (130°W-70°W) (5)	EP+NA+WA (150°W-50°W) (6)	P+NA+A (160°E-00) (8)	Average over all areas
WNA	low	3	2	0.27	0.25	0.23	0.25	0.12	0.22
	medium	6	4	0.43	0.36	0.30	0.23	-0.02	0.26
	high	11	6	0.52	0.47	0.39	0.24	0.02	0.33
WNA + CNA	low	6	4	0.35	0.32	0.29	0.26	0.19	0.28
	medium	12	7	0.50	0.42	0.36	0.27	0.10	0.33
	high	23	9	0.60	0.55	0.46	0.30	0.19	0.42
NA	low	9	5	0.44	0.43	0.40	0.32	0.21	0.36
	medium	18	9	0.55	0.48	0.41	0.31	0.09	0.37
	high	34	10	0.56	0.51	0.45	0.32	0.15	0.39
NA + E	low	13	8	0.37	0.38	0.36	0.28	0.24	0.33
	medium	26	14	0.49	0.47	0.43	0.31	0.20	0.38
NA + E + J	low	15	10	0.34	0.33	0.29	0.16	0.17	0.26
	medium	30	14	0.54	0.47	0.44	0.32	0.29	0.41
Average over all grids				0.46	0.42	0.36	0.26	0.13	0.33 (grand average)

atmospheric circulation are desired. The maximum value of adjusted explained variance (EV´) is obtained where the site location spans a longitudinal sector that is about twice as broad as the sea-level pressure area.

Figure 2.29: Explained variance (EV), adjusted explained variance (EV´), adjusted reduction of error (RE´), and reduction of error (RE) of January sea-level pressure for WNA + CNA, as a function of site density (site per million km^2; L = low, M = medium, H = high) in the corresponding grid (site location). Predictors at each site are January temperature and precipitation. The plotted values are averages of the respective dependent and independent data sets.

Fourth, the numerical values of reduction of error (RE) are often much lower than the values of adjusted explained variance (EV´). Compared to EV´ statistics, RE statistics provide more conservative and probably more realistic estimates of the expected accuracy of reconstruction of the pattern of sea-level pressure. If RE statistics are not available, RE´ statistics should be used.

Finally, the RE statistics are positive numbers in almost every experiment. This indicates that there is some skill in the reconstruction of large-scale patterns of sea-level pressure from information about temperature and precipitation at particular sites.

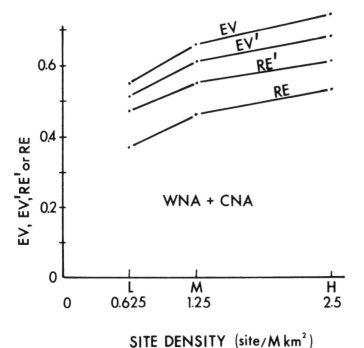

Figure 2.30: Adjusted explained variance (EV´), indicated by dots (.), and reduction of error (RE), indicated by crosses (x), of January sea-level pressure for two areas (WNA, P + NA + A) as a function of grid (location). Medium site density for each location. Predictors at each site are January temperature and precipitation. The plotted values are averages of the respective dependent and independent data sets.

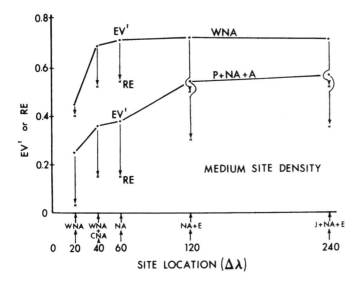

INTRODUCTION

The Editors

In Chapters 3 and 4, the current extent of the
dendroclimatological data base is reviewed and the
prospects for expansion in space and time are assessed.
The geographical areas referred to in these two chapters
reflect areas of research activity rather than climatically or dendrochronologically distinct regions. The
climatologists present at the Second International
Workshop on Global Dendroclimatology played an active part
in the discussions and research priorities were assessed
on both climatological and dendrochronological grounds.
This was the first time that the potential users of
dendroclimatic data had been directly involved at this
early stage of the reconstruction process and it proved of
great value. Knowledge of the characteristics of the
spatial patterns of climatic variation evident in instrumental records was used to guide sampling strategies,
alongside the more commonly used principles of dendrochronology.

Salinger describes the climatology of the Southern
Hemisphere. The small land area in the temperate zone and
its generally low relief, as well as the presence of the
ice-covered continent of Antarctica, promote a vigorous
year-round circulation. The strong but eccentric circumpolar vortex of low surface pressure is one manifestation of this. The associated extratropical depressions
provide the seasonal or year-round rainfall that supports
the forest and woodland vegetation of the southern parts
of the continents and New Zealand. This rainfall is
modified by rain-shadow effects in eastern New Zealand and
Argentina and in interior South Africa and Australia. The
strong and persistent subtropical high pressure belt may
influence the tracks of these depressions, and also plays
an important role in the seasonal climate regimes in lower
latitudes. The Southern Oscillation is a most important
feature in low latitudes of the Southern Hemisphere, and
it influences climatic variability at higher latitudes in
both hemispheres as well.

The paucity of land in the Southern Hemisphere
obviously limits both the development of a tree-ring data
base and the extent of comparative climatological data.
As a consequence, the approach to reconstruction of
climate parameters beyond the local and regional scales
will probably depart somewhat from the schemes developed
and successfully applied in parts of the Northern
Hemisphere. Even if tree-ring records could be developed
covering all of the land areas, the network size would
still be small in comparison with the total area of the
hemisphere. Therefore, the most effective approach may be
one specifically directed towards reconstructing the
behaviour of some of the more important features of the
general circulation, such as the circumpolar vortex, the
subtropical high pressure belts, and the Southern
Oscillation.

Tree-ring work in the Southern Hemisphere has lagged
greatly behind that in the Northern. Until very recently
few dendrochronological studies had been attempted, and
rarely were accurately dated and well-replicated
chronologies produced. The reasons for this lag include
the scarcity of experienced people, the apparent lack of
suitable species or habitats in some regions, a low level
of interest in climate variability within the broader
scientific community in some countries, and the lack of
suitable wood in archaeological or historical contexts
that might have led to development of an archaeological
dendrochronology. The situation changed dramatically about
six years ago, when local workers in South Africa,
Australia, and Argentina, as well as outside investigators, began comprehensive programmes of tree-ring
research. The relatively small size of the tree-ring data
base in the Southern Hemisphere reflects the recent
initiation of sampling and analysis programmes. The
chronologies are, however, of high quality because modern
sample size and replication requirements generally have
been met, site characteristics are well described, the
dating is reliable, and modern techniques have been used
in the processing and analysis of the data. The total of
less than 100 up-to-date chronologies does provide a firm
starting point for future work. This data base and future
prospects for its expansion are discussed by Dyer and
LaMarche (Southern Africa), Holmes and Boninsegna (South
America), and Ogden and Dunwiddie (Australasia).

The needs and prospects for further collection
differ from region to region. In the temperate zone, New
Zealand, Tasmania, and parts of Argentina and Chile have
fairly dense coverage, although gaps do exist. Future
sampling should include a greater variety of species and
climate response characteristics, as well as sampling for
densitometric and other analyses. Temperate areas that are
poorly covered include southern areas of South America,
mainland Australia, and South Africa. Recent collections
are available for parts of these areas, but dating
problems remain to be solved for some species. Areas with
virtually no coverage include the subtropical and tropical
parts of Africa, Australia, South America, and the islands
of Madagascar, Indonesia, New Guinea, New Caledonia, and
the smaller islands of the tropical South Pacific. One of
the reasons for the lack of progress in dendrochronology
in this vast region is the general lack of interest in the
past, although several groups have recently been working
on problems of warm temperate, subtropical, and tropical
dendrochronology in South Africa, South America, and
Australasia. A major obstacle has been the lack of clearly
defined rings in most species and the presence of obvious

intra-annual growth bands in others; these problems lead
to major difficulties in dating. Several promising species
have been identified, however, and future work could
emphasise the study of trees in taxa that have proven
their worth in extratropical regions of the Southern
Hemisphere.

CLIMATIC CONTEXT

M.J.Salinger

INTRODUCTION

Many early writers worked under the misconception
that the climatology of both hemispheres was very similar
but, in some respects, the Southern Hemisphere is unique.
Major divergences stem from the features that influence
the atmospheric circulation, especially the latitudinal,
longitudinal, and altitudinal distribution of land, sea,
and ice (Taljaard, 1972a). While much has been published
concerning the climatology of the Northern Hemisphere and
will not be repeated in this volume, the Southern
Hemisphere has received comparatively little attention,
mainly due to limited data availability. A brief summary
of the climate of the Southern Hemisphere is, therefore,
presented here. General climatological points of relevance
to dendroclimatology are discussed later by Kelly (this
volume). Further information on Southern Hemisphere
climatology can be found in Schwerdtfeger (1976) for South
America, Gentilli (1971) for Australia and New Zealand,
and Orvig (1970) for Antarctica.

The Southern Hemisphere consists of less than
one-fifth land, of which about 60% lies north of $30^{\circ}S$, and
30% south of $60^{\circ}S$ (Table 3.1). Land is completely absent
between 56 and $63^{\circ}S$ with very little between 40 and $65^{\circ}S$,
coinciding with the zone of southern westerlies. The other
80% of the hemisphere comprises oceans with the Pacific
occupying the greatest area, then the Indian, followed by

Table 3.1: Percentage zonal oceanic surface in each
hemisphere.

LATITUDE ZONE ($^{\circ}$)	HEMISPHERE	
	NORTHERN	SOUTHERN
0 - 10	77.2	76.4
10 - 20	73.6	78.0
20 - 30	62.4	76.9
30 - 40	57.2	88.8
40 - 50	47.5	97.0
50 - 60	42.8	99.2
60 - 70	29.4	89.6
70 - 80	71.3	24.6
80 - 90	93.4	0.0
0 - 90	60.6	80.9

the Atlantic. The water masses influence the tropical and
subtropical weather systems. In contrast, about 40% of the
Northern Hemisphere is land-covered and the greatest
percentage area of ocean per latitude band occurs north of
$70^{\circ}N$ and south of $30^{\circ}N$. The Arctic is dominated by ocean,
albeit covered with a thin layer of ice. The eastern
Northern Hemisphere is dominated by land and the western
by the North Pacific Ocean.

Each of the Southern Hemisphere continents has
different topographical characteristics. The Australian
land surface is mostly less than 350m in height which
promotes important seasonal thermal contrasts between
ocean and land. South Africa is a moderately high plateau
rising to between 1,000m and 2,000m above sea level and is
located where subsidence predominates from the anticyclonic circulation. The Andes rise abruptly from the
Pacific coast and run the length of South America
(8,000km). The remainder of South America to the east
consists of lowlands or low altitude plateaux. The Andes
are the most important topographical obstruction to the
hemispheric circulation in tropical and middle latitudes;
the very persistent anticyclone of the southeast Pacific
is one of the consequences of this mountain barrier.
Antarctica, with a total area of 14 million square
kilometres, accounts for over 90% of the Earth's ice. The
eccentricity of its location with respect to the South
Pole and its high relief are very significant to the mid-
and high-latitude circulation. West Antarctica, or that
part which adjoins the Pacific and the southwest Atlantic
Oceans, has diverse topography averaging 850m above sea
level. East Antarctica faces the Indian and southeast
Atlantic Oceans and is a lofty ice dome averaging 2,450m,
making the hemispheric circulation eccentric. The westerly
vortex is displaced towards the equator in the southeast
Atlantic and Indian Ocean sectors.

TEMPERATURE

Summer. The highest mean air temperature at the
surface occurs over the continents in summer, especially
in the northwest of Australia. High temperatures also
occur in South Africa and in central Argentina and Brazil,
just east of the Andes. By adjusting altitudinal temperatures to sea level, longitudinal temperature contrasts
reveal these as areas of positive anomalies, by 4 to $8^{\circ}C$
in South Africa and exceeding $8^{\circ}C$ in Australia and the
Argentine. Over the southern oceans, the large expanse of
the tropical western Pacific and a small area just east of
Mozambique record the highest air temperatures.
Longitudinal comparisons show negative anomalies in the
eastern parts of oceans centring on $25^{\circ}S$ with the greatest
departures along the western coasts of South Africa and
South America. The equivalent western sides of the ocean
demonstrate smaller positive departures. Except for the
land areas mentioned above, hemispheric temperatures
decrease polewards outside tropical latitudes, with the

lowest values occurring in East Antarctica. In middle latitudes, the Pacific is warmer than the southeast Atlantic and Indian Ocean sectors, with differences as high as 8°C at 50°S. The rate of temperature drop with latitude is greatest in the Atlantic and Indian oceans with the strongest gradients at mid-latitudes. South of the Antarctic Convergence, thermal gradients are weak. At 20°E the temperature drops 12°C between 40 and 50°S in January but only by 2°C between 55 and 65°S (van Loon, 1972a).

Winter. The temperature distribution (van Loon, 1972a) shows a poleward temperature decrease outside tropical latitudes and near landmasses. The same landmasses which exhibit the highest summer temperatures are relatively cooler in winter, especially the Australian interior. If these temperatures are adjusted to sea level, then they are much lower than those of South Africa and the Argentine. There is a very strong temperature gradient over the frozen areas of the Antarctic ocean where the summer gradient is weak. Again, temperatures are lowest on the plateaux of East Antarctica. Longitudinal temperature departures resemble the summer pattern over the oceans; the Pacific is warmer when compared with the cooler Atlantic and Indian regions of the hemisphere in mid-latitudes. The summer to winter temperature change is largest (about 5°C) between 30 to 40°S.

SYNOPTIC CLIMATOLOGY

Many of the above features are related to the synoptic systems of the south (Taljaard, 1972b; Streten, 1980). The seasonal differences in air pressure are not considered here because they are small when compared with the Northern Hemisphere, but these changes are very significant in producing seasonal patterns in the circulation (for further details, see Taljaard et al., 1969).

Anticyclones. In summer, anticyclones are most frequent in the zone 23 to 42°S (Taljaard, 1972b; Streten, 1980) and are very common in three areas; the southeast Pacific, the South Atlantic to the west of South Africa, and the Indian Ocean. Smaller frequency maxima (van Loon, 1972b) occur in the Great Australian Bight and to the west of North Island, New Zealand. Of the three anticyclones, the Indian ocean system shows the least latitudinal extent because of the restricted zone along which it can occur. The areas of highest frequency are also those areas where old anticyclonic cells dissipate and slow down off the west coasts of the continents. New anticyclones are born between 30 and 40°S just to the east of South America and Africa. Thus the two continents provide breaks in the core of high anticyclonic frequency circumscribing the hemisphere. Anticyclones are almost absent from 45°S to the coast of Antarctica except in the southwest Pacific. The strongest systems in the Southern Hemisphere occur well south of the subtropical ridge and about three to five degrees poleward of the core of high anticyclonic frequency. There are two types south of 40°S: systems of cold air which form in the intensifying ridges following a cold outburst in the rear of a cyclone, and the extension of the warm blocking anticyclonic cells from the subtropical ridge. The usual movement is east and slightly northward although southward components do occur off the southwest Australian and South African coasts. These systems also propagate eastward across the mountain barrier of the Andes and the west coasts of South Africa and Australia.

In winter, the core of maximum frequency (Taljaard, 1972b; van Loon, 1972b; Streten, 1980) is displaced by about four degrees towards the equator compared with the summer. Anticyclones in this season are most common in the southeast Pacific and near southeast Australia. They are also frequent in the southeast Atlantic, the Indian Ocean, to the east of Queensland, and the east of the South Island of New Zealand. East of Australia and over New Zealand a bifurcation of the anticyclonic belt occurs in winter. Strong anticyclones are more frequent at this time of year than in other seasons.

Cyclones. Cyclones or depressions (Taljaard, 1972b; Streten, 1980) are most frequent in all seasons over the seas between 60 and 75°S surrounding Antarctica. Of the seven regions of recurrent high occurrences, four coincide with embayments on the Antarctic coast, with the eastern part of the Ross Sea having the highest frequencies. Depressions become stationary in the bays, which can be considered the graveyards of cyclones in the Southern Hemisphere.

In summer, heat lows occur in the Argentine, over South Africa, Madagascar, and northern Australia. A unique feature of the circulation which is prominent in winter is two bands of high cyclone frequency which spiral to the southeast. One extends from 22°S east of the Andes, and another from the northern part of the Tasman Sea to Antarctica. The latter band is associated with the bifurcation of the anticyclonic belt during winter and here depressions are frequent in all seasons but especially winter, when they tend to be cut-off lows. Cyclones on average move towards the east-southeast and usually breed between 35 and 55°S except to the east of South America and in the eastern Pacific where the northern limits are 30 and 25°S respectively. The Tasman Sea is a much favoured birthplace of Southern Hemisphere depressions. Tropical cyclones form over the tropical oceans in the southwest Pacific and Australian region and the southwest Indian Ocean. Annual frequencies vary markedly (Raper, 1978).

THE SOUTHERN OSCILLATION

With continuing research (Bjerknes, 1969; Kidson, 1975a,b), the Southern Oscillation (Walker & Bliss, 1932, 1937; Troup, 1965; van Loon & Madden, 1981) is considered to play an increasingly important role in the operation of the general circulation. It affects tropical and subtropical areas from longitudes between Indonesia to South America in the Pacific with the main centres of action located in the anticyclone of the southeast Pacific and the area of lower pressure over Indonesia. The oscillation involves a pulsating of these two centres: when pressure is higher than usual in one, it is often lower than usual in the other.

The atmospheric and oceanic circulations of the Pacific produce upwelling of very cold water along the South American coast and thus the sea-surface temperature is $10^{\circ}C$ warmer to the east of Australia than off the Peruvian coast. This temperature gradient drives a west-east air exchange, known as the Walker Circulation (Walker & Bliss, 1937). When the Southern Oscillation is strong, the temperature gradient increases across the Pacific because of more upwelling of cold water and a more vigorous Walker Circulation. When it weakens, so does the temperature gradient and the Walker Circulation as less upwelling of polar water occurs.

PATTERNS OF CLIMATIC VARIATION

A brief resumé is now given of some of the patterns of Southern Hemisphere climatic variation. For details of longer term fluctuations, see Pittock et al. (1979) and Bowler (1981). Trenberth (1981) has described hemispheric trends in the general circulation over the last 20 years.

South Africa has no uniform pattern of variation. Instead, areas of the country which have differing precipitation regimes experience different climatic fluctuations on different timescales (Tyson et al., 1975). Four rainfall areas can be defined: the southwestern Cape with a winter maximum (Mediterranean type), the all-seasons belt of the southern coast, the arid western deserts with equinoctial maxima, and the larger eastern area with precipitation mainly in summer. Each of these has characteristic long period oscillations of which the quasi-20-year period of eastern South Africa is the most strongly developed. The southern coast has a weak 10-year oscillation, the southwestern Cape regime is more complicated, with periods greater than 20 years, and the desert experiences a quasi-biennial fluctuation.

Precipitation variation in Australia shows two main patterns (Pittock, 1975, 1978). The first is more important in the east of the continent, especially in the interior of New South Wales, coinciding with areas of summer precipitation maximum. The second acts along the south coasts and in Tasmania in the belt of winter maximum rainfall and, in an opposite sense, along the New South Wales coast. The first pattern is strongly correlated with the Southern Oscillation and hence the activity of the Walker Circulation in the Pacific to the east, and the second with the latitude of the surface high pressure belt over eastern Australia which controls the penetration of westerly rainfall along the southern coasts. These two patterns explain about half of the total variance in time.

New Zealand patterns are complex because of the axial relief almost at right angles to the mean circulation flow from the west-southwest (Salinger, 1979, 1980). There is little seasonal differentiation in precipitation amounts. Rainfall on the west coast and far south of South Island is very significantly correlated with positive anomalies of westerly winds over the country when the westerlies are strong. The east and northeast of North Island show highly significant negative correlations with weak or non-existent westerly circulation over New Zealand. Higher precipitation in these areas occurs with easterly winds from the Pacific Ocean. Meridional indices suggest that months and years of high precipitation occur in areas exposed to the north in the northern half of North Island and the northern extreme of South Island when excess northerly winds occur. Above normal rainfall in the south of South Island is triggered by anomalous southerly winds. The Southern Oscillation correlates significantly with northerly values of the meridional index (Trenberth, 1976): when flow from the northeast is enhanced, precipitation occurs in areas exposed to the north, associated with more frequent blocking anticyclones to the east of South Island and with cut-off lows in the Tasman Sea. The circulation trends have been towards this state since 1950, with a associated $0.5^{\circ}C$ increase in regional temperatures (Salinger, 1979).

The Andes effectively isolate any influence of the Pacific anticyclone and its oscillation on eastern Argentina and the South Atlantic (Sturman, 1979). When the east Pacific anticyclone is strong (Pittock, 1980a), precipitation is decreased along the Chilean coast from 20 to $36^{\circ}S$ with also a slight increase around $45^{\circ}S$. The transpolar index (Pittock, 1980a) represents the eccentricity of the westerly vortex about the South Pole with a tendency for it to be either displaced towards South America or eastern Australia. When it is displaced towards South America precipitation is increased in latitudes near $40^{\circ}S$ over Chile but reduced in central Argentina from 26 to $40^{\circ}S$ consistent with more westerly activity in the South American area.

In tropical areas, the most important departures of pressure, temperature, and rainfall are related to the Southern Oscillation (Kidson, 1975a,b). A global analysis suggests that when the South Pacific high is weak, pressures are higher over Indonesia and northern Brazil. The southern westerlies in the Australasian quadrant are also stronger.

The patterns described cover about one-fifth of the Southern Hemisphere as only 20% of this hemisphere is

land. Regrettably, the patterns over the other 80% of
water are virtually unknown.

SOUTHERN AFRICA
T.G.J.Dyer
INTRODUCTION

As southern Africa constitutes a large landmass
extending over a range of latitudes and longitudes, it is
subject to various climatic controls. The major upper-air
circulation controls are semi-permanent high pressure
cells which dominate most of southern Africa (Trewartha,
1961). For example, the combined effect of a stable,
relatively persistent, South Atlantic oceanic anticyclone
and the cold Benguela current, gives rise to arid
conditions along the west coast. In contrast to the small
irregular longitudinal fluctuations experienced by this
anticyclone, the South Indian oceanic anticyclone
undergoes a significant eastward displacement during
summer with a corresponding westward movement toward South
Africa during winter (Vowinkel, 1955). These differences
in stable anticyclonic control are responsible for a
decline in precipitation amounts received from east to
west across southern Africa. Whilst most of the
subcontinent receives summer rainfall, intensification of
the plateau high pressure cell together with its northward
displacement permits the presence of the mid-latitude
westerlies over the southern tip of Southern Africa with
accompanying winter rainfall (Taljaard et al., 1961).
Superimposed on the large-scale influences on rainfall due
to the general circulation patterns are the effects of
relief and distance from the warm Indian Ocean (Cole,
1961). The decline in rainfall amounts toward the interior
is associated with decreased reliability. As a result,
large areas of the subcontinent have been declared drought
prone by the Department of Agriculture (de Swardt &
Burger, 1941).

It has been shown that in the drier areas where
moisture stress is greatest, the climate signal within the
ring series is enhanced (Fritts, 1967). Under temperate
conditions, the climate signal is likely to be weak and
new techniques may be required to ensure the reliability
of the reconstructed series. Such modifications of the
established tree-ring analysis approach may necessitate
the collection of much larger samples and the more
extensive use of statistical cross-dating. Unfortunately,
although computerised cross-matching is a valuable aid,
its use imposes a limitation in that ring samples have to
be assessed in the laboratory. This removes the
possibility of on-the-spot evaluation in the field and
makes the collection of suitable samples a time-consuming
task.

Difficulties inherent in tree-ring analysis are
associated not only with temperate regions but with summer
rainfall areas as a whole. Under temperate conditions it
is the amount of rainfall that weakens the climate signal

present in the tree-ring series, while in summer rainfall
areas it is the time at which the rain falls that weakens
the clarity of the rings themselves. This has been shown
by Glock & Agerter (1962) who related specific growth-ring
anomalies to different regimes. Where rain was received
prior to the first flush of growth, as in Mediterranean
climates, growth rings had sharp boundaries and
discontinuous growth layers seldom occurred. In summer
rainfall areas, however, where most water becomes
available only after growth has been initiated, ring
margins were often diffuse and multiple, and discontinuous
rings were common. The tree-ring analyst who is not
working in an arid environment or a Mediterranean region,
is, therefore, likely to be hindered not only by a weak
climate response in tree growth but also by poor clarity
of ring boundaries.

PROBLEMS WITH SOUTH AFRICAN INDIGENOUS TREES

In order that attention be focussed on dendroclimatologically usable tree species, it is necessary to
exclude unsuitable species that exhibit features such as
poor ring structure, short life, or scarcity. These
characteristics, which are fairly easy to determine, serve
to limit the field of investigation prior to a microscopic
assessment of ring structures. The longevity of a tree is
the most obvious and important characteristic to consider
when selecting trees for dendroclimatological study.
Obviously, to extend rainfall records back in time one
needs to work with the longest living trees possible. In
the first instance, age must be assessed on a priori
grounds of size, girth, height, and so on. In South
Africa, these factors have not always proved helpful.
Only after careful dendrochronological study can tree age
be established with an acceptable degree of certainty.
Small size inevitably excludes a number of the 960 woody
species recorded in the National Tree List, many of which
are not, however, trees in the generally accepted sense
(de Winter & Vahrmeijer, 1972).

Related to the criteria of site selection are the
principle of sensitivity and the law of limiting factors.
These principles are directed at the minimisation of
non-climatic influences and the enhancement of macroclimate information. Locally, the ideal is to select trees
that favour extreme locations where water is the critical
factor controlling growth. Unfortunately, the ideal is not
easily attainable in South Africa. Because the life
expectancy criterion is of paramount importance, the
principles of site selection cannot be adhered to
strictly. In South Africa, there is a general scarcity of
large indigenous trees. Impressive specimens are mostly
located along river beds and in tropical forests. There
are a few exceptions. For example, the Acacia and Ficus
genera are medium-tall trees which occur in drier
localities (van Wyk, 1972). In general, however, large
trees are very sparse in semi-arid areas. As regards tree

species located in remnants of tropical forests, again restrictions and limitations are encountered. There are very few virgin forests left in South Africa and those that do exist are located along the southern and eastern coasts, and in patches in the mountain ranges. These are areas favoured with mild temperatures and well distributed rainfall, 762 to 1,524mm/yr (von Breitenbach, 1974; Palmer & Pitman, 1972). This limited natural forest resource, which covers less than 2% of South Africa (Palmer & Pitman, 1972), has, moreover, been exploited by wood-cutters early this century. According to von Breitenbach (1974), timbermen selected the best trees, and thereby practically depleted the forests of large indigenous genera such as Podocarpus and Widdringtonia. Today these species are protected which causes difficulties in obtaining suitable large samples.

Whereas in North America Pinus longaeva lives for more than 4,000 years and pine trees in general are numerous and widely distributed, in South Africa Podocarpus falcatus reaches a maximum age of about 700 years (von Breitenbach, 1974) and the indigenous conifers are protected, while large trees are generally sparsely distributed. South Africa's forest relics also contrast strongly with those in the Northern Hemisphere. They are different from the temperate, deciduous forests found in Europe and the coniferous forests species is common in South African forests. Unfortunately, in accordance with the law of limiting factors, competition in a forest tends

to mask any climate response in tree growth. In areas without extremes in temperature or aridity and with great competition between species, the climate signal in the growth-ring series is likely to be weak.

Implied in the principle of replication is the use of core samples. Obviously if large samples are required, coring is the practical solution when collecting wood samples for a chronology. Unfortunately the standard Swedish increment borer which extracts a core approximately 42cm long and 0.4cm in diameter cannot penetrate timbers with a density greater than $0.85g/cm^3$. This is, therefore, an additional factor which limits the number of species to be investigated.

In order to facilitate the evaluation of ring structures in the 108 woods that have been collected, a rating system has been devised (Table 3.2). Using these criteria five species seemed to show promise (Table 3.3). In fact, Podocarpus and Widdringtonia have proved to be most suitable trees. Most work has been carried out on Podocarpus but the results have not been very encouraging owing, mainly, to poorly defined rings, missing rings, and extreme areas of ring convergence. The work described is a very brief outline of the research carried out by the Climatic Research Group at the University of Witwatersrand (Director: P.D.Tyson). A more detailed record of the work done has been given by Lilly (1977).

COMMENT
V.C.LaMarche Jr.

Southern Africa probably presents greater problems in the application of dendrochronological techniques, and in the use of tree-ring data for palaeoclimatic studies, than any other region in the Southern Hemisphere. As Dyer emphasises, the effects of general circulation patterns combined with topography result in patterns of precipitation amount and seasonal distribution that severely restrict the extent of forests. Furthermore, particularly in the large areas of summer-dominant, equinoctial, or erratic rainfall distribution, the semi-desert, savannahs, or woodland vegetation is composed exclusively of angiosperm species.

One of the first problems that Dyer points out - that of diffuse ring boundaries and multiple intra-annual

Table 3.2: Criteria used in rating dendrochronological potential.

CRITERION RATING	CRITERIA
7	Slow growth rate (6-14 rings per cm)
6	Boundary parenchyma
5	Ring porosity
4	Denser latewood fibres
3	Semi-ring-porous structures
2	Growth-related variations in parenchyma patterns
1	Evidence of a ring-structure
0	Diffuse porosity
-1	Sapwood/heartwood differentiation
-2	Banded parenchyma
-3	False or disjointed rings
-4	Missing or discontinuous rings
-5	Indistinct boundaries
-6	Fast growth rate (3 rings per cm)
-7	Macroscopically deceptive ring pattern

Table 3.3: Dendrochronologically promising indigenous trees of South Africa.

SPECIES	RATING
Albizia forbesii	6
Burkea africana	6
Ekebergia capensis	7
Fagara davyi	13
Vepris undulata	13

growth bands - may not be so closely tied with macro-climate as he seems to believe. As I have emphasised elsewhere in this volume (Chapter 1), genetic potential can also play an important role. Thus, suitable species probably exist even in summer rainfall areas of southern Africa. A second major point is the need for screening a large number of tree and shrub species to assess their dendrochronological potential. The pioneering efforts of the Witwatersrand group to develop a rating system, as reported by Lilly (1977), represent a sound basic approach that is highly appropriate for the anatomically complex angiosperms. However, application of this kind of semi-objective numerical method should be supplemented by the judgment of a trained and experienced dendrochronologist.

As Dyer points out, the conifers Widdringtonia and Podocarpus seem to have immediate potential. An accurately dated and well-replicated ring-width index chronology, 413 years in length, has been developed from Widdringtonia cedarbergensis at a site near Cape Town (Dunwiddie & LaMarche, 1980a). The main problem with Podocarpus is extreme circuit variability, manifested in gross variations in ring width around the tree's circumference and in the frequent "wedging-out" of individual rings and groups of rings. Their proper study requires complete transverse sections, which are difficult to obtain because of highly appropriate protectionist measures. However, I believe that Podocarpus latifolius, a small but very slow-growing tree of the Mediterranean climate region in the southwest Cape, would yield accurate chronologies with intensive study. The large and long-lived podocarps of the southern and eastern forests also show great potential. But, in both cases, densitometric or possibly even isotopic data are more likely than ring widths to yield a meaningful climate signal. Although not well explored, it also seems likely that such parameters would be much less sensitive to the competitive affects mentioned by Dyer as characterising warm temperate to subtropical forests.

It should also be mentioned that, despite logistic and regulatory problems, it has been possible to obtain large sample collections in South Africa. Several hundred increment cores and cross-sections were collected during the period 1976 to 1978 from trees of nine species at some 24 sites, mainly in southern Cape Province. These collections and the Widdringtonia chronology are documented in LaMarche et al. (1979e). Overall, I am more optimistic than Dyer appears to be about the potential for dendroclimatology in southern Africa, but recognise that its development will require a sustained long-term research effort.

ARGENTINA AND CHILE

R.L.Holmes

INTRODUCTION

Modern tree-ring studies in South America were begun only in the mid-1970s when the Laboratory of Tree-Ring Research of the University of Arizona started a programme of tree-ring sample collection for the reconstruction of Southern Hemisphere palaeoclimates. This is a summary of the dendrochronological aspects of recent work on the reconstruction of climate for temperate latitudes in South America. Climatic analyses have been published by Pittock (1980a,b).

PREVIOUS DENDROCHRONOLOGICAL WORK

R.T.Patton joined a timber survey party in 1946-47 and took extensive core samples on the Hornopirén Peninsula south of Puerto Montt in Chile. His collections, mostly Fitzroya cupressoides, are at the Laboratory of Tree-Ring Research (University of Arizona). During the austral summer of 1949-50, E.Schulman of the Laboratory of Tree-Ring Research made a reconnaissance field trip to Chile and Argentina (Shulman, 1956). He obtained cores from trees of several conifer species, three of which seemed to offer promise. Schulman found Austrocedrus chilensis to be similar to western North American conifers in clarity of rings and in good cross-dating qualities. He found Araucaria araucana to have ill-defined rings in most cases, and was able to date only half of his samples of this species. Schulman recognised the longevity of Araucaria araucana, and obtained a ring count of 1,060 years in a log at a Chilean lumber mill. In a third and very long-lived conifer, Fitzroya cupressoides, Schulman found some cross-dating, but, because of circuit variability, reliable cross-dating over more than a few decades could not be established.

In early 1974, V.C.LaMarche Jr. (Laboratory of Tree-Ring Research, University of Arizona) and others made exploratory collections in central Chile. Intensive sampling was done at El Asiento near San Felipe, Chile, of the northernmost known stand of Austrocedrus chilensis. They derived a well-replicated site chronology which contains the oldest core yet dated from South America, dating from 1007 AD, and developed a reconstruction of annual rainfall at Santiago, Chile (LaMarche, 1975).

SAMPLE COLLECTIONS AND CHRONOLOGIES

Study of the available literature and reworking of the sample collections of Schulman and Patton provided a basis for tentatively judging the suitability of several arboreal species for dendrochronology. Tortorelli's (1956) excellent book on Argentine forests was particularly helpful. This systematically describes each species by habitat, geographical distribution, size, and the appearance of the wood and the rings, and includes photomicrographs of wood sections.

Since the trees from tropical and subtropical latitudes generally lack well-defined annual rings, often produce several growth layers per year, and show poor circuit uniformity and cross-dating (Fritts, 1976, pp.12 and 95-96), a decision was made to restrict the study to the area south of 30°S. Sample collection was undertaken during the three austral summers from September 1975 through February 1978. Exploratory cores were taken of all species that might be of use for dendrochronology and, after assessment in the field laboratory, intensive sampling was conducted throughout the ranges of the four species showing greatest promise (Holmes, 1978). In most cases, three or four cores per tree were taken to aid in dating and to strengthen the statistical analysis. A few cross-sections were obtained from lumber mills, but these served the purpose of investigating the characteristics of the species rather than of chronology development since the sections were few and their exact provenance unknown.

Figures 3.1, 3.2, and 3.3 show the location of sites from which chronologies were derived and sites of other exploratory collections (LaMarche et al., 1979a,b). In all, our South American collections include eight species of conifers in three families and 29 species of dicotyledons in 17 families (Table 3.4). There are 51 collection sites in Argentina and 21 in Chile, which so far have yielded 21 good site chronologies in Argentina and 11 in Chile (Table 3.5a and b). Of the 32 site chronologies, 18 are <u>Araucaria araucana</u>, 13 are <u>Austrocedrus chilensis</u> and one is <u>Pilgerodendron uviferum</u> (LaMarche et al., 1979a,b).

Figure 3.1: Map of west-central Argentina showing sites of collections from which chronologies were derived (solid circles) and other collections (open circles). Sites are identified by a three-letter code (see Table 3.5a). From LaMarche et al. (1979a).

DESCRIPTION OF SUITABLE SPECIES

Araucariaceae. <u>Araucaria</u> <u>araucana</u> (Mol.) C. Koch, (pehuén), is a large tree, up to 35m tall and 2m in diameter, of striking and characteristic form, which grows on the Cordillera de Nahuelbuta in Chile and in the semi-arid Andean foothills of both Argentina and Chile. The northernmost extent of <u>Araucaria</u> in Argentina is at 37°20′S in the foothills west of Chos Malal in Neuquén Province and at about the same latitude in Chile. Its southernmost extent is about 40°20′S on both sides of the Andes (Boninsegna & Holmes, 1978a). In addition, <u>Araucaria</u> grows in the higher latitudes of the Cordillera de Nahuelbuta around 37°45′S near the Chilean coast.

The Andean stands of <u>Araucaria</u> are very open, devoid of understory, and the ground cover is usually sparse tussocks of a tough bunch grass. Stands are usually on rocky ground, often on knolls of large rocks which are the first areas free of snow in the spring. Mature trees typically have an umbrella shape, with a clear trunk up to some 80% or more of their height. These characteristics lend themselves to a technique developed for site selection at the Argentine Institute of Nivology and Glaciology (I.A.N.I.G.L.A.) in Mendoza. Inspection of stereoscopic pairs of aerial photographs at a scale of 1:50,000 enabled us to preselect promising stands, saving considerable time in the field through advance knowledge of the size of the stand, surrounding topography, and means of access. <u>Araucaria</u> growing in the Nahuelbuta range appears to be different from that growing in of the Andes. The Nahuelbuta type has a slimmer trunk and branches, and grows in much more humid areas, often in company with large trees of <u>Nothofagus</u> <u>dombeyi</u> and tall, dense

Figure 3.2: Map of southern Argentina showing sites of collections from which chronologies were derived (solid circles) and other collections (open circles). Sites are identified by a three-letter code (see Table 3.5a). From LaMarche et al. (1979a).

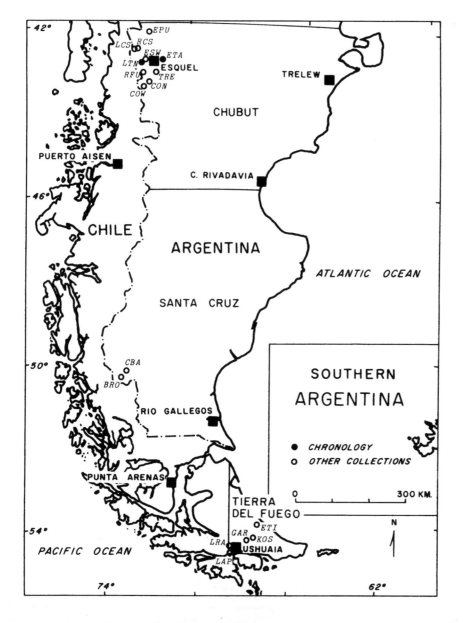

understory, a situation not seen in the Andes. The
Nahuelbuta type is generally slower growing radially,
displays vague ring boundaries with greater frequency, and
produces occasional rings of white wood which aid in
cross-dating.

On well-prepared surfaces, <u>A</u>. <u>araucana</u> generally
exhibits well-defined ring series. Ring boundaries are
often not obviously demarcated by colour change or by
marked thickening of cell walls as in many conifer
species, but in nearly all cases they can be determined by
a change in tracheid cell size. Occasional series will
have such vague boundaries as to be undecipherable;
usually these are "tight" series consisting of very narrow
rings. Surface preparation is more critical with <u>Araucaria</u>
than with any other conifer species I have seen. Schulman
(1956) was able to date only half of his samples of this
species. Examination of his samples shows that he prepared

his surfaces by slicing with a razor blade at a 60° angle
to the tracheids. We prepared sample surfaces progressing
through four or five grades of sandpaper from medium to
fine to obtain a highly polished surface at right angles
to the tracheids. With samples prepared this way the ring
structure is much clearer, and thus we were able to date
90% of our <u>Araucaria</u> samples.

<u>Araucaria</u> reproduces partly by root sprouting.
Several times we saw small trees growing from superficial
roots of a larger tree up to 10m away. One tree was
completely surrounded by root sprout offspring forming a
fence 5m in radius. This habit may affect the growth rates
of all the interconnected trees in an unpredictable
manner; a biological factor influencing the climate signal
from such trees. The hypothesis has not been tested for
<u>Araucaria</u>. Also, <u>Araucaria</u> is dioecious and female trees
reportedly tend to produce heavy seed crops in alternate

Figure 3.3: Map of central Chile showing sites of collections from which chronologies were derived (solid circles)
and other collections (open circles). Sites are identified by a three-letter code (see Table 3.5b). From LaMarche
et al. (1979b).

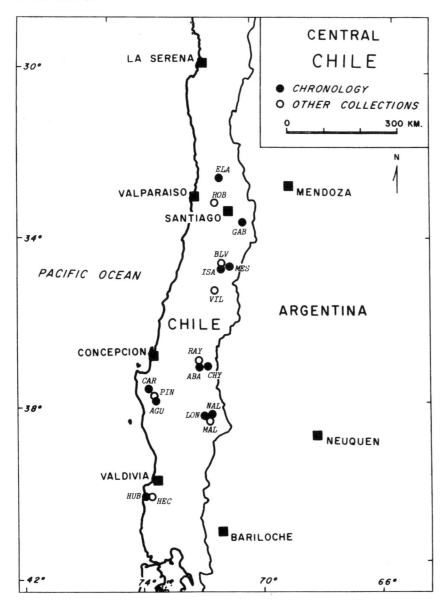

years. We saw no evidence that this affects ring growth.

Gale force winds blew during our sampling at some Araucaria sites. Evidence (bark erosion and rock polishing on the windward side) points to a near constancy of such winds, yet the trees lean only one to two degrees downwind and show very little "flagging", with branches nearly as long and as numerous toward the wind as away from it. This is in marked contrast to Austrocedrus chilensis which shows marked flagging (branches to leeward only) under conditions of considerably less wind velocity and constancy.

Cupressaceae. Austrocedrus chilensis (D.D Don) Endl., (ciprés), is a medium-size conifer of up to 25m height found on sites ranging from very steep, rocky, and arid slopes to gently sloping mesic hillsides. Its northernmost extent in Chile is $32^\circ 40'S$ at El Asiento, and in Argentina just north of Huinganco at $37^\circ 04'S$. In both countries the southernmost limit of Austrocedrus is about $44^\circ S$ near the Río Corcovado.

In the northernmost Austrocedrus sites, El Asiento and Huinganco, we found mostly old individuals, but at other sample sites the majority of the individuals were young, 70 years or younger, with only a few old trees (Boninsegna & Holmes, 1978b). In searching for appropriate sites for sample collection, it was much more common to find stands in which all the trees were young. For this reason we conclude that Austrocedrus is extending its range. Tortorelli (1956, p.256) also believes this, and we saw further evidence of extension in a series of photographs taken every few years since early in this century at San Martín de Los Andes (Neuquén), showing a hillside near the town. In the early photos this hillside is bare, but in subsequent photos Austrocedrus is becoming established, until today a very dense forest covers the hillside and all the surrounding terrain.

In some environments Austrocedrus grows mainly with multiple trunks, a growth habit that may be undesirable for dendroclimatology for the same reasons as root sprouting in Araucaria. Another problem is rot, which attacks from the centre outward in a great many Austrocedrus, destroying the oldest rings and often making it extremely difficult to extract the increment borer. Austrocedrus is sensitive to the wind, exhibiting strong flagging in areas of steady winds. In many localities, Austrocedrus is subject to severe disturbance by cutters of firewood, particularly near villages. We found only one site, Lago Rucachoroi in Neuquén Province, with both old Austrocedrus and old Araucaria together. The site was sampled with special intensity in order to study the possibly different responses of the two species to identical climatic and edaphic conditions (LaMarche et al., 1979a, pp.32-35).

Pilgerodendron uviferum (Don) Florin, (ten; ciprés de las Guaitecas), is a medium-size conifer reportedly attaining a maximum height of 20m and a maximum diameter of 100cm (Tortorelli, 1956, p.263), although all individuals we found were much smaller than this. It is found from about $40^\circ S$ latitude south to Tierra del Fuego. Its range in Chile extends somewhat further north than in Argentina. Pilgerodendron grows in the Valdivian and subantarctic forests as sparse open groups in swampy ground, as scattered individuals in humid closed-canopy parts of the forest, and occasionally as pure stands on the higher slopes of some Chilean islands (Heusser, 1966). We sampled Pilgerodendron on swampy ground in the Valdivian forest near the Chilean coast and derived a short chronology (LaMarche et al., 1979b, pp.32-33).

Pilgerodendron exhibits very clear and fairly uniform annual rings, and although complacent, it cross-dates well. Its potential for dendrochronology would be very good if sufficient individuals could be found of several hundred years' age. The wide latitude range of the species makes it especially promising for future work.

Table 3.4: Number of sites collected by species. Adapted from LaMarche et al. (1979a,b).

SPECIES	NO. OF SITES COLLECTED	SPECIES	NO. OF SITES COLLECTED
Aextoxicon punctatum	1	Araucaria araucana	23
Austrocedrus chilensis	23	Azorella nucamentacea	1
Baccharis salicifolia	1	Berberis grevilleana	1
Bulnesia retama	1	Cercidium australe	1
Colletia spinossisima	1	Dacrydium fockii	1
Drimys winteri	1	Eucryphia cordifolia	1
Fitzroya cupressoides	3	Guevina avellana	1
Larrea divaricata	1	Larrea nitida	1
Laurelia philippiana	1	Laurelia sempervirens	1
Lomatia hirsuta	2	Maytenus boaria	4
Monttea aphylla	1	Myrtus luma	1
Nothofagus alpina	1	Nothofagus antarctica	3
Nothofagus betuloides	2	Nothofagus dombeyi	4
Nothofagus obliqua	4	Nothofagus pumilio	6
Persea lingue	1	Pilgerodendron uviferum	1
Podocarpus andinus	1	Podocarpus nubigenus	1
Proustia cuneifolia	1	Saxegothaea conspicua	1
Schinus molle	1	Schinus polygamus	1
Weinmannia trichosperm	1		

FUTURE PROSPECTS

Early contacts made by V.C.LaMarche Jr. with several institutions in Argentina and Chile proved very valuable in our later work there and have helped to generate growing interest in dendrochronology in both countries. The Botanical Section of the Department of Natural Resources of the Fundación Bariloche (Bariloche and Buenos Aires) and the Department of Ecology of the Universidad Católica de Chile (Santiago) have begun independent dendrochronological projects since 1976.

The Argentine Institute of Nivology and Glaciology in Mendoza collaborated closely with us throughout our field work, due to the interest of A.E.Corte, Director of I.A.N.I.G.L.A. J.A.Boninsegna, Acting Director, accompanied us on most of our field trips. He has established there a Laboratory of Dendroclimatology, which operates under his leadership and has begun work on some of the more difficult but potentially useful species. A great deal of interest developed in dendrohydrology in Argentina and Chile during the preparation for publication of an article on the extension of riverflow records using tree-ring chronologies (Holmes et al., 1978, 1979; this volume). The application of tree rings to hydrological reconstruction may attract even greater interest than their application to primary climatic factors.

In South America, tree rings may be able to assist in reconstruction of volcanic events and their effects. Methods may be developed to use additional species for dendrochronology in the tropical and subtropical areas of nothern Argentina, Bolivia, Paraguay, and Brazil, the relict forests of Fray Jorge y Talinay in northern Chile, and the cultivated windrow trees on estancias toward the dry edge of the humid Pampa. Finally, some of the species growing south of $44°$S latitude may prove to be workable and of sufficient age, thus permitting the southward extension of dendroclimatic coverage by over 11 degrees of latitude.

Table 3.5a: Summary of South American chronologies: Argentina. Adapted from LaMarche et al. (1979a).

IDENT*	SPECIES**	STARTING AND ENDING YEAR	MEAN SENSITIVITY	STANDARD DEVIATION	AUTO-CORRELATION
ANG	AR	1717-1974	0.14	0.31	0.83
CAV	AR	1444-1974	0.13	0.23	0.70
CHP	AR	1246-1974	0.15	0.22	0.65
CLL	AU	1539-1974	0.17	0.26	0.62
COP	AR	1640-1974	0.13	0.18	0.63
CYM	AU	1543-1974	0.23	0.33	0.62
ELM	AU	1690-1974	0.20	0.30	0.62
ETA	AU	1540-1974	0.20	0.32	0.70
HNG	AU	1418-1975	0.17	0.25	0.60
KIL	AR	1700-1974	0.22	0.28	0.48
LAL	AR	1306-1974	0.15	0.24	0.65
LTN	AU	1700-1974	0.18	0.23	0.54
MML	AR	1690-1976	0.12	0.17	0.56
MOQ	AR	1601-1974	0.13	0.26	0.80
PRA	AR	1486-1974	0.13	0.18	0.62
PRI	AR	1140-1974	0.16	0.25	0.71
PRP	AR	1459-1974	0.14	0.24	0.77
RAH	AR	1483-1974	0.13	0.18	0.56
RUC	AR	1392-1976	0.14	0.18	0.55
RUC	AU	1572-1976	0.19	0.32	0.74
TRO	AR	1617-1976	0.12	0.17	0.62

* See Figures 3.1 and 3.2   ** Species Identification: AR = Araucaria araucana, AU = Austrocedrus chilensis

Table 3.5b: Summary of South American chronologies: Chile. Adapted from LaMarche et al. (1979a).

IDENT*	SPECIES**	STARTING AND ENDING YEAR	MEAN SENSITIVITY	STANDARD DEVIATION	AUTO-CORRELATION
ABA	AU	1733-1975	0.16	0.19	0.40
AGU	AR	1242-1975	0.10	0.16	0.65
CAR	AR	1440-1975	0.09	0.19	0.79
CHY	AU	1641-1975	0.20	0.30	0.73
ELA	AU	1011-1972	0.18	0.25	0.57
GAB	AU	1131-1975	0.17	0.25	0.60
HUB	PL	1868-1975	0.14	0.19	0.55
ISA	AU	1568-1975	0.14	0.22	0.60
LON	AR	1664-1975	0.11	0.16	0.63
MES	AU	1796-1975	0.17	0.24	0.59
NAL	AR	1386-1975	0.13	0.23	0.72

* See Figure 3.3   ** Species identification: AR = Araucaria araucana, AU = Austrocedrus chilensis,
PL = Pilgerodendron uviferum

COMMENT

J.A.Boninsegna

Translation: R.L.Holmes

Holmes has described the chronology development
carried out by V.C.LaMarche Jr. and his group in Argentina
and Chile from about 32 to 44°S latitude. They used almost
exclusively two coniferous species: Austrocedrus chilensis
and Araucaria araucana. The possibility exists of
extending the above-mentioned area by studying the
characteristics of other species south of 44°S, there are
nearly pure stands of Nothofagus pumilio (lenga).
Exploratory sampling and examination of cross-sections
indicate that Nothofagus pumilio possesses annual rings
visible with some difficulty. By improving ring visibility
through the use of stain and/or ultraviolet light, we have
cross-dated this species from a site at 50°S latitude.
Maximum age at the site is 280 years. Two preliminary
chronologies have been developed from this site from the
lower limit and near the upper limit of the forest. They
exhibit characteristic features which are being
investigated. Nothofagus pumilio extends to the southern
limit of the South American continent at 55°S. Further
study is needed to explore the dendroclimatological
potential of this species.

In northern Argentina, within the vegetation type
known as the Tucuman-Bolivian forest, several species of
interest are found at different altitudes, such as
Podocarpus parlatorei, Cedrela lilloi, Juglans australis,
and Phoebe porfiria. Podocarpus parlatorei has clearly
visible rings, but exhibits a great frequency of "wedging"
which makes cross-dating nearly impossible. Cedrela
lilloi, Juglans australia, and Phoebe porfiria have
clearly defined rings that show high correlation
(averaging 0.75) among radii within a single tree. The
maximum ages attained by these trees is under study, as is
the possibility of cross-dating among individuals of the
same species. The range of these species extends from
northern Argentina, through the Bolivian rainforest, as
far as the eastern part of the Cordillera of the Andes in
Peru. Their wide distribution makes them of special
interest.

Further site sampling of Araucaria araucana and
Austrocedrus chilensis has been carried out, especially
along the eastern limit of these species at the edge of
the Patagonian plateau, in order to fill out the coverage
of the area studied by members of the University of
Arizona team. In parallel with these studies we are, at
I.A.N.I.G.L.A., conducting a major effort to gather all
possible meteorological information from the areas where
chronologies have been derived. For various reasons this
information is widely scattered. It includes records from
estancias, government institutions, railroads,
agricultural experiment stations, and similar bodies.
Owing to the diversity of sources, special care must be
exercised in screening data series. At the same time we

have commenced historical climatology, mainly by searching
for documentary references in the archives of the Spanish
Colonial period, approximately 1550 to 1820. Knowledge
derived from these records will be important in the
verification of reconstructions from tree rings.

AUSTRALASIA

John Ogden

GEOGRAPHICAL SETTING

Location and physical setting. The area being
considered here comprises the Australian mainland, and the
associated islands of Tasmania to the south and New Guinea
to the north, and the isolated islands of New Zealand
situated some 1,600km to the east of Australia.
Continental Australia, Tasmania, and New Guinea are part
of the same crustal plate and have been connected by land
at times of low sea-level during the Pleistocene
(Jennings, 1972). Consequently, they share elements of
their flora and fauna, and comprise a single, if diverse,
biogeographical region. New Zealand has been isolated from
the other southern continents at least since the upper
Cretaceous (60-80Myr BP; Weissel et al., 1977), and its
flora and fauna are predominantly endemic at the species
level.

Australia has a land area of 7,682,000km^2 lying
mostly between 15 and 35°S and 115 and 150°E. This vast
area is characterised by plains and plateaux; the average
elevation of the surface is only 300m and almost 90% of
the continent is less than 500m in elevation. The three
main structural components are the Eastern Uplands, rising
to over 2,000m altitude at Mt. Kosciusko, the gently
warped Central Basin and the predominantly low-lying,
semi-desert Western Shield (Mabbutt, 1970). Tasmania is
basically a southern extension of the Eastern Uplands, and
is Australia's smallest, wettest, and most mountainous
state. Much of its area is rugged uplands exceeding 1,000m
in height, dominated by a high central plateau. Within
Tasmania, there are strong west-east contrasts in climate,
landforms and vegetation. The island of New Guinea lies
north of Australia and comprises the countries of Papua
New Guinea (Papua Niu Gini) and Irian Jaya (Indonesia).
This tropical island is predominantly mountainous, with
extensive intermontane highlands and many peaks in excess
of 4,000m. The main divide is orientated southeast to
northwest and is drained by several large rivers which
have built up extensive coastal plains. New Zealand
(267,000km^2), situated between latitudes 34°S and 47°S
and longitudes 167°E and 178°E, is narrow and mountainous,
with its major dividing range reaching over 3,600m, and
running southwest to northeast. South Island is
predominantly mountainous with extensive glaciers.
Alluvial outwash plains occur along the eastern flank of
the mountains. North Island takes the form of an arcuate
cross, with a northern extension of the alpine system
meeting the volcanic plateau which dominates the centre of

the island. Both islands are geologically complex and structurally diverse (Cochrane, 1973).

Biogeographical background. The geological background to the biogeography of the region is described by Dietz & Holden (1970), Fleming (1962, 1979), and Weissel et al. (1977). It is convenient to simplify the biotic diversity of the region with a threefold division into:

    a) an ancient element derived from continental Gondwanaland, with generic affinities throughout the southern continents and generally moist temperate (oceanic) ecological requirements;

    b) a tropical element, colonising from the Malaysian region as the distances between this area and Australia and New Zealand declined; and

    c) an endemic element including some large genera such as _Eucalyptus_ in Australia and _Hebe_ in New Zealand, evolving _in situ_ in response to oscillating climatic conditions and extensive topographic changes during the Quaternary.

Forest vegetation. General accounts of the Australian forest vegetation can be found in Leeper (1970) and Carnahan (1976), the latter with the most recent vegetation map. Information on a regional basis can be found in state floras and in Specht et al. (1974). A structural comparison of the rain forests of Australia, New Zealand, and New Guinea is given by Webb (1978). The vegetation of New Guinea is described by Paijmans (1975, 1976) and the Tasmanian vegetation by Jackson (1965). A general description of the forest vegetation of New Zealand, with notes on the most important trees, is given by Godley (1976), and a comprehensive summary and literature review by Wardle (1973a). Maps of the contemporary forest areas, and of forest vegetation _circa_ 1840 in New Zealand are given by Wendelken (1976).

The Gondwanic floristic element is characterised by the tree genera _Nothofagus_, _Phyllocladus_, _Podocarpus_, _Dacrydium_, the family Araucariaceae, and numerous genera of ferns and lower plants (Schuster, 1972). This element is particularly prominent in New Zealand, where montane forests are dominated by various evergreen species of _Nothofagus_, and lowland forests commonly have an emergent stratum of members of the Podocarpaceae with smaller hardwood trees of tropical or subtropical affinity, _Weinmannia_, _Elaeocarpus_, _Metrosideros_, forming the subcanopy. Where they remain undisturbed, these forests commonly form altitudinal continua, with the diverse podocarp-hardwood forests of the lowlands gradually giving place to floristically simpler _Nothofagus_ forests with increasing altitude, although relatively sharp discontinuities are sometimes found. At lower latitudes, in New Guinea, the Gondwanic element is restricted to the montane zone, where _Nothofagus_ has undergone considerable speciation. The New Guinean lowlands have been colonised

predominantly by forest genera from Malaysia, including members of the Dipterocarpaceae. This contrast between the tropical lowland flora, derived from the north, and the southern origin of the montane flora is stressed by Schuster (1972), but its reality has not been objectively assessed (Walker & Guppy, 1976).

The tropical Malaysian floristic element extends from New Guinea into the wetter mountainous regions of eastern coastal Queensland from Cape York southwards where it is mixed with species of the Australian endemic element. On the slopes of the Atherton Tablelands, over 100 tree species per 0.4ha may be found. Prominent families include Lauraceae, Rutaceae, Myrtaceae, Meliaceae, Myristicaceae, Proteaceae, Sapindaceae, and Rubiaceae. Rain-forest areas in Queensland are discontinuous amongst the ubiquitous _Eucalyptus_-dominated associations. Further south, in New South Wales, rain-forest patches become smaller, less diverse, and restricted to sites where orographic rainfall or fertile soils make conditions exceptionally suitable. With the inclusion of _Nothofagus_ _moorei_ (in New South Wales) and _Nothofagus_ _cunninghamii_ (in Victoria) they acquire a Gondwanic element, which gradually preponderates further south; the temperate rain forests of western Tasmania have close generic and structural affinity with similar _Nothofagus_- and _Dacrydium_-dominated associations in New Zealand. The island has several endemic conifers, most notable being two species of _Athrotaxis_ (Taxodiaceae) (Ogden, 1978b).

In Tasmania, as elsewhere in Australia, the balance between rain-forest and sclerophyll communities dominated by _Eucalyptus_ is determined by the subtle interplay of soil fertility and fire frequency (Jackson, 1968). Fire has been a potent force in the evolution of the widespread endemic element which gives the unique character to the Australian landscape (Singh et al., 1980). This element is characterised by families and genera largely restricted to Australia, including some sections of the Protoeaceae, Epacridaceae, _Eucalyptus_, _Acacia_, and _Callitris_.

Climate. Regional climates in Australasia are dominated by the seasonal latitudinal migration of the major pressure systems, locally modified by oceanic and topographic effects. In the north, there is a belt of migrating anticyclones and the thermal equator or intertropical convergence zone (ITCZ). These dominate the humid tropical climate of highland New Guinea and the strongly seasonal subtropical rainfall regime of northern mainland Australia. In the south, the cold fronts associated with low pressure cells in the belt of the westerlies bring characteristically variable and frequently wet weather to Tasmania, the southern mainland, and New Zealand. Comprehensive summaries of the Australian climate can be found in Gentilli (1971) and Linacre & Hobbs (1977). Gentilli (1972) gives details on a regional basis and

Langford (1965) describes the climate of Tasmania.

The climate of Papua New Guinea is summarised by Ryan (1972). The climate of New Guinea can be broadly described as humid equatorial, but is greatly influenced by topography. Rainfall maxima generally occur in summer, when the ITCZ is south of the island and the northwest "monsoon" winds prevail. The southeast trades penetrate the windward lowlands during much of the remainder of the year, but only locally influence the extensive mountainous areas. Northern Australia shows a marked alternation between wet and dry seasons. Short, wet summers with overcast skies and heavy convective rainfall are related to inflow of moist tropical air when the ITCZ is at its southernmost position, followed by long, warm winters with generally clear skies as the ITCZ moves north. The southern part of the continent tends to have dry summers and wet winters, reflecting seasonal shifts in the normal tracks of cyclonic and anticyclonic cells, and a Mediterranean climate regime is characteristic of much of southwestern and southern Australia.

Large regions of central eastern Australia have two distinct wet seasons – a major one in winter. From the point of view of tree growth, however, it may be reasonable to regard this region as having one prolonged humid period, from December to June (Gentilli, 1972). Tasmania lies south of the normal anticyclone track, and is predominantly influenced by a progression of cold fronts that bring heavy precipitation to its mountainous western side. Precipitation decreases and is more variable to the east, where semi-stationary anticyclones in the Tasman Sea can also contribute to flow patterns leading to drought conditions. A significant feature of the Australian climate is its high variability, especially in the drier regions. Erratic rainfall and disastrous droughts are recurring themes over most of Australia.

Excellent accounts of the climate of New Zealand are given by Garner (1958), Maunder (1970), and Tomlinson (1976). Ecological aspects are discussed by Coulter (1973) who also supplements the bibliography of Sparrow & Healy (1968). Mountain climates are described by Coulter (1967, 1973) and by Mark & Adams (1973). Data sources are listed by Crawford (1975) in an account of a symposium published by the New Zealand Meteorological Service, and an annotated list is given by Maunder (1970). The climate of New Zealand is governed by its position in the path of migrating anticyclones and depressions of the westerly belt and its high local relief. The oceanic influence provides relatively mild climates, without marked seasonal extremes, over much of the country. However, the barrier of the high Southern Alps brings heavy rainfall to the southwest coast of South Island, and creates a "rain-shadow" effect in central Otago.

There is an extensive literature relating to the Pleistocene and Holocene climates of the Australian region (Moar, 1966, 1971; Bowler et al., 1976; Australian Academy of Science, 1976; McGlone & Topping, 1977; Walker, 1978; Macphail, 1979; Soons, 1979; Burrows & Greenland, 1979), as well as to climatic variability during the periods of historical and instrumental records (Salinger, 1976; Pittock et al., 1978; Hessell, 1980). Such information may be useful as a basis for calibration of tree-ring sequences and for independent verification of dendro-climatic reconstructions.

DENDROCHRONOLOGY AND DENDROCLIMATOLOGY

Australia. The indigenous peoples of Australia did not build extensive wooden structures. Moreover, the predominant tree genera of the continent (Eucalyptus, Acacia) do not usually form clear annual rings and are short-lived (Ogden, 1978a; Wellington et al., 1979). For these reasons, archaeologists and ecologists have paid scant attention to tree rings until their value as proxy climate indicators become more widely appreciated during the 1970s. Consequently, tree-ring research has a short history in Australia, and emphasis has been on climatic interpretation. A review of research prior to 1978 and an appraisal of the dendrochronological potential of Australian trees is given by Ogden (1978a). The most extensive sampling is that of LaMarche and co-workers (LaMarche et al., 1979d).

The widespread genus Eucalyptus is largely unsuitable for dendroclimatological work due to diffuse ring boundaries and intra-annual bands. In subalpine species with relatively clear annual rings, ring width may be chronically influenced by the large phytophagous insect population on the trees (Morrow & LaMarche, 1978; Readshaw & Mazanec, 1969). However, some progress has been made in cross-dating Eucalyptus pauciflora from timberline stands in the mountains of southeastern Australia (Banks, personal communication, 13.10.1979).

The genus Callitris (Cupressaceae) is widely distributed in the semi-arid regions of the continent and there is evidence to suggest that the width of the annual increment in such areas is determined primarily by rainfall (Lange, 1965; Pearman, 1971; Johnston, 1975). LaMarche and co-workers sampled 350 individuals from four species in this genus (and the related Actinostrobus) at 38 sites throughout the southern half of Australia (see Figures 3.4a,b,c; site details are given in Lamarche et al., 1979d). However, owing to difficulties in determining annual ring boundaries, only two short chronologies have so far been developed for sites with a highly seasonal rainfall distribution in the vicinity of Perth. The negative autocorrelation for the combined chronology (Table 3.6) may imply a biological rhythm overriding the climate signal.

The greatest potential for dendrochronology in Australia lies with the long-lived endemic Tasmanian conifers in the genera Athrotaxis, Phyllocladus, and Dacrydium. Ogden (1978b) outlines research on the two

Athrotaxis species, maps sites, tabulates his collections, and illustrates good cross-dating between widely separate sites. LaMarche et al. (1979d) present four chronologies for this genus, 10 for Phyllocladus aspleniifolius and one for Nothofagus gunnii (Figure 3.4d, Table 3.6). The longest chronology, for Athrotaxis cupressoides, spans the period 1028 AD to 1975 AD, but several extend back to the 1500s or earlier. The marked contrast between the mean chronology statistics for the relatively insensitive, highly autocorrelated, Athrotaxis species, and the more sensitive, less strongly autocorrelated, Phyllocladus aspleniifolius is also illustrated by Ogden (1978a). Nothofagus gunnii is intermediate in these respects, but trees of this species probably rarely exceed 300 years in age.

These Tasmanian species have different chronologies with only exceptionally narrow rings in common. In the case of the deciduous Nothofagus gunnii a lagged response may be involved, but the difference between Athrotaxis and Phyllocladus is likely to be due to control by different limiting factors. In the case of timberline Athrotaxis cupressoides and montane Athrotaxis selaginoides, summer temperatures may determine growth, while submontane Phyllocladus aspleniifolius may be responsive to summer droughts and mild winters. Until analyses of response functions and climatic calibrations are available, further speculation is unwarranted.

Papua New Guinea and the Australian tropics. A review of dendrochronological studies and tree age determination in the Australian tropics (including Papua New Guinea) is given by Ogden (1981). Although rings are commonly present in trees in non-seasonal environments, there is difficulty in determining their annual nature. There is a basic need for more research on wood anatomy, phenology, and the climatic responses of the trees.

Figure 3.4a: Site map, western Australia.

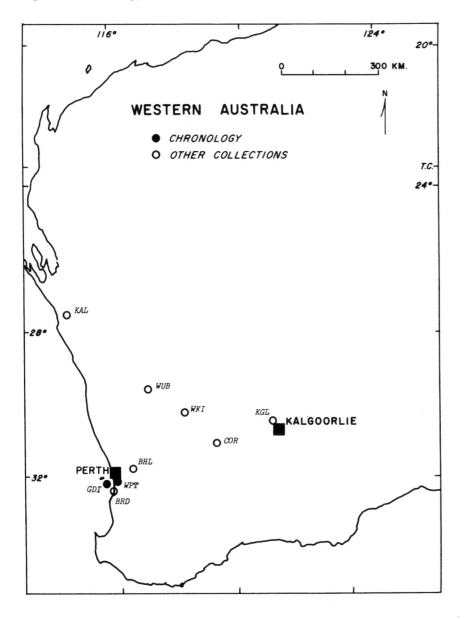

In open woodlands in the summer rainfall zone some progress has been made with species of <u>Eucalyptus</u> (Mucha, 1979), <u>Callitris collumellaris</u> (Hammer, personal communication, 6.3.1979), and <u>Diospyros ferrea</u> (Duke et al., 1981). However, all of these species are short-lived and have diffuse ring boundaries with abundant false rings. In all cases the evidence indicates that diameter increment occurs during the rainy season. Long-lived trees (age > 1,000yr) occur in the closed rain-forests of Queensland and montane Papua New Guinea, and some of the genera present are known to be suitable for dendrochronology elsewhere (for example, <u>Agathis</u>, <u>Araucaria</u>, <u>Podocarpus</u>, <u>Dacrydium</u>, and <u>Nothofagus</u>). Enright (1978) provides some evidence that the rings of <u>Araucaria hunsteinii</u> are sometimes approximately annual, but Ash (personal communication, 13.10.1979) has found discrepancies between radiocarbon ages and ring-counts for older individuals of several rain-forest species.

The only thorough anatomical growth study to have been made on a rain-forest tree in Australia appears to be that of Amos & Dadswell (1950) on <u>Beilschmiedia bancroftii</u> (Lauraceae), which becomes dormant during the summer wet season. Likewise ring width in <u>Pisonia grandis</u> (Nyctaginaceae) may be negatively related to wet season rainfall, but this species is short-lived and has a limited distribution so that it is of little dendrochronological value (Ogden, 1981). Reduced growth during the summer wet season may be due to raised temperatures and reduced average light intensities (increased cloudiness) adversely affecting the tree's carbohydrate balance.

New Zealand. Studies using tree rings in New Zealand have recently been reviewed by Dunwiddie (1979) and are summarised by species in Table 3.7. A photographic survey of the wood anatomy of New Zealand species is given by

Figure 3.4b: Site map, south-central Australia.

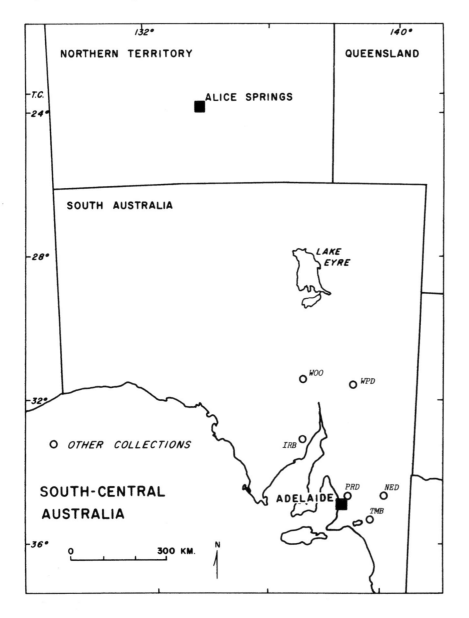

Meylan & Butterfield (1978).

Although both Druce (1966) and Topping (1971) used ring-width sequences for dating purposes, in general, tree-ring studies in New Zealand have centred on the ages of living trees rather than on the development of cross-dated ring-width chronologies. First attempts at constructing chronologies for archaeological dating were mainly unsuccessful (Bell & Bell, 1958; Cameron, 1960; Scott, 1964). Consequently, while many of the studies listed in Table 3.7 provide valuable background information about ring characteristics, they are of limited relevance to the main theme of this review.

Preliminary studies by Carter (1971) produced a short dated chronology for the montane species _Nothofagus solandri_ var. _cliffortioides_. This showed a positive correlation with _previous_ summer temperatures. In contrast both Scott (1972) and Norton (1979) obtained positive correlations between _current_ summer temperatures and ring widths for this species. Scott (1972) attempted cross-dating and a form of response function analysis using _Nothofagus solandri_ var. _cliffortioides_, _Podocarpus hallii_, _Phyllocladus alpinus_, _Discaria toumatou_, and two introduced species of _Pinus_ from sites at Mt. Ruapehu and Lake Takepo. His conclusions were largely negative: "The present study confirms earlier conclusions (Bell, 1958; Bell & Bell, 1959; Cameron, 1960; Scott, 1964) that because of difficulties of ring recognition, lack of radial uniformity, difference between adjacent trees, and difference in pattern between species and areas, dendro-chronological techniques will be extremely difficult to apply in New Zealand" (Scott, 1972). However, in addition to the positive response to early summer temperature in _Nothofagus solandri_, mentioned above, he obtained a significant negative response to summer temperature in _Phyllocladus alpinus_, but no significant correlations with either temperature or precipitation in _Podocarpus hallii_.

Figure 3.4c: Site map, southeast Australia.

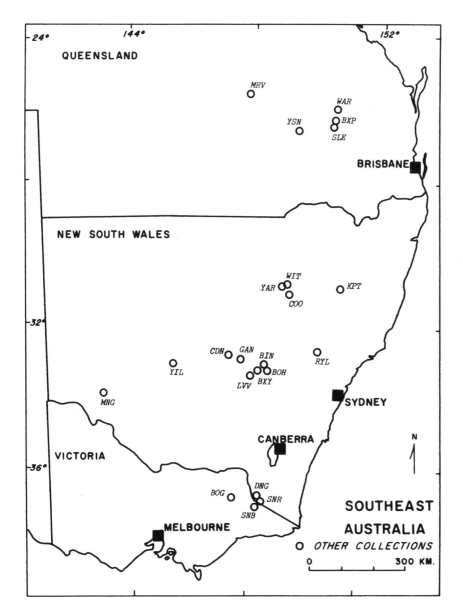

Wells (1972) investigated the palaeoecology and dendroclimatological potential of the latter species on the Pisa Range of Central Otago. Her study indicated the presence of ancient wood in the area and provided some useful phenological data, but in general she agreed with Scott (1972) and Carter (1971) in suggesting that Podocarpus hallii was not a sensitive climatic indicator.

The largely negative appraisals of the dendro-climatic potential of the species they studied reached by Scott (1972), Wells (1972), and some earlier workers may be criticised for small sample sizes, poor site selection and replication, and in some cases lack of adherence to strict cross-dating techniques. In view of these weaknesses, Dunwiddie's (1979) survey, based on cores and discs from 20 New Zealand tree species collected from 45 sites throughout North Island and South Island, supercedes all previous work.

Dunwiddie (1979) illustrates wood cross-sections and

representative portions (1850 to 1976) of nine chrono-logies from the 20 he developed. Seven species are represented in these chronologies, the longest of which is for Libocedrus bidwillii and covers the period 1256 to 1976. Details of sites and species are given in Dunwiddie (1978, 1979) and in greater detail by LaMarche et al. (1979c; Figures 3.4e,f) who also provide index chronologies. A synopsis is presented in Tables 3.8 and 3.9. These results reveal a contrast in response between the relatively insensitive and strongly autocorrelated Libocedrus bidwillii, Dacrydium biforme, and Dacrydium colensoi, and the more sensitive and weakly autocorrelated Phyllocladus trichomanoides and Phyllocladus glaucus. This contrast mirrors that between the two Athrotaxis species and Phyllocladus asplenifolius in Tasmania. In part it may be accounted for by the different altitudinal ranges of the species involved. Thus, if the New Zealand chronologies are examined by altitude (excluding

Figure 3.4d: Site map, Tasmania.

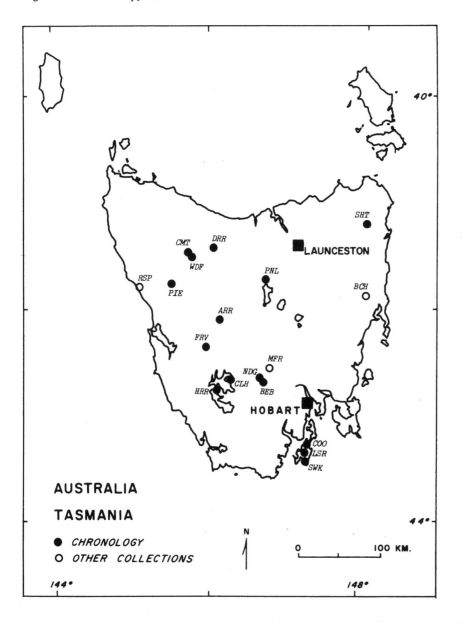

Phyllocladus glaucus) mean sensitivity is found to
decrease (r = 0.56; n = 18; P < 0.02) and autocorrelation
to increase (r = 0.50; n = 18; P < 0.05) with the altitude
of the sample. Phyllocladus glaucus, in common with some
samples of other Phyllocladus species, has negative
autocorrelation, indicating a marked biennial alternation
in ring width. This biennial oscillation has been noted in
other New Zealand species (for example, Agathis australis,
Lloyd 1963; Ogden, unpublished observations) and is likely
to be phenological rather than climatic in origin. Its
prevalence appears to limit climatic interpretations based
on Phyllocladus glaucus despite the apparently good
analysis of variance and cross-correlation results for
this species (Table 3.9). In general, the analysis of
variance results (LaMarche et al., 1979c) are similar to
those of Scott (1972), who found that his chronologies
accounted for from 9%, Podocarpus hallii, to 37%,
Nothofagus solandri var. cliffortioides, of the total
variance.

No climatic interpretation of these chronologies has
been published, although the presence of synchronous
narrow rings between some North and South Island sites of

Libocedrus bidwillii indicates control by relatively
large-scale synoptic weather patterns. Response functions
have been obtained for all chronologies and a more
detailed analysis is in progress (LaMarche, personal
communication, 21.6.1979). In view of the montane location
of most of the sites it is not surprising to find that
significant responses to temperature variables are more
common than those involving precipitation, but apparently
few sites show an indication of the temperature increase
which may have occurred throughout New Zealand during the
last half century (Salinger, 1979; Hessell, 1980). The
altitudinal difference between the tree-ring sites and the
long-term weather records may be implicated, but the
possibility that the indexing procedure has removed the
long-term climatic influence must be borne in mind.

In contrast to the Northern Hemisphere, the forests
in many areas of New Zealand were largely undisturbed by
man until colonisation by Europeans and introduced
browsing mammals began to have an impact during the 19th
century. Consequently, it has been supposed that the age
structure of some tree species within these forests
reflects changes in climatic conditions influencing their

Figure 3.4e: Site map, New Zealand, South Island.

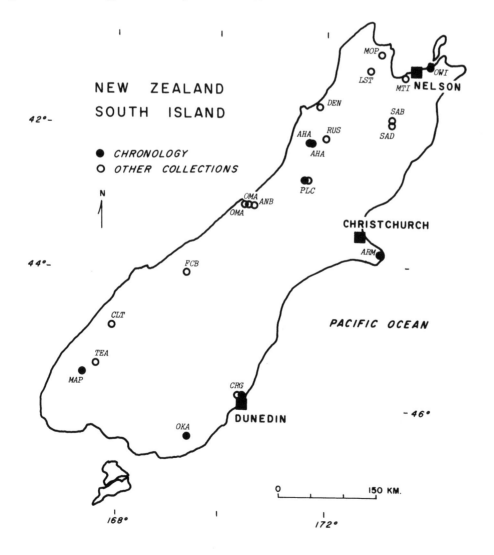

regeneration. Since Holloway (1954) suggested that the apparently anomalous size-class frequency distributions and disjunct spatial patterns of Dacrydium cupressinum and Nothofagus spp. could be accounted for by postulating climatic changes, the subject has been a dominant theme in the New Zealand ecological literature and several reviews are available (Wardle, 1963a, 1978; Molloy 1969; Burrows & Greenland, 1979). Wardle (1963b) found that fluctuations in growth rate (ring widths measured in decades) were consistent between nine Dacrydium cupressinum trees on Secretary Island. A more comprehensive study of this species in the Longwood Ranges, where Holloway (1954) obtained most of his evidence, is currently in progress. Preliminary results indicate that the decadal chronology, which is based on discs from 100 trees for the first 800 years and extends to about 1000 BP, will provide a palaeotemperature record (Bathgate, personal communication, 17.9.1979). Franklin (1969) found that he could cross-date particular years marked by dense latewood in Dacrydium cupressinum and that these years had cool autumn temperatures. Against these apparently promising

results with this species must be set others (Dunwiddie, 1979; Ogden, unpublished observations) recording problems with indistinct ring boundaries and lobate growth.

Until more is known about the regeneration of New Zealand tree species it is unlikely that age-structures alone can shed much light on climatic oscillations. However, consideration of closed forests where several distinct generations exist concurrently (for example, Athrotaxis in Tasmania, Ogden, 1978b; Libocedrus in New Zealand, Clayton-Green, 1977) raises some important considerations for the dendroclimatologist. For example, a cohort of old emergent trees may experience a different set of limiting factors to the cohort of subcanopy individuals which will eventually succeed it. During this replacement process competitive hierarchies develop within each cohort, so that biological interactions may suppress the climate signal for a majority of individuals. Moreover, if each generation has developed under different climatic or ecological conditions and the extensive mortality associated with this manner of regeneration has been selective, then different generations within a

Figure 3.4f: Site map, New Zealand, North Island.

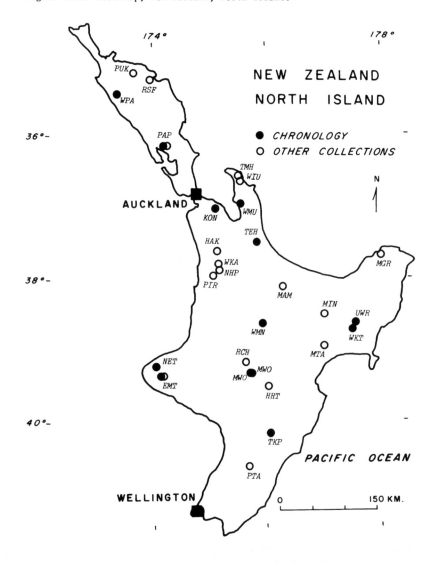

species may have different response functions. These considerations suggest that a policy in which samples from different age-classes are analysed separately is to be preferred; averaging together ring-width sequences from younger and older generations may blur the climatic signals of the latter.

REVIEW OF SUITABLE GENERA AND SOME FUTURE PROSPECTS

In this section, I attempt to summarise some of the main conclusions about the suitability for dendro-climatological research of some of the most widespread, and apparently suitable, genera in the Australasian region, and make some comments on possible future developments.

The genus Phyllocladus (Podocarpaceae) has representatives in New Zealand, Tasmania, and montane Papua New Guinea. Nothing is known of its ring characteristics in the latter country, but elsewhere it appears to reach ages of 500 to 900 years and to have clear annual rings,

which are sensitive. There is an indication that sensitivity is a response to moisture stress, but the presence of long biennially alternating sequence of wide and narrow rings contributes to low (sometimes negative) autocorrelation coefficients, and is a puzzling feature of the genus.

The genus Dacrydium (Podocarpaceae, including the related Dacrycarpus in Papua New Guinea) has a similar distribution to Phyllocladus. It includes some very long-lived trees (age > 2,000yr, Dacrydium franklinii) in habitats varying from lowland temperate rain forest on flood plains to alpine timberlines. Some of the lowland genera are complacent, and/or have markedly lobate growth, but long chronologies have been developed for two New Zealand montane species (Dacrydium bidwillii, Dacrydium colensoi, LaMarche et al., 1979c). Dacrydium franklinii in Tasmania can also be cross-dated in some cases. Preserved logs of this species have recently been sampled for carbon isotope analysis and one has recently been radiocarbon

Figure 3.4g: Site map, recent collections, North Island.

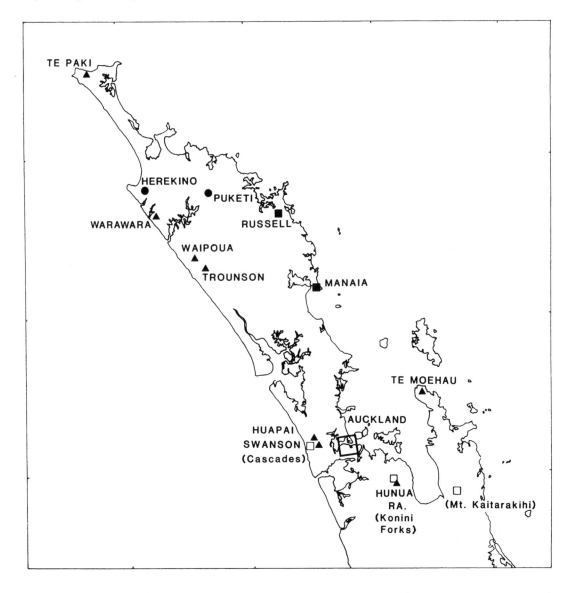

dated at > 6000 BP (Francey, personal communication, 28.3.1981).

Dacrydium cupressinum occurs in a wide range of habitats throughout New Zealand (Franklin, 1968). Despite several attempts no annual ring-width chronologies have yet been developed for the species but the results of Franklin (1969) suggest that a densitometric approach may be more profitable. X-ray densitometry equipment has been in use at the Forest Research Institute (Rotorua) in New Zealand for several years (Ellis, 1971) and excellent facilities are currently available.

The ecologically similar genera Libocedrus (Papuacedrus in Papua New Guinea) (Cupressaceae) and Athrotaxis (Taxodiaceae) also contain timberline representatives from which long chronologies have been constructed. The chronologies obtained so far from Dacrydium, Libocedrus, and Athrotaxis (LaMarche et al., 1979c,d) are alike in being relatively insensitive and highly autocorrelated. Where species in these genera occur with Phyllocladus they appear to respond differently to climate. As most of the species studied are found at timberline or in montane forests, summer temperature may be found to be an important component of their response functions. In the case of Athrotaxis selaginoides in Tasmania, preliminary results indicate a closer relationship between mean summer temperatures and ^{13}C

content than between mean summer temperatures and either total ring width or latewood width (Pearman et al., 1976; R.J.Francey, personal communication, 20.4.1977).

Several species in this group have regeneration strategies which generate even-aged stands, and two or three distinct generations may sometimes be present in an area. Interference within and between generations will add to the "noise" in the chronology, and should be minimised by careful site and cohort selection.

The genus Podocarpus is widespread in the Southern Hemisphere but although some cross-dating has been reported (Bell & Bell, 1958), in general, attempts to use the genus have had little success (Dunwiddie, 1979). Preserved logs are known in archaeological contexts in New Zealand.

The most widespread and abundant genera of the Australian mainland, Eucalyptus (Myrtaceae) and Acacia (Leguminosae), usually appear to be difficult or impossible to cross-date, and may be unsuitable for dendroclimatology in most cases. However, the possibility that new methods will change this conclusion must be borne in mind. LaMarche and co-workers collected many samples of the almost equally widespread genus Callitris (Cupressaceae), but have so far had little success with cross-dating on most sites. It appears that these characteristically Australian genera are adapted to

Table 3.6: Synopsis of chief characteristics of some Australian chronologies. Data of LaMarche et al. (1979d).

| SPECIES | TOTAL NUMBERS | | | CHRONOLOGY CHARACTERISTICS | | | | | |
	SITE CHRONOLOGIES	TREES	RADII	MAXIMUM LENGTH (YRS)	MEAN AUTO-CORRELATION	AVERAGE MEAN SENSITIVITY	% ABSENT RINGS (RANGE)	% VARIANCE ACCOUNTED FOR BY CHRONOLOGY	MEAN CORRELATION BETWEEN TREES (RANGE)
Athrotaxis cupressoides	1	43	91	947	0.65	0.13	0.00	28-37	0.35_0.42
A. selaginoidus	1	14	33	777	0.62	0.14	0.08	23	0.27
Callitris robusta	2	12	37	63	0.06	0.21	0.00	37	0.40
Nothofagus gunnii	1	4	14	243	0.52	0.17	0.00	37	0.40
Phyllocladus asplenifolius	10	75	184	664	0.26	0.29	0.0-3.96	24-49	0.26-0.54
COMPOSITE CHRONOLOGIES (SITES)									
Phyllocladus asplenifolius BIT 159	3	23	70	433	0.03	0.26	-	26	0.36
Phyllocladus asplenifolius LYL 159	2	10	22	427	0.30	0.26	-	23	0.30
Athrotaxis cupressoides MTB 139	2	32	66	947	0.64	0.13	-	33	-
Callitris robusta PER 119	2	12	37	63	-0.08	0.20	-	24	0.20

erratic rainfall, which characterises so much of the continent, by possessing an opportunistic growth pattern not clearly tied to seasons. For this reason, even where individuals possess similar patterns of growth bands these may be difficult to date absolutely. Despite these problems Callitris can be cross-dated in some circumstances, and as it seems likely to reflect a valuable precipitation chronology for the semi-arid regions in which it grows it deserves considerable attention.

Nothofagus (Fagaceae) forms the timberline on many New Zealand mountains, and is an important component of the montane forests of New Zealand, Papua New Guinea, Tasmania and South America. The species are mostly ring-porous and have clear annual rings, at least in high latitudes. Chronologies have been constructed for the deciduous Nothofagus gunnii in Tasmania, and the evergreen Nothofagus solandri in New Zealand. The latter may be responsive to summer temperatures (Norton, 1979). The

Table 3.7: Listing by species of main publications dealing with dendrochronology, ring counts, phenology, and related subjects in New Zealand, and the location of the main extant core and/or disc collections. * Nomenclature follows Allan (1961).

SPECIES*	PUBLICATIONS ARRANGED CHRONOLOGICALLY	LOCATION OF COLLECTIONS
Podocarpus totara	Lockerbie (1950), Batley (1956), Bell & Bell (1958), Cameron (1960), Lloyd (1963), Dunwiddie (1979)	
P. hallii	Oliver (1931), Bell & Bell (1958), Carter (1971), Scott (1972), Wells (1972), Dunwiddie (1979)	T T V
P. nivalis	Wardle (1963c), Wells (1972)	Ch
P. ferruginens	Bell & Bell (1958), Dunwiddie (1979)	T BD
P. spicatus	Lockerbie (1950), Bell & Bell (1958), Wardle (1963a) Scott (1964), Dunwiddie (1979)	T Ak W
P. (Dacrycarpus) dacrydioides	Bell & Bell (1958), Cameron (1960), Lloyd (1963), Scott (1964), Dunwiddie (1979)	T Ch W
Dacrydium cupressinum	Lockerbie (1950), Cameron (1957, 1960), Wardle (1963a,b), Lloyd (1963), Franklin (1968, 1969), Jansen & Wardle (1971), Dunwiddie (1979)	T Ak Ch W
D. biforme	Wardle (1963c), Dunwiddie (1979)	T
D. colensoi	Dunwiddie (1979)	T
Phyllocladus alpinus	Wardle (1963c,1969), Scott (1972), Topping (1972), Dunwiddie (1979)	T Ch
P. trichomanoides	Lloyd (1963), Herbert (1977)	T Ak
P. glaucus	Dunwiddie (1979)	T
Libocedrus bidwillii	Wardle (1963a,c), Cunningham (1964), Scott (1964), Druce (1966), Clayton-Greene (1972), Wardle (1978), Dunwiddie (1979), Norton (1979)	T Ch BS ? W
L. plumosus	Dunwiddie (1979)	T
Agathis australis	Bell & Bell (1958), Jansen (1962), Lloyd (1963), Wilson & Grinsted (1976), Dunwiddie (1979)	T Ak FRI (R)
Nothofagus solandri	Wardle (1963c), Bussell (1968), Wardle (1970), Carter (1971), Scott (1972), Dunwiddie (1979), Norton (1979)	T Ch
N. menziesii	Williams & Chavasse (1951), Bell & Bell (958), Wardle (1963c), Bussell (1968), Herbert (1972), Dunwiddie (1979)	T
N. fusca	Bussell (1968), Ogden (1978c), Dunwiddie (1979)	Ak BD
Other species	Lockerbie (1950), Bell & Bell (1958), Wardle (1963c), Mark et al. (1964), Druce (1966), Bussell (1968), Scott (1972), Dunwiddie (1979)	T Ak FRI (R) BD
Abbreviations:	Ak = Auckland University, Botany Department ED = Botany Division, DSIR, Christchurch Ch = Canterbury University, Botany and/or Forestry Departments, Christchurch FRI (R) = Forest Research Institute, Rotorua T = Laboratory of Tree-Ring Research, Tucson, Arizona V = Victoria University, Wellington, Botany Department W = Waikato University, Hamilton, School of Biological Sciences	

genus deserves much more study.

The family Araucariaceae is represented in the region by the genera <u>Agathis</u> in northern New Zealand, Queensland, and New Guinea, and <u>Araucaria</u> in tropical montane forest in Queensland and New Guinea. Long-lived individuals of both genera occur (Ogden, 1978a, 1981) and may offer some prospect of extending dendroclimatology into the tropics, perhaps using wood characteristics other than simple ring widths.

One of the most exciting prospects in the long-term dendroclimatology of region is the possibility of constructing a very long chronology for <u>Agathis</u> <u>australis</u> (Araucariaceae) in New Zealand. Not only do individuals of this species live to great ages (<u>circa</u> 1,000 years), but also there is abundant subfossil wood preserved in peat bogs. Radiocarbon dates indicate that this preserved wood ranges in age from <u>circa</u> 1700 to <u>circa</u> 40000 BP (Wilson, personal communication, 3.6.1977). Stable carbon isotope studies (δ^{13}C) have been made on cellulose from a 1,000-year old <u>Agathis</u> <u>australis</u> tree by Wilson & Grinsted (1976) and the results are interpreted as a temperature chronology. The wood samples were dated by ^{14}C rather than tree rings. Although cross-dating of living trees is frequently difficult, in certain circumstances it can be achieved (Bell & Bell, 1958; Dunwiddie, 1979) giving prospect of a radiocarbon calibration curve covering a time span similar to those for the Northern Hemisphere. This is the subject of a cooperative research programme involving M.Barbetti

(University of Sydney), R.J.Francey (Division of Atmospheric Physics, C.S.I.R.O., Melbourne), and the author. Where conventional cross-dating cannot be achieved it is planned to attempt cross-correlation of density patterns. Sites sampled are shown in Figure 3.4g.

CONCLUDING REMARKS

This survey has covered a latitudinal range in some ways equivalent to the expanse between Norway and Saharan Africa, and required an attempt to integrate findings from many climatic zones. With our present rudimentary knowledge of the dendroclimatology of the region such integration is not possible, and the most that has been achieved may have been to emphasise our chief areas of ignorance. Some of these may reside in the scanty distribution of weather records across a hemisphere dominated by ocean, or in their short lengths, but others, more fundamental to palaeoclimate reconstruction from tree rings, stem from lack of basic ecological information about the species involved.

In view of the difficulties, LaMarche and co-workers have been notably successful in establishing a network of chronologies for the temperate parts of the region. The longest chronology covers almost 1,000 years. These data are currently being evaluated for climatic reconstruction, and seem to offer good prospects. However, I note that there are no obvious or simple correlations between longer-term (10 to 50-year) trends in the published index

Table 3.8: Synopsis of chief characteristics of some New Zealand chronologies. Data of (1) Dunwiddie (1979), (2) Scott (1979), and (3) Norton (1979).

SPECIES	TOTAL NUMBER			CHRONOLOGY CHARACTERISTICS				
	SITE CHRONOLOGIES	TREES	RADII	MAXIMUM LENGTH (YRS)	MEAN AUTO-CORRELATION	AVERAGE MEAN SENSITIVITY	% ABSENT RINGS (RANGE)	REFERENCE
Agatnis australis	1	11	33	264	0.00	0.21	0.7	(1)
Dacrydium biforme	1	7	25	409	0.75	0.10	0.1	(1)
Dacrydium colensoi	2	21	69	573	0.61	0.12	0.1-1.0	(1)
Libocedrus bidwillii	8	97	295	720	0.70	0.15	0.0-2.2	(1)
Phyllocladus alpinus	1	12	38	259	0.57	0.13	0.0	(1)
Phyllocladus glaucus	4	42	110*	441	−0.16	0.44	0.7-2.6	(1)
Phyllocladus trichomanoides	4	35	103*	312	0.16	0.25	0.3-0.7	(1)
Discaria toumatou	1	4	12	36	0.12	0.49	0.0	(2)
Nothofagus solandri var. cliffortioides	3	12	12	34	0.42	0.39	0.0	(2)
Phyllocladus alpinus	2	6	14	36	0.49	0.26	0.0	(2)
Podocarpus hallii	1	5	15	36	0.52	0.52	0.0	(2)
N. solandri var. cliffortioides	2	24	48	245	0.36	0.26	1.6	(3)

* In one chronology (WPA) these species were combined and the number of radii of each species is not separately recorded. See also LaMarche et al. (1979d).

chronologies and secular variation in climate parameters over the same period. This raises doubts about the use of indexing procedures designed to eliminate growth trends. At this stage in our understanding of the ecology of the species involved, a localised, site by site, approach to climatic calibration might be more prudent than a large scale overall analysis, despite the fact that the latter approach alone may lead to correlation with synoptic weather patterns. See, however, LaMarche & Pittock (this volume).

We are accustomed to think of "sensitivity" as a species and site concept; in the case of long-lived Tasmanian conifers it may also include a time component. LaMarche et al.'s (1979c,d) chronologies show some marked changes from "sensitive" to more "complacent" periods. In New Zealand, independent data on climatic oscillations over the relatively recent past can be obtained from glaciers, and a fairly reliable chronology seems to be emerging (Burrows & Greenland, 1979). Such independent proxy records will be valuable in calibrating the tree-ring sequences, and may suggest new forms of sampling and/or analysis which will allow genuine secular climatic trends to be identified in tree-ring sequences.

COMMENT
P.W.Dunwiddie

Many significant advances have been made in dendrochronological studies in Australasia in the 1970s. Ogden has documented the many areas and species on which tree-ring work has already been carried out. Although climate reconstructions from much of this material are not yet available, there is considerable evidence that many ring-width chronologies contain a substantial climate signal. Cross-dating between chronologies separated by as much as several hundred kilometres in New Zealand suggest a dominant climate influence. Response functions for the same species from different sites exhibit similar, coherent patterns. Principal component analyses of tree-ring chronologies in New Zealand show patterns

similar to those derived using climate data. See, also, Campbell (this volume) and LaMarche & Pittock (this volume).

The importance of expanding this dendrochronological base cannot be overemphasised. Expansion of the network must include other areas - most notably the Australian continent - as well as further points within the current range. The genus Callitris will require considerable effort to produce chronologies, but is the most likely prospect in much of Australia. I have matched ring patterns between trees in several areas of New South Wales and South Australia, but better time controls are needed to develop chronologies. Fire scars of a known date and logs from buildings when the time of construction is known can help in this regard. In New Zealand, Dacrydium biforme, Dacrydium colensoi, and several species of Nothofagus may be useful on the west coast and in the extreme south. Work by T.Bird (Laboratory of Tree-Ring Research, University of Arizona) suggests that long chronologies may be developed from Dacrydium franklinii in Tasmania, a species found on the west coast. Even where other chronologies exist, the use of different species may give additional climate information owing to varying climate-growth responses in the species.

Ogden has questioned the extent of the climate signal in trees with low or negative autocorrelation values, such as Phyllocladus. This "biennial signal" is extremely marked, compared with many Northern Hemisphere species, and deserves careful investigation. The synchronism of this biennial oscillation, as well as lower frequency variations, between sites in New Zealand suggests a degree of climate influence. The extent to which biological factors amplify the pattern, and the effect this will have on reconstructed climate series, remains to be seen. Filtering out, or creation of, long-term climatic trends is probably minimal in the oldest chronologies in Tasmania and New Zealand, where growth in most trees was largely linear for hundreds of years. The problem may be magnified in shorter

Table 3.9: Synopsis of analyses of variance and cross-correlation for New Zealand species. Data of LaMarche et al. (1979e).

SPECIES	NUMBER OF CHRONOLOGIES	% VARIANCE ACCOUNTED FOR BY MEAN CHRONOLOGY MEAN (RANGE)	CORRELATION BETWEEN TREES MEAN (RANGE)
Agathis australis	1	23	0.26
Dacrydium	1	16	0.20
D. colensoi	2	14 (9–18)	0.17 (0.14–21)
Libocedrus bidwillii	8	27 (18–34)	0.30 (0.22–0.38)
Phyllocladus alpinus	1	16	0.19
P. glaucus	4	41 (37–46)	0.45 (0.41–0.51)
P. trichomanoides	3	31 (29.32)	0.30 (0.23–0.37)

chronologies, and where evaluation is made using
polynomial curves. Nevertheless, the appearance of similar
low-frequency patterns in several New Zealand chronologies
indicate that such problems may be avoidable in some
sites.

The quality of chronologies, and inferred palaeoclimatic interpretation, must always be carefully
evaluated. The decadal chronology being developed as a
palaeotemperature record using a species (Dacrydium
cupressinum) with considerably vague ring and lobate
growth patterns must be viewed with caution. The use of
X-ray densitometry to achieve cross-dating for Dacrydium
cupressinum, Agathis australis, or other species, when
visual dating cannot be obtained remains to be
demonstrated in the Southern Hemisphere. In many cases, I
have observed that even quite clear ring series must be
visually dated before the densitometric patterns can be
similarly dated. The clues for cross-dating using the
human eye often exceed those provided by a graph of wood
density.

The study of age structure in New Zealand forests,
and interpretations of past climate based thereon, occur
frequently in the literature. I concur with Ogden that
such studies are unlikely to shed much light on climatic
oscillations, but for different reasons. Having examined
cores from hundreds of individuals, and having seen the
extent of heart rot, growth surges and suppression, the
difficulties with estimating distance to pith, age to
coring height, and ring uncertainties in undated
specimens, I can only regard most of these studies with
scepticism. The uncertainties in estimates would often
exceed the differences in age class.

INTRODUCTION

The Editors

The distribution of land and its relief differs markedly between the two hemispheres. This influences both the potential extent of the tree-ring data base and the nature of the climatic and atmospheric circulation features that are to be reconstructed. The atmospheric circulation of the Northern Hemisphere does not have the year-round vigour of the Southern Hemisphere. Pronounced seasonal changes in climate and the strength and position of the major circulation features occur, particularly in the continental interiors. Kelly discusses climatological points of relevance to dendroclimatic analysis.

The Arctic area, reported on by Jacoby, Brubaker and Garfinkel, and Lawson and Kuivinen, is of special interest as climatic variations in high latitudes are particularly marked and may indicate changes over a much wider region. As these authors show, there has been a considerable amount of work undertaken in high latitudes, particularly in Alaska where a clear climate signal in tree rings has been demonstrated. This region has its special problems related to access to sites and the properties of the tree-ring series themselves. It is an area with great potential and one especially requiring international cooperation.

Western North America, reported on by Brubaker, is one of the most heavily worked areas of the globe. It is the source of most climate reconstructions so far. The range of species, the altitudinal gradients, and the latitudinal range have all contributed to this success, although there is much still to be done. Many gaps still exist in the tree-ring network and many species have yet to be investigated. Its position at the eastern edge of the largest ocean of the hemisphere makes it particularly sensitive to climatic variation.

In contrast to the west, eastern North America, discussed by Cook, Conkey, and Phipps, was neglected in the early days of the science. Much of the native forest has been destroyed and what remains is dominated by angiosperm genera. Techniques for handling these genera have been more extensively developed in Europe than the U.S.A. In spite of these difficulties, recent progress, including the development of 48 long modern chronologies, shows the great potential of this area. Eastern North America will be vital to any attempt to determine the past climate of the North Atlantic sector; data from the eastern United States and Europe could be combined in a grid surrounding the North Atlantic Ocean. Such a reconstruction is rapidly becoming feasible as chronologies up to 250 to 300 years long with a strong climate

signal are becoming available from both areas. The value of density measurement in eastern North America, as well as in the Arctic and Alpine regions, is well illustrated.

In Europe, which is reported on by Eckstein, Pilcher, and Bitvinskas, many of the existing chronologies are not in a form suitable for dendroclimatology without reworking. The high quality of cross-dating and limited dendroclimatological study have shown that trees from the mesic hardwood forests have a strong common climate signal. The Alpine areas, reported on by Braker and Bednarz, although dealt with separately because of different methodologies, will form part of a single unit with lowland Europe for a climate reconstruction grid. As pointed out by several contributors, human influence on the forests is strong in the Northern Hemisphere but nowhere more so than in Europe and the Mediterranean region. Grazing by domestic animals is a particularly severe factor in the Mediterranean region. Nevertheless, as pointed out by Munaut and Serre-Bachet, a clear climate signal is present in tree-ring chronologies in this area. Density measurements on coniferous timber from a variety of areas in in Europe are valuable additions to the tree-ring data base. Europe has the potential for reconstructions extending back beyond 1000 AD using composite chronologies from archaeological and other non-living timber.

Dendroclimatology is apparently advancing in eastern Europe, although little firm information has been published in the West. As the U.S.S.R. forms a major part of the European landmass, its contribution to the hemispheric data base cannot be ignored. The Asian continent, reported on by Zheng, Wu and Lin, and Hughes, is the largest landmass of the hemisphere and perhaps the least studied. The impressive list of long-lived tree species in China highlights the potential of the region. It is in an area such as this that instrumental, documentary, and tree-ring records of climate may be fruitfully combined. International cooperation is again essential if this region is to be included in a hemispheric data base.

CLIMATIC CONTEXT

P.M.Kelly

INTRODUCTION

The climatology of the Northern Hemisphere has been discussed in depth by many authors. General summaries can be found in Flohn (1969), Rex (1969), Lamb (1972, 1977), Lockwood (1974, 1979), and Trewartha (1980). Regional descriptions have been given by Orvig (1970) for the Arctic, Wallén (1970, 1977) for Europe, Griffiths (1971) for Africa, Arakawa (1969), Lydolph (1977), and Borisov (1965) for Asia, and Bryson & Hare (1974) for the North American continent. The characteristics of the atmospheric circulation have been described by Palmén & Newton (1969), Newell et al. (1972, 1974), and Chang (1972), and a

straightforward account of the physical mechanisms
underlying the general circulation of the atmosphere is
given by Barry & Chorley (1976). Barry & Perry (1973)
review synoptic climatological research in depth, and
references to methodological practice will be found
therein. Mitchell et al. (1966) contains a concise summary
of the major statistical techniques used in climatology.

The study of climatic change, while a branch of
climatology, impinges on many disciplines. Lamb (1972,
1977), N.A.S. (1975), Gribbin (1978), and Berger (1981)
provide an introduction to the many facets of this field.
The relationship between society's actions and climatic
change is discussed by Tickell (1973), Schneider & Mesirow
(1976), Flohn (1980), Kellogg & Schware (1981) and W.M.O.
(1981). The literature concerning climatic change over the
Northern Hemisphere has been summarised by Lamb (1972,
1977); more recent data have been given by Berger (1980)
and Wigley (1981).

Studies of hemispheric temperature variations during
the last 100 years has been reviewed, most recently, by
Jones et al. (1982). There have been a number of analyses
of hemispheric-scale spatial patterns of variation in the
major climate variables: surface pressure (Kutzbach, 1970;
Kidson, 1975a,b); 500mb height (Craddock & Flood, 1969);
hemispheric temperatures (Barnett, 1978); and Arctic
climate variables (Walsh, 1977, 1978; Kelly et al., 1982).
Numerous regional analyses have been published. Little
attention has been paid to hemispheric-scale variations in
precipitation owing to the highly localised nature of
fluctuations in this variable.

THE AVAILABILITY OF CLIMATE DATA

Given that the major aim of dendroclimatology is to
extend the limited climate data base back in time, the
availability of instrumental climate data for calibration
and verification will appear to be a problem whatever
region is studied. It is, however, only an insurmountable
problem when attempts are made to extract more information
from the available tree-ring and climate data than is
statistically advisable (Lofgren & Hunt, this volume;
Gordon, this volume). The availability of climate data
(length and reliability of series, types of parameters,
and so on) must influence the design of any dendroclimatic
research project, from sampling strategy through response
and transfer function development and calibration to
verification. It is better to produce a small, well-
verified reconstructed data set than a large set of data
of untested veracity.

The length of climate time series available to
dendroclimatology varies greatly from region to region.
While a number of compilations of climate data are
available in digital form (Jenne, 1975), they are by no
means complete, particularly prior to 1900. Additional
data exist in meteorological agency archives and
elsewhere, and should be sought. If found, the later

inclusion of such series in existing climate data banks is
a valuable addition to the climate data set alongside
any dendroclimatic reconstructions produced.

Restrictions in the climate and tree-ring data bases
limit the geographical areas where large-scale
reconstruction of climate, such as achieved in North
America (Fritts, this volume), will be possible. While the
spatial continuity of climate patterns will enable
extrapolation over adjacent areas from these regions, the
verification of such extrapolation is difficult. In many
regions, and probably when reconstruction for the
hemisphere as a whole is attempted, it will be necessary
to take a more problem-orientated approach: to tailor the
design of the analysis to answer a particular climatological problem. This is also advisable in areas where
sampling is restricted for logistical, political, or
financial reasons.

RELIABILITY OF CLIMATE DATA

Even the inclusion of a particular time series in a
digitised data bank is no guarantee of reliability. Claims
that data have been homogenised should be checked
carefully. There are a number of different methods of
homogenising climate data. One tendency, particularly
prior to 1960, was to remove all long-term trends on the
basis of the belief that climate was invariant on decadal
and longer timescales. It is necessary, therefore, to
determine the method of homogenisation. Standard methods
have been described by Mitchell et al. (1966), Wahl
(1968), Bradley (1976), and Salinger (1979). In general,
climate data should not be corrected unless there is
first-hand evidence of error (for example, documentary
evidence of a change in site location or observational
practice, or marked inconsistency with neighbouring
records that cannot be explained by local effects). If
there is doubt, it is better not to use the record or, if
it must be used, to leave the data unchanged but to bear
the uncertainty in mind throughout the analysis.

REPRESENTATIVENESS OF CLIMATE DATA

The spatial patterns of climatic change are complex.
Whatever the timescale under consideration, climatic
changes vary in magnitude and timing from region to
region. Patterns of changes in precipitation are
particularly complex spatially. This needs to be borne in
mind when designing a dendroclimatic analysis and when
interpreting reconstructed data. The selection of data
representative of a particular region is not straightforward. It is often advisable to avoid the use of
single-station data and use spatial averages. Spatial
averaging tends to emphasise longer-term macroclimatic
variations as local effects are masked. The choice of
which region to average over can be guided by principal
component analysis, as can the assessment of the spatial
representativeness of any data for a single site or

limited geographical region. The publication of principal
component analyses of instrumental climate data for
previously unstudied regions, though primarily undertaken
in the context of a dendroclimatic analysis, is a valuable
addition to climatological knowledge. Such analyses can
also guide the formulation of indices to be used in areas
where limitations in the dendroclimatological data base
mean that large-scale reconstruction is not possible. For
example, principal component analysis of mean-sea-level
pressure data for the Northern Hemisphere shows that the
major spatial patterns of variation are related to changes
in the strength and position of the atmospheric "centres
of action" (the Iceland Low, the Siberian High, and so
on). The reconstruction of indices based on the location
and intensity of these features should be investigated
further for the Northern Hemisphere (see the Introduction
to Chapter 3).

CHOICE OF CLIMATE VARIABLES FOR RECONSTRUCTION

The selection of climate variables for reconstruction is primarily determined by the climate-growth
response of the particular trees sampled. It is, however,
worth bearing in mind that different types of reconstructed climate variables will be later employed for
different purposes, and the type of reconstruction
attempted could be influenced by the later use of the
reconstructed data. For example, reconstruction of
short-term drought is dependent on accurate modelling of
high-frequency component of the variance, whilst an
attempt to determine past variations in the average
temperature of the Northern Hemisphere to provide a
context for future carbon-dioxide-induced changes would be
dependent on accurate modelling of lower-frequency
variance. Given that the distribution of variance with
frequency (the variance spectrum) found in tree-ring data
is highly variable, the selection of dendroclimatic data
"tuned" to a particular problem is likely to be
advantageous. Similarly, it may well be desirable to
select climate variables for reconstruction on the basis
of their statistical characteristics (autocorrelation,
variance spectrum, and so on) being similar to those of
the available tree-ring data. In particular, parameters
which have a strong autoregressive or moving-average
nature, such as some drought indices, degree days, ocean
temperatures, and sea-ice extent, should be further
investigated. As more reconstructions become available, it
will be possible to compare in depth the statistical
characteristics of the reconstructed and the instrumental
data used in calibration. See, for example, Blasing (1981)
and Chapter 5 of this volume.

In the case of analyses involving principal
components of climate data, it may be advisable to reject
certain components on a priori grounds before attempting
calibration, reconstruction, and verification. If, for
example, the spatial pattern of a particular component

does not have a strong centre affecting the area of the
tree-ring grid (see Pittock, this volume), the likelihood
of successful reconstruction is low, and its inclusion in
the analysis will result in calibration and verification
statistics which are poorer than if it is excluded at the
outset. Post hoc selection of results must, however, be
avoided (Pittock, this volume).

A CHALLENGE

Pittock (this volume) suggests that climatologists
would be content with some additional climate data. I
would argue that, for certain purposes, climatologists
would be happy with any additional data (Kelly, 1979);
even a reconstruction of a single, and not necessarily the
first, principal component of a large-scale data set could
be a valuable aid in determining the causes of past
climatic change or assessing the frequency of occurrence
in the past of certain scenarios for future change (Wigley
et al., 1980). It is, however, absolutely necessary that
reconstructions be thoroughly verified. This is not only
to ensure scientific rigour, but also to convince potential users outside the field, who may take a justifiable
stance of scientific scepticism, that the data are to be
trusted and to indicate the extent to which they can be
relied on.

THE ARCTIC
G.C.Jacoby
INTRODUCTION

The Arctic regions are very important areas
climatologically, as long-term temperature variations
there are of greater magnitude than in lower latitudes
(Brinkmann, 1976; Kelly et al., 1982). Temperature is
usually the primary climate variable that influences tree
growth. Several of the species that grow in the Arctic
regions are long-lived and the preservation of dead
material is fairly good owing to the low temperature. The
potential for long-term reconstruction of past climates in
the Arctic is, therefore, very good.

The term Arctic frequently means the area north of
the latitudinal treeline. For this discussion, I will
extend the area covered to include the subarctic and
boreal forest zones. The subarctic is used to include the
forest-tundra ecotone and the northern boreal forest, that
is, the forest-and-barren region. The southern boreal
forest, a closed forest, is included although this last
area is not generally considered Arctic. A discussion of
these terms is given in Barry & Ives (1974). This section
will concentrate on the latitudinal forest-tundra ecotone
and on the forest-and-barren regions of the boreal forest.
Figure 4.1 shows the northern treeline and the zones of
northern forests in the western hemisphere. Within these
regions there are two treelines, the latitudinal and
altitudinal.

Most dendroclimatic studies in the Arctic emphasise

the temperature signal recorded by tree growth. However, even in trees that are stressed primarily by temperature variations, there often appears to be moisture stress towards the end of the growing season. This can be used dendroclimatically given proper site selection (Stockton & Fritts, 1971a).

GEOGRAPHIC SETTING

The actual northern treeline ranges from $55°N$ at Cape Henrietta Maria on the southern shore of Hudson Bay, Ontario Province, Canada (Ives, personal communication, 10.5.1979) to $76°N$ on the Taymyr Peninsula, U.S.S.R. (Isachenko, 1955). There is a strong asymmetry to the treeline. Nowhere in the western hemisphere does it extend above $70°N$. This asymmetry and the variations in the tree-growth patterns along the treeline indicate the inhomogeneity of the high-latitude climatic regions. During individual periods, different sectors of the Northern Hemisphere may be similar or opposite in climatic trend (Lamb, 1977, pp.400-402; Kelly et al., 1982).

Throughout this Arctic zone there are vast areas of low relief with an elevation of less than 1,000m. These areas include the Alaskan interior, central and eastern Canada, eastern Sweden, and Finland. In western Canada and Alaska, the region also includes the highest mountains in North America, over 6,000m. Norway and western Sweden have mountain ranges reaching over 2,000m. The effect of these topographic differences is that in the low-elevation, low-relief areas there is a real latitudinal forest-tundra

ecotone and treeline. In the mountainous areas, local climate is more predominant than in the low-relief areas. Slope orientation and local air drainage as well as elevation influence the distribution and growth of trees. These factors interfere with a well-defined latitudinal treeline. In the southern portion of the boreal forest, the treeline may extend to a few thousand metres in elevation. In areas like the Brooks Range in Alaska, the latitudinal and altitudinal treelines coincide at some locations. There are extensions and outliers beyond the regional treeline. These extensions and outliers are often in local climatic zones such as river valleys in Alaska and Canada, fjords and coastal valleys in Scandinavia; or they can be relict stands of trees which if destroyed, would probably not re-establish themselves by seed (Nichols, 1975).

There is a wide variety in soil types. The three major groups are tundra soils, podzols, and azonal soils (Strahler, 1960, p.245). In sites on the well-drained azonal soils one would expect moisture stress to become important to tree growth.

One of the unique features of the northern forest zone is permafrost. Continuous or discontinuous permafrost underlies much of the Arctic and subarctic region. In western Canada and Alaska, the forest-tundra ecotone is subparallel to and within the zone of continuous permafrost. Because of the permafrost and freeze-thaw cycles in the forest and barren area and the forest-tundra ecotone area, one of the primary erosional mechanisms is

Figure 4.1: Boreal forest regions of North America after Rowe (1977) for Canada and Viereck (1972) for Alaska.

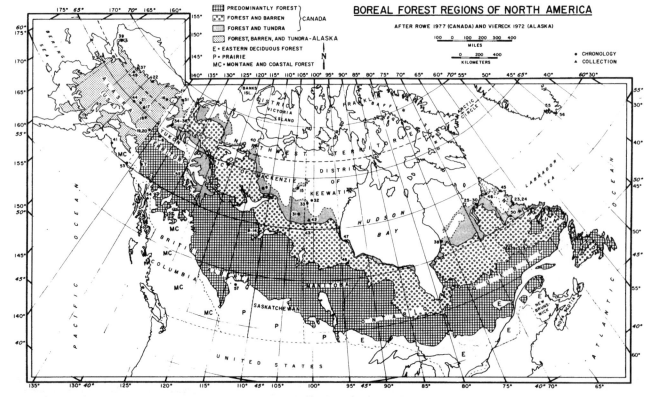

solifluction. This can take place on slopes of 15 to 35 degrees, but also even in areas where the slope is approximately 2 to 3 degrees (Leopold et al., 1964, p.346). Owing to solifluction, there are often what are called "drunken forests" where trees may stand at various angles from vertical. This is an important phenomenon because it is difficult to obtain good dendroclimatological samples in these areas owing to reaction wood and abnormal growth in the trees. Micro-site changes due to cryogenic alterations can occur even on stable areas. Small ice lenses can affect the growth of individual trees and complicate ring-width patterns (Viereck,1965). Another effect of the permafrost is interference with drainage in many areas. The permafrost forms an impermeable layer beneath the surface soil material. This poor drainage can create treeless bogs in areas where the climatic conditions are actually suitable for tree growth.

The boreal forest lies mostly in the "cool snow forest climate" as described by Köppen (Petterssen, 1969). I have included in this discussion the high-latitude coastal forests (Rowe, 1972; Viereck & Little, 1972) because of their proximity to the Arctic areas. These cool-rainy areas occur along the southeastern coast of Alaska, northwest coast of Canada and Scandinavia.

The present location of the northern treeline approximates to the mean summer position of the Arctic front and it has been theorised that this frontal position is what controls the limit of tree growth (Bryson, 1966). A good discussion of the relationships between climate parameters and the location of treeline is given in Larsen (1974). There are locations where the forests, if

destroyed by fire or some other phenomena, would not regenerate (Nichols, 1975). There are also areas that appear to be in equilibrium with the northern limit of growth, and areas where tree growth is moving north. It has been suggested that there may be places where the forests have not yet reached their survival limit owing to migration lag (Elliott, 1979). In addition, because wood is a scarce commodity in some Arctic regions, there may be anthropogenic modifications of the treeline.

SPECIES USEFUL IN ARCTIC DENDROCLIMATOLOGY

In the northern region, there is a somewhat limited number of tree species that may be useful for dendroclimatology. In the forest-tundra ecotone and the forest-and-barren areas, there is low species diversity relative to temperate or tropical regions. Of the species present, not all achieve the needed longevity or arborescent form. A life span of at least 200 years is necessary for a species to be of much value. In the southern portion of the boreal forest there are several additional important species. Individual species have their own northern limits. A map of the northern limits of Picea, Abies, Pinus, and Larix is given in Larsen (1974, p.342). Table 4.1 lists the species which have been used or may be useful in the northern regions.

White spruce, Picea glauca (Moench) Voss, is an especially good species because it grows throughout the entire northern forests of the western hemisphere and living specimens can exceed 500 years in age, although heart-rot is a problem in many older trees (Jacoby & Cook, 1981). Eastern larch or tamarack, Larix laricina (Du Roi) Koch, is also widely distributed and has been used to develop long chronologies (Fritts, 1976). Larch does not appear to attain the age of white spruce, but on sites where the two species are mixed the larch appears to have more ring-width variation. Long chronologies have been developed from black spruce, Picea mariana (Mill.) B.S.P. Its distribution is similar to white spruce but it is not recorded as being as long lived. Owing partly to the occurrence of black spruce in wetter and less stable areas, it tends to present more problems of reaction wood interfering with normal growth and a clear response to climatic variation.

The oldest Scandinavian chronologies are from Scots pine, Pinus sylvestris L. (Siren, 1961). It is widely distributed in the northern areas of the eastern hemisphere and can grow on a great variety of soils (Dallimore & Jackson, 1954, p.493). Norway spruce, Picea abies (L.) Karst, has also been used (Mikola, 1962). In the zone of continuous boreal forest and the mountains of western North America there are two firs that might be of use. The balsam fir, Abies balsamea (L.) Mill., extends to 57°N but is not reported to achieve great age (Fowells, 1965, p.12). The longer-lived (Fowells, 1965, p.37) subalpine fir, Abies lasiocarpa (Hook) Nutt., grows in

Table 4.1: Species for northern dendroclimatology.

COMMON NAME	SPECIES
balsam fir*	Abies balsamea (L.) Mill.
subalpine fir	Abies lasiocarpa (Hook.) Nutt.
Alaska cedar	Chamaecyparis nootkatensis (D.Don) Sprach
Larch, tamarack	Larix laricina (Du Roi) K.Koch
Norway spruce	Picea abies (L.) Karst.
white spruce	Picea glauca (Moench) Voss
black spruce	Picea mariana (Mill.) B.S.P.
sitka spruce	Picea sitchensis (Bong.) Corr.
jack pine*	Pinus banksiana Lamb.
lodgepole pine	Pinus contorta Dougl.
Scots pine	Pinus sylvestris L.
western hemlock	Tsuga heterophylla (Raf.) Sarg.
mountain hemlock	Tsuga mertensiana (Bong.) Carr.
northern white cedar (eastern arborvitae)	Thuja occidentalis L.
western red cedar (giant arborvitae)	Thuja plicata Donn.
European birch*#	Betula pubescens Ehrh. (Betula alba L.)
paper birch*#	Betula papyrifera Marsh.

* These species are relatively short-lived and it is uncertain how useful they may be for dendroclimatological purposes.

Betula papyrifera has a wide geographic distribution and B. pubescens is one of the few trees native to Greenland and Iceland.

mountainous areas and has been sampled (Table 4.2). Of
the North American pines, only lodgepole pine, Pinus
contorta Dougl., and jack pine, Pinus banksiana Lamb.,
occur at high latitude. Old chronologies have been
developed from the former but the latter seldom achieves
sufficient age to be of much use.

In the south-central region of the boreal forest,
the northern white cedar, Thuja occidentalis L., may have
potential. South of the true boreal forest, but still at
high latitude, the southeast coast of Alaska contains old
growth stands of western hemlock, Tsuga heterophylla
(Raf.) Sarg., and mountain hemlock, Tsuga mertensiana
(Bong.) Carr. These hemlocks have been found to be over
500 years old on two or more locations (Jacoby, 1979;
Cropper & Fritts, 1981) Sitka spruce, Picea sitchensis
(Borg.) Carr., has been found to live to over 700 years
near Icy Bay, Alaska (Jacoby, 1979). In these same western
coastal areas of North America, the Alaska cedar,
Chamaecyparis nootkatensis (D.Don) Sprach, and the western
redcedar, Thuja plicata Donn, are long lived and, although
the ring structure is not uniform, it may be possible to
produce good chronologies from these species (Parker,
personal communication, 15.11.1980). The above coastal and
montane species are included because they could play a
role in regional studies such as that reported by Blasing
& Fritts (1975).

It is difficult to say if there is much potential
for angiosperms. The widespread Alnus, Populus, and Salix
do not have any species that reach sufficient age.
Certain birch, Betula, are widespread, growing in
Greenland and Iceland, in addition to North America and
Scandinavia. However, few Betula appear to reach any great
age.

DATING OF ARCTIC TREE SPECIMENS

The dating process in the north presents two
problems. There are locally absent rings and, partly owing
to strong serial persistence, the ring-width variation is
less than in some lower forest-border areas. In white
spruce, there are locally absent rings in the periods of
extremely low growth which correspond to the Little Ice
Age and a cold period in the middle of the 19th century.
In spruce specimens from the forest-tundra ecotone, there
are often long series of micro-rings of the order of a few
hundredths of a millimetre wide. Some of these rings
consist of only two cells in the radial direction. Even
during less severe climatic periods there can be missing
rings. Dating is complicated by the lack of variation in
some tree ring-width series so that defining signature or
notable rings for skeleton-plot type of dating may be
difficult. For cross-dating, longer sequences may be
needed than are necessary in other areas where there is
greater sensitivity and ring-width variation.

In white spruce from the forest-tundra ecotone we
have not had problems with false or double rings. They

have been reported to occur in other species in Norway.
Most cases appear to be attributed to moisture stress in
southeastern Norway (Hoeg, 1956). There is a real problem
with very light latewood that is barely discernible on
many occasions in spruce from the treeline area in North
America. This light latewood appears to correspond to low
temperatures and/or reduced insolation. In spite of being
a problem in ring definition, light latewood is often
consistent within a site and can be a useful aid in
cross-dating. Until now, most dating in North America has
been done by the standard skeleton-plot method (Stokes &
Smiley, 1968) but Parker (personal communication,
15.11.1980) is testing computerised dating methods and
feels they will be successful.

DENDROCLIMATOLOGY

There has been a substantial amount of work in
dendroclimatology for the Arctic regions. Much of this
work has centred on inter-relationships between tree
growth and temperature (Giddings, 1941; Schove, 1954;
Mikola, 1962). It has been shown that tree-ring series can
be used to gain insight into past pressure systems in the
northern areas (Blasing & Fritts, 1975). Also, quantitative reconstructions of temperature parameters have been
made (Jacoby & Cook, 1981).

Figure 4.2 shows a response function for a white
spruce ring-width chronology from the forest-tundra
ecotone in Yukon Territories (Y.T.), Canada, and monthly
temperature and precipitation from Dawson. As this figure
indicates, temperature is usually an important variable
influencing tree growth. It should be noted that moisture
can also play a significant role towards the end of the
growing season. The general form of this response function
seems to occur often in various sites along the northern
treeline in North America. It is probable that with
appropriate site selection the moisture signal present in
tree-ring series could be amplified.

Owing to the severe cold and long winters, the
"climatic window" (Fritts, 1976, pp.238-239) of northern
trees may be narrower than in lower latitudes. During the
short season of radial growth in the subarctic there are
days when the actual air temperature may be 25 to 30°C
under bright cloudless skies. A general dendroclimatic
assumption in the north is that warmer summer temperatures
lead to increased radial growth. With the generally low
precipitation of around 250 to 300mm/yr, a combination of
moisture stress and high temperature could reduce
photosynthesis on many days during especially warm
summers. The negative effect of temperature above optimum
combined with moisture stress on photosynthesis of
ponderosa pine, Pinus ponderosa Laws., has been documented
by Hadley (1969). A similar effect could complicate
dendroclimatic interpretation of northern tree growth.

Although there is a similarity in the form of some
response functions, the Arctic region is not a climat-

ically homogeneous region. An indication of the variations that can occur along the Arctic treeline is given in Figure 4.3. These are plots of annual ring-width indices from three sites along the northern treeline from Yukon Territories (Canada), Quebec (Canada), and Finland.

OTHER METHODS: DENSITOMETRY AND ISOTOPES

In addition to ring-width analyses, it appears that there may be great opportunities for using densitometric methods in dendroclimatology in northern areas. There are considerable variations in latewood density that are readily discernible by microscopic examination and that can be quantified by use of X-ray densitometry. It is probable that this relatively new technique will find much application in Arctic tree-ring studies. Parker & Henoch (1971) have shown the potential for obtaining better parameters of tree growth in western Canada and preliminary investigations at Lamont-Doherty Geological Observatory (University of Columbia) of white spruce from northwestern Alaska confirm the potential value of densitometry. However, owing to the narrowness of many rings near the treeline, there will have to be high resolution capability. The problems of specimen geometry discussed by Schweingruber et al. (1978a) will be crucial for successful application of X-ray densitometry. Even with a real resolution of 0.01mm, an extremely narrow ring would yield only a few density values. These few points may be inadequate for a good density profile.

Analysis of the isotopic composition of wood is currently being investigated as a dendroclimatic tool (Jacoby, 1980b; Long, this volume; Wigley, this volume). Briefly, analyses of the stable isotope ratios of

hydrogen, oxygen, and carbon may provide further information about growth environments. However, these analyses are much more costly and time-consuming than tree-ring width or densitometric analyses. A study of the I.A.E.A. isotopic data from precipitation indicates that the higher latitudes of western North America may be better sites for these techniques than lower latitudes (Lawrence, 1980). Near Edmonton, Canada, good results were achieved relating $^{18}O/^{16}O$ ratios to temperature parameters (Gray & Thompson, 1976).

FUTURE PROSPECTS

The Arctic region has been less densely sampled over the last few decades than some other areas. Except for Scandinavia, the meteorological records for calibrating tree-ring series in this area are quite short. It is desirable, therefore, to develop chronologies that reach the present time to increase the time span for calibration purposes. The excellent early work by Giddings (1941) and others, referred to in Mikola (1962), clearly demonstrates the potential for long records from tree-ring material in the Arctic areas by the combination of archaeological materials, driftwood, and living trees. Giddings (1948) developed one of the longest tree-ring series available for the Arctic area. Several Scandinavian chronologies exceed 500 years in length (Schove, 1954; Hoeg, 1956; Siren, 1961).

Recent sampling of the Arctic areas is rather widely spaced (Figure 4.1, Tables 4.2, 4.3, 4.4). Some of these samples are yet to be rigorously developed into chronologies and some of the extant chronologies lack good replication. In spite of these drawbacks, it could be said

Figure 4.2: Typical response function for white spruce at the northern tree line.

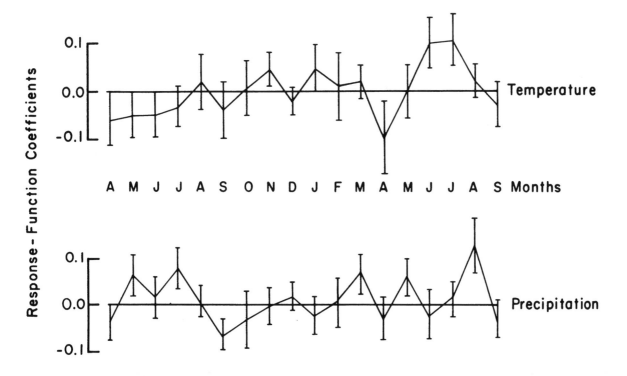

that reconnaissance of the Arctic areas is well under way. I believe the importance of the northern region to climate studies is established. The need now is for researchers to utilise modern techniques and also to innovate in order to develop good chronologies from Arctic material.

In addition to the new techniques of measurement, such as densitometry and stable isotopes, there is potential for improving the extraction of climate signals from tree-ring series for the northern area. Improved numerical analyses may allow amplification of the climate signal from northern trees. Various types of mathematical filters could be used or certain deconvolution techniques applied to the time series. In order to apply these techniques, however, it is necessary to gain greater understanding of the physiological processes that cause such effects as the persistence which tends to dampen the higher-frequency climate signals in the northern tree-ring series. To this end, there is a need for further studies such as those described in Tranquillini (1979). To gain more insight into the physiological response of different species to climatological factors, there should be careful observations during entire yearly cycles for several years. With this information it would be possible to apply different models to the dendrochronological time series in order to improve the reconstruction of climatological variables.

Another approach which is just beginning to receive more attention is the use of subfossil wood in Arctic tree-ring studies. There is subfossil wood in alluvium that may be only a few centuries old (Giddings, 1938), and subfossil wood with ^{14}C dates ranging to almost 5,000 years (Lamb, 1977, plate IV). In the literature relating to the early northern explorations, there are reports of logs in alluvium in many of the Arctic Islands, for example, the note of logs in alluvium several metres above the beach on the north shore of Banks Island (Miertsching, 1854). There are other areas where the burial of wood material appears to have been an episodic phenomenon and there may be a strong possibility that rigorous sampling of these layered alluvial areas could produce a cross-dated, long-term time series (Parker, personal communication, 15.11.1980). With floating chronologies that cannot be anchored to living material (where reliance has to be placed on ^{14}C dating), if the provenance is accurately determined it would be possible to infer climatic variations during those approximate time periods.

In addition to the standard types of meteorological parameters, that is, monthly precipitation and temperature, new parameters should also be examined for the purpose of dendroclimatological analyses. Such new parameters should include degree days above a threshold temperature (for example, about 4°C) as a variable affecting tree growth. Perhaps a parameter similar to the Palmer Drought Severity Index that combines the effects of temperature and moisture could be developed for northern treeline investigations.

Tree-ring series from broad regions should be incorporated into an overall network which could lead to a more comprehensive climatic analysis such as that by Blasing & Fritts (1975). More studies of this type will

Figure 4.3: Tree-ring index chronologies for three longitudes at the northern tree line: (1) data by Sirén (1961) in Lamb (1977); (2) data from Fritts (1978); (3) data from Jacoby & Cook (1981).

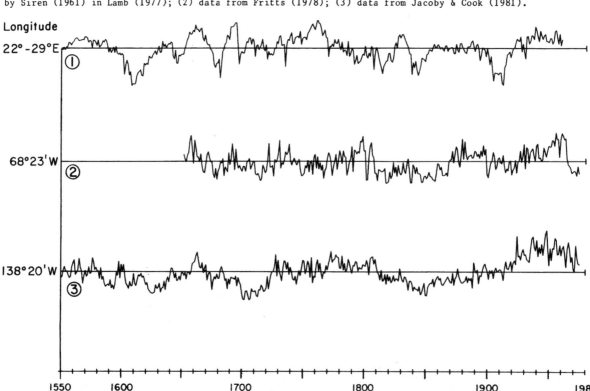

Table 4.2: Recent tree-ring chronologies or collections in the Boreal Forest regions of North America. Map reference is for Figure 4.1. Compiled by L.D.Uland & G.C.Jacoby with the cooperation of J.P.Cropper.

REFERENCE	SITE NAME	SPECIES	TIME SPAN	LAT. N	LONG. W	REFERENCE
1	Barlow Dome	Picea glauca	1820-1966	63° 49′	137° 32′	Drew (1975b)
2	Chapman Lake	P. glauca	1710-1966	64° 51′	138° 19′	Drew (1975b)
3	Swede Creek	P. glauca	1800-1966	64° 08′	139° 43′	Drew (1975b)
4	Dawson Fire Lookout	P. glauca	1870-1966	64° 04′	139° 20′	Drew (1975b)
5	Sixty Mile	P. glauca	1790-1966	64° 08′	140° 35′	Drew (1975b)
6	Gold Creek	P. glauca	1750-1966	64° 06′	140° 49′	Drew (1975b)
7	Bullion Creek	P. glauca	1690-1966	61° 01′	138° 37′	Drew (1975b)
8	Lake Beniah	P. glauca	1747-1970	63° 29′	112° 71′	Drew (1975b)
9	Athabasca River	P. glauca	1708-1970	58° 22′	111° 32′	Stockton & Fritts (1971a)
10	Quatre Fourches	P. glauca	1765-1970	58° 47′	111° 27′	Stockton & Fritts (1971a)
11	Revillon Coupe	P. glauca	1783-1970	58° 52′	111° 18′	Stockton & Fritts (1971a)
12	Peace River I.	P. glauca	1804-1970	58° 58′	111° 55′	Stockton & Fritts (1971a)
13	Peace River II.	P. glauca	1698-1970	58° 59′	111° 26′	Stockton & Fritts (1971a)
14	Claire River	P. glauca	1760-1970	58° 53′	111° 53′	Stockton & Fritts (1971a)
15	Thelon Game Sanctuary	P. glauca	1574-1969	63° 50′	104° 12′	Drew (1975b)
16	Procrastination Cr.	P. glauca	1633-1962	67° 40′	142° 30′	Blasing & Fritts (1975)
17	Twelve Mile Summit	P. glauca	1650-1962	65° 20′	146° 00′	Blasing & Fritts (1975)
18	Salcha River Headwaters	P. glauca	1650-1962	64° 55′	144° 00′	Blasing & Fritts (1975)
19	Dawson Junction	P. glauca	1443-1962	64° 10′	141° 30′	Blasing & Fritts (1975)
20	Mount Fairplay	P. glauca	1680-1962	63° 50′	142° 00′	Blasing & Fritts (1975)
21	Hermann's Cabin	P. glauca	1750-1962	65° 20′	147° 30′	Blasing & Fritts (1975)
22	Chandalar Lake	P. glauca	1785-1962	67° 30′	148° 30′	Blasing & Fritts (1975)
23	Nain Forest (A)	P. glauca	1802-1973	56° 33′	62° 00′	Fritts (1976, 1978′)
24	Nain Forest (B)	P. glauca	1769-1973	56° 33′	62° 00′	Fritts (1976, 1978′)
25	Ft. Chimo 1-S	P. mariana		58° 22′	68° 23′	Fritts (1976, 1978′)
26	Ft. Chimo 2-S	P. mariana	1700-1974	58° 22′	68° 23′	Fritts (1976, 1978′)
27	Ft. Chimo L	Larix laricina	1650-1974	58° 22′	68° 23′	Fritts (1976, 1978′)
28	Ft. Chimo L	L. laricina	1753-1974	58° 22′	68° 23′	Fritts (1976, 1978′)
29	Ft. Chimo L	L. laricina	1677-1974	58° 22′	68° 23′	Fritts (1976, 1978′)
30	Ft. Chimo L	L. laricina	1641-1974	58° 22′	68° 23′	Fritts (1976, 1978′)
31	Nicol Lake	Picea mariana	1873-1974	61° 35′	103° 29′	Kay (1978)
32	Slow River	P. mariana	1810-1974	63° 02′	100° 47′	Kay (1978)
33	Dubwant River	P. mariana	1740-1974	62° 37′	101° 17′	Kay (1978)
34	Twisted Tree- Heartrot Hill	P. glauca	1459-1975	65° 00′	138° 20′	L-DGO#
35	River Crag	P. glauca	1635-1975	65° 40′	138° 00′	L-DGO
36	Cat Track	P. glauca	1696-1975	65° 57′	137° 15′	L-DGO
37	Arrigetch	P. glauca	1586-1975	67° 27′	154° 03′	L-DGO
38	Gulf Hazard	P. glauca	1681-1976	56° 10′	76° 34′	L-DGO
39	412-Noatak	P. glauca	1515-1977	67° 56′	162° 18′	L-DGO
40	Coppermine River (several sites)	P. glauca	1340-1977	66°-68°	114° 117	L-DGO
41	Herring-Alpine	Tsuga heterophylla	1422-1972	60° 26′	147° 45′	Fritts (1976, 1978′)
42	Ennadai Lake (several sites)	Picea glauca P. mariana Larix laricina	1814-1977 1793-1977 1638-1977	61°	101°	Elliot (1979, 1980′)
43	Kasba Lake	P. mariana	1776-1977	60° 06′	101°53′	Elliot (1979, 1980′)
44		Picea glauca P. mariana Populus balsamifera	1768-1977	65° 67° 67°	148° 144° 144°	Juday (1979′)
45	Napaktok Bay	Picea glauca	1781-1978	57° 56′	62° 35′	Elliot (1979, 1980′)
46	Koroc River	P. mariana Larix laricina	1666-1978 1690-1978	58° 04′	65° 02′	Elliot (1979, 1980′)
47	Churchill	P. glauca	1650-1978	58° 43′	94° 04′	L-DGO
48	Sky Pilot Creek	P. glauca	1725-1978	56° 24′	94° 22′	L-DGO
49	Walker Lake (several sites)	P. glauca	1627-1977	67° 00′	154° 00′	Brubaker (1978′)
50	Border Beacon	P. glauca	1660-1976	55° 20′	63° 15′	Fritts (1978′)
51	Spruce Creek	P. glauca	1570-1977	68° 31′	138° 40′	Church (1978′)
52	Mt. Sheldon	P. mariana Abies lasiocarpa	-1977 -1977	62° 62°	130° 130°	Noel (1978′)
53	Dwarf Trees (Icy Bay)	Tsuga mertensiana and T.heterophylla	1100-1979	60° 01′	141° 49′	L-DGO
54	Juneau	Tsuga sp.				Posmentier (1979′)
55	Narssarssuaq (Greenland)	Betula pubescens	1870-1977	61° 09′	45° 02′	Lawson (1980′)
56	Quinquadalen (Greenland)	B.pubescens	1876-1977	60° 16′	44° 30′	Lawson (1980′)
57	Edmonton	P. glauca	1882-1969	53°	113°	Gray & Thompson (1976)
	Okak Bay Area*	Picea glauca Larix laricina		57°	62°	Elliot (1980′)
	Desulo Lake*	P. glauca		56° 03′	63° 48′	Elliot (1980′)
	Sukakpak Mt.*	P. glauca		67° 36′	149° 48′	L-DGO
	Sheenjek River*	P. glauca		68° 37′	143° 40′	L-DGO

* Not shown on map   # Lamont-Doherty

add greatly to the value of the individual climate
reconstructions of local areas. For such network analyses
and to maximise the use of data, there should be cooperation and data exchange, either on an individual basis
or through a central source such as the International
Tree-Ring Data Bank. Collection of northern material can
be very expensive. After accepting the human inclination
to keep proprietary data for a reasonable time, there
should be encouragement to release these data into the
general scientific domain.

COMMENT

L.B.Brubaker & H.Garfinkel

As Jacoby has pointed out, the Arctic holds great
potential for dendroclimatic research because of its
diversity of extreme macro- and microclimate environments
and an abundance of relatively long-lived trees. Several
recent studies (Kay, 1978; Garfinkel & Brubaker, 1980;
Jacoby & Cook, 1981; Cropper, 1981) have demonstrated
conclusively that Arctic trees contain significant
information for reconstructing climates. Probably the most
obvious conclusion which can be drawn from Jacoby's
discussion is that dendroclimatic research has only begun
to explore the potential of Arctic trees. Of the numerous
topics deserving consideration in Arctic dendroclimatology, five seem especially important to us. These are
discussed below, but not necessarily in order of their
importance.

Updating chronologies. The early work of Giddings (1938,
1941, 1943, 1947, 1948), Oswalt (1952, 1958), and others
(Marr, 1948; Hustich, 1956) in Alaska and Canada is
impressive. Jacoby lists nearly 40 chronologies which were
collected before 1950. The statistical properties of these
chronologies have been examined by Cropper & Fritts (1981)
and those which are judged well suited for climate reconstructions should be updated for use in current research.

Further analysis of existing chronologies. Although
numerous chronologies have been collected in the North
American Arctic during the past two decades, few have been
entered into response function analyses and fewer yet have
been used for climate reconstruction. These analyses
should be routinely carried out by Arctic dendroclimatologists.

A potential problem in such studies, however, is the
scarcity of long-term instrumental climate records. It may
not be possible in some Arctic areas, for example, to find
climate stations within 300km of a given tree-ring site.
Long distances such as this can weaken statistical relationships between tree-ring and climate series. Nevertheless, Garfinkel & Brubaker (1980) have found strong
climate-growth correlations between climate and tree-ring
sites 350km apart in central Alaska (tree-ring data from
white spruce, Picea glauca, on the southern flank of the
Brooks Range and climate data from Fairbanks, Alaska).
These correlations are strong enough to permit the
reconstruction and verification of climate data from
Fairbanks, Alaska. This study also shows that verifiable
climate reconstructions can be obtained from chronologies
which exhibit relatively low mean sensitivity, low
standard deviation, and high autocorrelation. Such
findings suggest that Arctic researchers should perform
climate analyses, even though their chronologies do not
meet statistical criteria which have been established for
selecting suitable sites in the arid southwest United
States (Fritts & Shatz, 1975). Finally, this study
demonstrates that chronologies within rather small
geographical areas in the Arctic can produce accurate
reconstructions of single-station climate data.

It is important to emphasise that the major research
goal in the Arctic, as in other regions, should be to establish large spatial grids of chronologies for synoptic-
scale reconstructions. Cropper (1981) has established a
very good network for Alaska and western Canada by
assembling and calibrating chronologies which have already

Table 4.3: Chronologies in Scandinavia.

SITE NAME	SPECIES	TIME SPAN	LAT. N	LONG. W	REFERENCE
Nokutusvaara (near Kiruna)	Pinus sylvestris	1578-1950	68°	20°	Schove (1954)
Steigen-Sorfold	P. sylvestris	1396-1950	67°35	16°	Schove (1954)
Kuovhonvopio	Picea albies	1464-1950	68°15'	19°15'	Schove (1954)
Muddus Nat'l Park Muddus Site A	Pinus sylvestris	1572-1971	*67°	*21°	Fritts (1976)
Muddus-2	P. sylvestris	1532-1971	*67°	*21°	Fritts (1976)
Ostersund	P. sylvestris	1671-1971	*63°	*15°	Fritts (1976)
Arosjak	P. sylvestris	1614-1971	*68°	*19°	Fritts (1976)
Arjeplog	P. sylvestris	1553-1974	*65°30'	*19°	Fritts (1976)
Lapland	P. sylvestris	1181-1960	68°-70°	22°-29°	Lamb (1977)
Utsjoki Kevo	P. sylvestris		#69°30'	#27°	Mikola (1962)
Inari Naatamojoki	P. sylvestris		#69°20'	#27°30'	Mikola (1962)
Inari Inarinjarvi	P. sylvestris		#68°40'	#27°30'	Mikola (1962)
Inari Lemmenjoki	P. sylvestris		#68°30'	#26°	Mikola (1962)
Inari Ivalojoki	P. sylvestris		#68°20'	#25°50'	Mikola (1962)
Kittula Pallastunturi	P. sylvestris		#68°10'	#23°40'	Mikola (1962)

* Estimated from Fritts (1976), Figure 3   # Estimated from Mikola (1962), Figure 16-2

Table 4.4: Tree-ring chronologies of North America, pre-1960. The latitudes and longitudes are estimated from maps in the various references and should only serve as a guide to general locations.

SITE NAME	SPECIES	TIME SPAN	LAT. N	LONG. W	REFERENCE
White Mt.*	Picea	1590-1937	66°	146°	Giddings (1941)
Alaska Range	Picea	1540-1938	64°	146-149°	Giddings (1941)
Noatak (comb.)	Picea	1640-1940	67-163°	162-163°	Giddings (1941)
Cleary Summit (comb.)	Picea	1790-1938	65°	147°	Giddings (1941)
Hope Creek	Picea	1590-1939	66°	146°	Giddings (1941)
12-mile Summit	Picea	1540-1937	66°	146°	Giddings (1941)
Hogatza River	Picea	1640-1938	67°	154°	Giddings (1941)
Hunt River	Picea	1640-1940	67°	159°	Giddings (1941)
Dahl Creek	Picea	1590-1940	67°	157°	Giddings (1941)
Squirrel River	Picea	1690-1940	67°	161°	Giddings (1941)
Ft.Yukon-White Eye	Picea	1690-1939	63-67°	141-148°	Giddings (1941)
Stephens Village (sic)	Picea	1770-1909	66°	149°	Giddings (1941)
Salcha Bluff	Picea	1850-1938	64°	145°	Giddings (1941)
Globe Creek Bluff	Picea	1640-1938	65°	148°	Giddings (1941)
Lower Goldstream	Picea	1640-1938	65°	148°	Giddings (1941)
Driftwood-Series A	Picea	1500-1860			Giddings (1941)
Driftwood-Series B	Picea	1550-1849			Giddings (1941)
Haycock	Picea	1600-1939	65°	161°	Giddings (1941)
Koyuk (Norton Bay) (comb.)	Picea	1640-1940	65°	161°	Giddings (1941, 1953)
Great Whale R.	Picea glauca P. mariana*	1765-1939	55°	77°	Marr (1940)
Gulf Hazard	P. glauca	1700-1939	56°	76°	Marr (1940)
Moose River	P. mariana	1800-1946	52°	81°	Hustich (1956)
Great Whale R. Hudson Bay	P. mariana	1800-1946	55°	77°	Hustich (1956)
Knob Lake	P. mariana	1800-1946	55°	67°	Hustich (1956)
Knob Lake	P. glauca	1800-1946	55°	67°	Hustich (1956)
Mecatina	P. mariana	1817-1946	53°	3°	Hustich (1956)
Nulato	Picea	1597-1947	65°	158°	Oswalt (1950)
Kaltag	Picea	1710-1947	64°	159°	Oswalt (1950)
Anvik	Picea	1733-1947	63°	160°	Oswalt (1950)
Holy Cross	Picea	1622-1947	62°	160°	Oswalt (1950)
Russian Miss.	Picea	1737-1947	62°	161°	Oswalt (1950)
Marshall	Picea	1703-1947	65°	162°	Oswalt (1950)
Shaktoolik	Picea	1700-1949	65°	160°	Oswalt (1950)
Paxson	Picea	1750-1950	63°	146°	Oswalt (1952)
Menasta Group (SIC)	Picea	1755-1950	63°	144°	Oswalt (1952)
Slana Group	Picea	1700-1950	63°	144°	Oswalt (1952)
Gakona	Picea	1780-1950	62°	145°	Oswalt (1952)
Chitina	Picea	1723-1950	63°	145°	Oswalt (1952)
Moody	P. glauca	1840-1952	64°	149°	Oswalt (1958)
Cantwell	P. glauca	1762-1952	63°	149°	Oswalt (1958)
Chulitna	P. glauca	1861-1952	63°	149°	Oswalt (1958)
Talkeetna	P. glauca	1854-1952	62°	150°	Oswalt (1958)
Eureka	P. glauca	1791-1952	62°	147°	Oswalt (1958)
Atlasta	P. glauca	1781-1952	62°	146°	Oswalt (1958)
Beaver Groups	Picea	1802-1941	67°	147°	Giddings (1943)
Purgatory Group	Picea	1740-1941	66°	148°	Giddings (1943)
Stevens Village Gr.	Picea	1675-1941	66°	149°	Giddings (1943)
Rampart North Gr.	Picea	1730-1941	65°	150°	Giddings (1943)
Tanana Group	Picea	1730-1941	65°	152°	Giddings (1943)
Birches Group	Picea	1700-1941	65°	153°	Giddings (1943)
Kokrines Groups	Picea	1720-1941	65°	155°	Giddings (1943)
McGrath,Kuskokwim Gr.	Picea	1700-1941	63°	156°	Giddings (1943)
Nulato Group	Picea	1700-1941	65°	158°	Giddings (1943, 1953)
Circle Group	Picea	1740-1941	66°	144°	Giddings (1943, 1953)
Mackenzie River Delta	Picea	1357-1945	67-68°	134-135°	Giddings (1947)
Kobuz-Kotzebue Series#		1978-1947			Giddings (1948)
Cape Darby	Picea	1527-1949	65°	163°	Giddings (1951)
Kobuk River-Amber Is.##		1700-1760			Giddings (1942)

*   These are named groups of trees used to form series of ring-width data. The groups are often different from what is termed a site in current literature. The ring-width data may not extend to the age of the oldest trees. There are other collections mentioned by Giddings but no associated ring-width information published. Beginning of time spans are estimates by Giddings.
#   Includes archaeological material and driftwood.
Other archaeological material was collected in this area, but not absolutely dated.

in which reconstructed climates are extreme may be periods of sparse tree establishment. Similarly, reconstructed mild periods may correspond to periods of high tree establishment. Such comparisons can be made, however, only in stands where disturbance events (for example, due to fires or landslides) have not been important.

Expedient sampling strategies. Dendroclimatic research in the North American Arctic has lagged behind that of temperate areas primarily owing to the logistical problems and expense of working in remote areas. Since most of Arctic North America is roadless, it is difficult to sample evenly spaced sites across large areas. It will probably be necessary to piece together a network of sites from collections donated by a number of scientists, not all of whom are dendrochronologists. Many botanists, zoologists, geologists, palynologists, and archaeologists are currently conducting research in the Arctic. Their field parties often spend long periods at single locations. It is reasonable to assume that some of these scientists, if carefully instructed, would be willing to collect increment cores for dendroclimatic research. been collected by other workers. However, it will take a long time to achieve this goal in other parts of the Arctic where existing chronologies are sparse owing to the logistical difficulties of field research. In the meantime, much useful climate information can be obtained by using relatively small tree-ring data sets to reconstruct single-station climate records.

Verification. Few climate records in the Arctic are long enough to provide adequate data for both the calibration and rigorous verification of climate reconstructions. This is an important problem, because dendroclimate reconstructions can not be accepted unless some portion of them is checked against independent climate data (Gordon, this volume).

While dendroclimatic reconstructions cannot be rigorously verified without instrumental data, it may be possible to obtain some qualitative confirmation of reconstructions by comparing them with records of neoglacial activity in areas such as the Brooks and Alaska Ranges. The terminus positions of many glaciers in these mountains have changed dramatically during the past few centuries producing moraines which are currently being dated by lichen measurements and tree-ring counts. Records of glacier movements during the past 500 years in some areas may, therefore, provide useful supportive evidence for dendroclimatic reconstructions. It may also be possible to use age-structure data from forests stands to support dendroclimatic inferences. For example, periods

Cooperation with other scientific disciplines. Dendroclimatic research techniques are especially useful to other scientific disciplines in the Arctic. For example,

response function analyses can provide information to tree physiologists who are studying growth-limiting factors in boreal forests and treeline environments. Some of the response function elements presented by Jacoby and in Garfinkel & Brubaker's (1980) study suggest that freeze-desiccation may be an important process at treeline sites, that precipitation may be more limiting in the Arctic than biologists have traditionally believed, and that various non-growth season conditions affect radial growth rates. Goldstein (personal communication, 1981) reports that response function results agree well with the results of direct ecophysiological measurements of trees at Arctic sites. In the future, therefore, response function analyses may prove to be valuable interpretive tools for tree physiologists, especially in areas where field measurements of climate-growth relationships are difficult.

COMMENT

M.P.Lawson & K.C.Kuivinen

Dendrogeomorphology. Verification with independent data is crucial to the scientific foundations of dendro-climatological research. The paucity of meteorological time series in the Arctic challenges our ability to verify independently growth response models and evaluate climate reconstructions. Thus, we turn to proxy evidence such as pollen assemblages, varves, oxygen-isotope profiles derived from ice cores, and glacier mass budgets. In looking to such evidence, we often compromise proximity to the chronology site or depend on the serendipity of applicable research performed by scientists in other environmental disciplines. Perhaps we should exploit the valuable chronological and environmental data obtainable by employing the principles being developed in dendrogeomorphology.

Many geomorphic processes, climato-genetic in origin, affect a diversity of growth responses (Alestalo, 1971; Shroder, 1978, 1980). Generally, response to environmental events such as (a) inclination, (b) corrosion, (c) burial of stemwood, (d) exposure of rootwood, (e) inundation, and (f) shear of rootwood can be detected and accurately dated. The typical resultant structural and morphological alteration to tree growth includes: (a) growth suppression; (b) growth release; (c) reaction-wood growth; (d) sprouting; or (e) ring termination. Normally the dendroclimatologist, employing traditional procedures, attempts to identify and reject specimens exhibiting such aberrations in the field. Unfortunately, much information, potentially useful as independent verification of past climatic regimes, often goes uncollected by field survey teams.

The potential for deriving climatic information from process-event-response systems in the Arctic has been demonstrated (Shroder, 1980) with respect to (a) fluvial activity; (b) mass wasting; (c) periglacial activity; and

(d) glaciation (Table 4.5). Matthews (1976) suggests that supplemental investigations of this nature become more critical if the calculation of standardised tree-growth indices results in the elimination of low-frequency climatic variance. Using a unique approach, he calibrates Little Ice Age palaeotemperatures in Norway by relating tree growth to glacier fluctuations. The application of similar approaches in the subpolar Arctic should prove extremely valuable to advances in both dendrogeomorphology and dendroclimatology.

Opportunities for dendroclimatic studies in Greenland. A survey of the literature reveals that very limited tree-ring research has been undertaken in Greenland. A study by Beschel & Webb (1963) demonstrated that stems of dwarf willows, Salix glauca, and of juniper, Juniperus communis, proved to be reliable indicators of age in western Greenland for the past 200 years. Olsen (personal communication, 4.8.1978) collected samples of birch, Betula pubescens, mountain oak, Sorbus decora groen-landica, and willow, Salix glauca callicarpaea, from sites in southern Greenland, but his attempt to compare fluctuations in animal populations with the tree-ring measurements was abandoned "because of the lack of concordance between the trees". Reconnaissance of stands of birch, Betula pubescens, growing in a uniquely fertile, isolated valley, Qinguadalen (60°16'N, 40°30'W), convinced the authors that, although the birch suffered considerable

Table 4.5: Subpolar geomorphological processes having potential for deriving climatic information. Compiled from Shroder (1980).

EARTH SURFACE PROCESS	SELECTED REFERENCES
1. Fluvial Activity	
a. Streamflow	Stockton & Fritts (1973)
b. Flood Frequency	Helley & LaMarche (1968)
c. Ice Jams	Smith (1974)
d. Channel Change	Becker (1978)
2. Mass Wasting	
a. Slope Denudation	Hueck (1951); LaMarche (1961)
b. Soil Creep and/or Snow Creep	Phipps (1974)
c. Landslippage	Moore & Mathews (1978)
d. Rock Glacier Movement	Shroder (1978)
3. Periglacial Activity	
a. Snow-Avalanche	Potter (1969)
b. Ice-Lens Growth	Viereck (1965)
c. Earth-Hummock Growth	Zoltai (1975)
d. Solifluction	Alestalo (1971)
e. Permafrost Fluctuation	Thie (1974)
4. Glaciation	
a. Historical Variations	Heusser (1964)
b. Equilibrium Displacement	Matthews (1976)

geomorphic stress, conventional dendroclimatic procedures would successfully yield valuable climatological information. Using standard dendrochronological techniques, a final chronology was produced based on 20 trees (20 paired radii). In the response functions as much as 42% of the variance could be explained by climate and 28% by prior growth, giving 70% of the total variance.

The response function for Qinguadalen, Betula pubescens, as might be expected from trees growing at the northern treeline, contains more weights significant for temperature than for precipitation, and weights for temperature during January to June are generally positive and significant. The weights for precipitation are usually negative and less significant with the exception of the positive and significant current May and June. Although the Qinguadalen birch chronology is relatively short, it does contain considerable information concerning the relationship between climate variables and tree growth for the area. This is significant not only for the establishment of a new climatically sensitive chronology within the Arctic region, but also, because of the close chronological control inherent in dendroclimatological investigations, for the useful supplement it provides to the ongoing studies of ice cores from the Greenland ice sheet (Dansgaard et al., 1975).

Comparative analysis of proxy data. Critical to an understanding of both the Arctic climate-growth relationship and the climate history of the Arctic region is the investigation of relationships among the established Arctic tree-ring chronologies and between the chronologies and regional historical climate reconstructions based on other sources of proxy data. For the southern Greenland area, there exist proxy data in various forms from which can be made inferences of climatic fluctuations. Koch (1945) provides a survey of year-by-year records of sea ice off the west coast of Greenland since 1820. The work of Smed (1946-67) describes monthly anomalies of sea-surface temperature in the northern North Atlantic Ocean. The fluctuations of Greenland's animals, sea birds, fish, and vegetative communities, and the effect upon these of climatic change, have been described by several investigators (Jensen, 1939; Vibe, 1967; Mattox, 1973). Climatic fluctuations for the region have also been traced through records of early Viking settlements (Simpson, 1966), archaeological remains (Vibe, 1967), and glacier fluctuations (Ten Brink & Weidick, 1974; Weidick, 1975).

Our own investigations have compared the chrono-logies and climatic inferences of two climate surrogates: the Qinguadalen birch tree-ring chronology and the time series of oxygen isotope values from ice core collected from the south dome of the Greenland ice sheet (W. Dansgaard, Geophysical Isotope Laboratory, University of Copenhagen; Polar Ice Core Analysis Program, State

University of New York at Buffalo). We found a poor
correlation between the two time series for the period
1875 to 1976, but visual comparison of the chronology
plots revealed good cross-dating of extreme values (that
is, low tree-ring indices corresponded with extremely
negative $\Delta^{18}O$-values). It must be realised, however, that
the tree rings reveal mainly the local growing season
temperature conditions, whereas the ice core records
reveal mean annual temperatures at best (Dansgaard et al.,
1975; Johnsen, 1978).

WESTERN NORTH AMERICA
L.B.Brubaker
INTRODUCTION

Tree-ring collections and dendroclimatic research in
western North America, from approximately 20 to 65°N latitude and 100 to 170°W longitude, are reviewed. Since this
region encompasses diverse geographical, climatic, and
forest zones and has been studied by dendrochronologists
for several decades, it is possible to consider only major
research accomplishments. In addition, discussion will be
limited to studies dealing with ring sequences of
coniferous species.

GEOGRAPHIC SETTING

General geography. The dominant geographical feature
of western North America is the Cordillera mountain system
(Atwood, 1940; Paterson, 1979). This system is composed of
two parallel ranges extending from northern Alaska to
southern Mexico (Figure 4.4). Between these ranges lies a
series of plateaux which are separated by irregular
transverse mountains. The eastern Cordillera range
consists of the Brooks, Rocky, and Sierra-Madre Oriental
Mountains. Peaks throughout this range rise above 3,500m,
but the highest elevations occur in central portions of
the United States. The western Cordillera range, composed
of the Alaska, Cascade, Sierra Nevada, and Sierra-Madre
Occidental ranges, is an extremely rugged mountain system
with numerous volcanic cones, some towering more than
4,000m. Bordering the Pacific Ocean is a smaller mountain
range consisting of rolling hills in Mexico, California,
and Oregon, rugged peaks in Washington, and a series of
islands along the coast of British Columbia and Alaska.

The plateaux between the Cordillera ranges are
typically greater than 1,000m in elevation. The landscape
is relatively flat in the southernmost plateau, the
Central Plateau of Mexico, but becomes irregular to the
north, where the Colorado river and its tributaries have
eroded large portions of the Colorado Plateau. The Great
Basin, covering extensive areas of western United States,
is a complex system of smaller basins divided by irregular
mountain blocks. The Columbia Plateau to the north is
smaller due to a westward trend in the Rocky Mountains.
This plateau is extremely flat and covered by numerous
lava flows, some as much as 700m thick. The Central

Plateau of British Columbia is broken up into small basins
by several transverse mountain systems. The Arctic Circle
passes through the northernmost Cordillera Plateau, an
extensive basin in central Alaska covered with thick loess
deposits of late-Wisconsin age.

Climate. Western North America is influenced by
surface airstreams originating over the Arctic Ocean,
Pacific Ocean, and Gulf of Mexico (Bryson & Hare, 1974).
The Arctic and Pacific systems dominate in winter, while
the Gulf of Mexico system is relatively strong in summer.
The boundaries between these air masses as well as their
effects on local climate are greatly modified by the
Cordillera mountain system.

The Arctic maritime airstream brings cold air into
the northern Cordillera Plateaux to approximately 60 to
65°N in summer and 40°N in winter. Areas influenced by
this airstream are usually cool and cloudy in summer and
cold, but clear, in winter. Occasionally winter systems
are strong enough to cross the western Cordillera
mountains, causing exceptionally cold, dry conditions in
Pacific coastal areas. Most parts of western North America

Figure 4.4: Major geographic features of western
North America: (1) Central Plateau of Mexico; (2)
Colorado Plateau; (3) Great Basin; (4) Columbia
Plateau; (5) Central Plateau of British Columbia;
(6) Central Basin of Alaska.

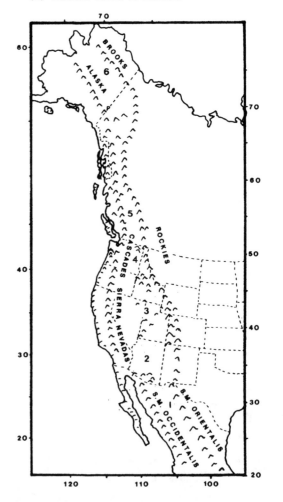

are dominated by the Pacific maritime airstream throughout the year. This airstream is composed of a cool northern portion that dominates during winter and a warm southern portion that is more prominent in summer. The northern portion consists of air which is deflected northward by the Cascade and Alaska ranges, causing cool, moist conditions throughout the year in northern British Columbia and southeast Alaska. The southern portion is air deflected southward by the Sierra Nevada and Sierra-Madre Occidental ranges, causing warm, dry conditions throughout the year in southern California and western Mexico. The boundary between these systems changes seasonally between 35 and 50°N, so that northern California is wet only in winter and Oregon, Washington, and southern British Columbia are dry only in summer. In contrast to surface airstreams, upper-elevation Pacific air passes across the Cordillera Mountains into the Cordillera Plateaux and the Great Plains. This extremely dry air causes desert or near desert conditions in the plateaux and western Great Plains.

Figure 4.5: Monthly distribution of total precipitation and average temperature for selected climate stations in western North America. Vertical units represent 10mm precipitation and 10°C temperature. (a) Fairbanks, Alaska; (b) Vancouver, Canada; (c) Edmonton, Canada; (d) Lander, Wyoming; (e) Los Angeles, California; (f) Guaymas, Mexico. Data from Bryson & Hare (1974).

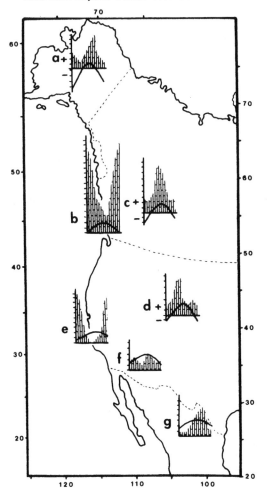

In general, the interaction of Pacific maritime airflows with mountain ranges results in strong precipitation and temperature gradients across short distances. Maritime conditions occur in coastal lowland areas and continental conditions prevail east of the mountain systems. High elevations are always colder and wetter than nearby low-elevation areas. Western Cordillera sites experience more precipitation and milder temperatures than those at comparable elevations and latitudes in the eastern Cordillera. Moist and unstable tropical air from the Gulf of Mexico invades the Central Plateau of Mexico, the Colorado Plateau, and the southern Great Basin during summer. Thunderstorms associated with this airflow cause a peak in summer precipitation. Two distinct rainy seasons typically occur in southwest United States and northwest Mexico where the Gulf of Mexico and Pacific airflows overlap. Figure 4.5 summarises geographical trends in precipitation and temperatures across western North America.

Forests. Forests of western North America are dominated by coniferous species (Fowells, 1965; Preston, 1966; Harlow et al., 1979). Little (1971) lists 56 species in this region. The names and approximate life spans of these are given in Table 4.6. Six major forest zones can be distinguished in the area: Coastal, Sierra, Rocky Mountain, Subalpine, Sierra-Madre, and Boreal (Figure 4.6). The distribution of these zones follows to some extent the location of major mountain ranges and airstream boundaries.

The Coastal zone corresponds to the northern portion of the Pacific airstream in Alaska, Canada, Washington, Oregon and California. Forests in this zone are dominated by very large, long-lived species, but the species diversity is low within individual stands. The most important species, from north to south, are Sitka spruce, _Picea sitchensis_, western hemlock, _Tsuga heterophylla_, western redcedar, _Thuja plicata_, Douglas-fir, _Pseudotsuga menziesii_, and coastal redwood, _Sequoia sempervirens_. The Sierra zone, corresponding to the southern portion of the Pacific airstream in California and Mexico, has a rich species composition. White fir, _Abies concolor_, California red fir, _Abies magnifica_, incense cedar, _Libocedrus decurrens_, western white pine, _Pinus monticola_, Jeffrey pine, _Pinus jeffreyi_, ponderosa pine, _Pinus ponderosa_, sugar pine, _Pinus lambertiana_, lodgepole pine, _Pinus contorta_, and Douglas-fir are common in forests throughout most of this zone. The Rocky Mountain zone occupies large areas of western North America where Pacific air penetrates into the Cordillera mountain system. Ponderosa pine and Douglas-fir are common throughout this zone. Lodgepole pine, western white pine, western larch, _Larix occidentalis_, and grand fir, _Abies grandis_, are most abundant in northern portions, while white fir, blue spruce, _Picea pungens_, and Colorado pinyon pine,

Table 4.6: Coniferous species of western North America. Data on life spans come from Peattie (1953) (P) and Fowells (1965) (F).

SCIENTIFIC NAME	COMMON NAME	LIFE SPAN (YRS)
Cupressaceae		
Libocedrus		
Libocedrus decurrens (Torrey) Florin	Incense cedar	500(P)
Chamaecyparis		
Chamaecyparis lawsonlana (Murray) Parlatore	Port orford cedar	500(P)
Chamaecyparis nootkatensis (D.Don) Spach	Alaska cedar	350(P) (F)
Juniperus		
Juniperus californica Carriere	Californian juniper	250(P)
Juniperus communis L.	common juniper	-----
Juniperus deppeana Steudal	alligator juniper	-----
Juniperus flaccida Schlectendahl	drooping juniper	-----
Juniperus monosperma (Engellmann) Sargent	cherrystone juniper	400(P)
Juniperus occidentalis Hooker	Sierra juniper	800-1000(F)
Juniperus osteosperma (Torrey) Little	Utah juniper	>3000(P)
Juniperus scopulorum Sargent	Rocky mountain juniper	3000(F)
Juniperus texensis van Melle	Texas juniper	-----
Juniperus virginiana Linnaeus	eastern juniper	300(F)
Thuja		
Thuja plicata Donn ex. D.Don	western redcedar	1000(P) 375(F)
Pinaceae		
Abies		
Abies amabills (Douglas) Forbes	Pacific silver fir	540(P) 540(F)
Abies balsamifera (Linnaeus) Miller	eastern balsam fir	200(P) 200(F)
Abies concolor (Gordon and Glendenning) Hoopes	white fir	>350(P) 360(F)
Abies grandis (Douglas) Lindley	grand fir	200(P) 280(F)
Abies lasiocarpa (Hooker) Nuttal	alpine fir	>250(F)
Abies magnifica Murray	California red fir	>500(F)
Abies procera Rehder	noble fir	600-700(F)
Abies venusta (Douglas) K.Koch	Santa Lucia fir	-----
Larix		
Larix laricina (DuRoi) K.Koch	tamarack	355(P) 230-240(F)
Larix lyallii Parlatore	alpine larch	-----
Larix occidentalis Nuttall	western larch	700-900(P) >700(F)
Picea		
Picea breweriana Watson	weeping spruce	-----
Picea engelnannii Parry	Engelmann spruce	>600(P) >500(F)
Picea glauca (Moench) Voss	white spruce	>200(P) >350(G)
Picea mariana (Miller) Britton, Sterns & Paggenberg	black spruce	250(P) 250(F)
Picea pungens Engelmann	Colorado blue spruce	-----
Picea sitchensis (Bong.) Carr.	Sitka spruce	700-800(P) 700-800(F)
Pinus		
Pinus albicaulis Engelmann	whitebark pine	426(P)
Pinus aristata Engelmann	bristlecone pine	5000(P)
Pinus attenuata Lemmon	knobcone pine	-----
Pinus balfouriana Greville & Balfour	foxtail pine	-----
Pinus cembroides Zuccarini	Mexican pinyon pine	-----
Pinus contorta Douglas	lodgepole pine	500-600(P) 450-600(F)
Pinus coulteri D.Don	Coulter pine	-----
Pinus edulis Engelmann	New Mexican pinyon pine	250(P) 973(F)
pinus flexilis James	limber pine	>300(P)
Pinus jeffreyi Greville & Balfour	Jeffrey pine	600(F)
Pinus juarezensis Lanner	Sierra Juarez pinyon pine	-----
Pinus latifolia Sargent	Apache pine	-----
Pinus lambertiana Douglas	sugar pine	-----
Pinus leiophylla var. chihuahuana (Engelmann) Shaw	Chihuahua pine	-----
Pinus monophylla Torrey & Fremont	single-leaf pinyon pine	200(P)
Pinus monticola Douglas	western white pine	615(P) 615(F)
Pinus muricata D.Don	Bishop pine	-----
Pinus quadrifolia Parry	four-leaved nut pine	-----
Pinus ponderosa Laws	ponderosa pine	1047(P) 726(F)
Pinus radiata D.Don	Monterey pine	150(P)
Pinus remorata Mason	Island pine	-----
Pinus reflexa Engelmann	Mexican white pine	-----
Pinus sabiniana Douglas	Digger pine	-----
Pinus toreyana Parry	Torrey pine	200(P)

Pinus edulis, occur in southern areas. This zone, also includes juniper, Juniperus sp., and juniper-pinyon pine woodlands which occupy intermontane basins in the United States.

The Subalpine zone consists of high elevation forests of the Cascades, Rockies, and Sierra Nevadas. Subalpine fir, Abies lasiocarpa, Engelmann spruce, Picea engelmannii, whitebark pine, Pinus albicaulis, and limber pine, Pinus flexilis, are important in the Rocky Mountains. Lodgepole pine and limber pine are most important in the Sierra Nevadas. The Cascades, where the snow pack is exceedingly high, are dominated by subalpine fir, Pacific silver fir, Abies amabilis, noble fir, Abies procera, and mountain hemlock, Tsuga mertensiana. The Sierra-Madre zone consists of open woodlands which occur in much of northern Mexico. Important species include Douglas-fir, ponderosa pine, Mexican pinyon pine, Pinus cembroides, Chihuahua pine, Pinus leiophylla, Apache pine, Pinus engelmannii, drooping juniper, Juniperus flaccida, and alligator juniper, Juniperus deppeana. The Boreal zone occupies areas of Alaska and northern Canada which are dominated by Arctic airstreams during four to ten months of the year. White spruce, Picea glauca, black spruce, Picea mariana, and tamarack, Larix laricina, are common in both Canada and Alaska, while jack pine, Pinus banksiana, and balsam fir, Abies balsamea, are abundant only in Canada.

DENDROCHRONOLOGY

Coniferous species in western North America are well-suited for dendroclimatic research because they typically:

a) produce one distinct layer of wood each year;

b) are free of severe ring-width asymmetries; and

c) respond to climatic variations for some measurable ring characteristic.

However, exceptions do exist. While virtually all species have well-defined annual rings, missing and intra-annual

latewood bands can occur at sites throughout the region. As a result, cores are routinely cross-dated to ensure the accurate dating of all rings. Missing rings are most common in semi-arid, lower treeline sites, while intra-annual bands can make dating difficult at extreme southern sites, where midsummer rains cause a resumption of earlywood formation at the end of the growing season. Cupressaceous species tend to form fluted trunks due to irregular circumferential growth in the lower portions of the stem. Because of this, ring-width patterns in different radii of the same tree are often too variable for cross-dating and subsequent use in dendroclimatic research. Nevertheless, several high quality chronologies have been obtained from incense cedar, indicating that members of Cupressaceae should be evaluated for dendroclimatic studies on a case-by-case basis. Apart from ring asymmetries, these species have several desirable characteristics. For example, many are widespread in extremely dry habitats of the intermontane plateaux where other conifers are rare. Since several species are long-lived (> 1,000 years) and highly resistant to rot, they have good potential for yielding very long chronologies.

The ranges of several western North American species lie entirely within regions of moderate climate (for example, most species of the Coastal Forest zone). Ring widths in such species commonly exhibit little year-to-year variation, and, therefore, may, at first, appear poorly suited for dendroclimatology. However, recent dendroclimatic studies of ring-width (Brubaker, 1980), oxygen-isotope (Gray & Thompson, 1976, 1977; Burke, 1979), and wood-density variation (Parker & Henoch, 1971) indicate that, in some cases, climate information can be extracted from the growth records of species growing in relatively mild climates.

Collections. Collections of tree-ring samples are extensive in western North America owing to the continued

Table 4.6: Continued.

SCIENTIFIC NAME	COMMON NAME	LIFE SPAN (YRS)
Pseudotsuga		
Pseudotsuga macrocarpa (Vasey) Mayr	bigcone Douglas-fir	-----
Pseudotsuga menziesii (Mirb.) Franco	Douglas-fir	1375(P) >1000(F)
Tsuga		
Tsuga heterophylla (Rafinesque)Sargent	western hemlock	>500(P)
Tsuga mertensiana (Bongard) Carriere	mountain hemlock	>500(P) 400-500(F)
Taxodiaceae		
Sequoiadendron		
Sequoiadendron giganteum (Lindley) Buchholz	giant sequoia	4000-4500(P) >4000(F)
Sequoia		
Sequoia sempervirens (Lambert) Endlicher	coastal redwood	220(P)2200(F)

efforts of dendrochronologists since the early part of
this century (Stokes et al., 1973; Drew, 1974, 1975a,b,
1976). While early workers typically sampled few trees and
seldom replicated cores within trees, today most
researchers collect two cores from each of 10 to 20 trees
of the same species at a site (Stokes & Smiley, 1968).
The replicate cores are from opposite sides of the tree in
a direction parallel to the slope contours. Most
collections to date have been made by scientists at the
Laboratory of Tree-Ring Research (University of Arizona).
Collections from this laboratory currently number more
than 1,000 sites. Approximately 150 of these have been
contributed to the International Tree-Ring Data Bank. An
additional 50 or more sites have been contributed by other
scientists working in western North America.

Figure 4.7 shows the locations of western North
America chronologies on file at the Laboratory of
Tree-Ring Research and the International Tree-Ring Data
Bank which satisfy the following criteria:

a) 10 or more trees of the same species;

b) two cores per tree; and

c) 100 or more years.

Table 4.7 summarises these chronologies according to
species, latitude, and elevation. Most samples are from:

a) Douglas-fir (35.6%), ponderosa pine (21.7%), and
   pinyon pine (15.4%);

b) latitudes 30 to 35°N (17.9%) and 35 to 40°N (43.0%);
   and

c) elevations between 1,500 to 2,000m (19.3%) and 2,000
   to 2,500m (38.7%).

The spatial coverage of these samples is not uniform
across the region because of the preponderance of
semi-arid sites from southern areas. This high sample
density reflects the proximity of these sites to the
Laboratory of Tree-Ring Research in Arizona and the
initial research emphasis on arid tree-ring sequences.

The statistical characteristics of a large number of
tree-ring collections have been evaluated by Fritts &
Shatz (1975). The purpose of their study was to choose an
evenly spaced grid of high quality chronologies for
dendroclimatic reconstructions. They considered only
chronologies which:

a) consisted of 10 or more trees with two replicate
   cores from opposite sides of the tree;

b) were precisely dated, measured, and standardised to
   form ring-width indices;

c) displayed high mean sensitivity, low first order
   autocorrelation, high standard deviation and high
   percent variance in mean yearly values; and

d) were strongly correlated with neighbouring
   chronologies for both low- and high-frequency
   components of chronology variance.

The statistical criteria (c) and (d) suggest strong growth
limitations by climate. When climate is the primary
limiting factor, for example, there should be high

year-to-year variations in ring-width indices, few
long-term trends due to disturbances such as logging and
fire, and strong correlations among tree chronologies
within a site and among site chronologies from nearby
locations.

The final network of Fritts and Shatz includes 102
sites (average mean sensitivity by age class, 0.35 to
0.39; serial correlation, 0.38 to 0.45; standard
deviation, 0.37 to 0.40). Sixty-five of these have
subsequently been chosen by Fritts et al. (1979) to
reconstruct climate variations in western North America.
As a part of this research, Fritts et al. verified that
site chronologies with the above characteristics produce
the best climate reconstructions. These criteria for
chronology selection, therefore, should be used as a
guideline in future research.

It is important to note, however, that tree-ring
sequences from less arid parts of North America can
contain significant climate information even though
chronology statistics do not compare well with those of
Fritts and Shatz. Tree-ring chronologies from the Pacific
Northwest (Brubaker, 1980) and Alaska (Garfinkel, 1979)
display average mean sensitivity, serial correlation, and

Figure 4.6: Distribution of forest zones in western
North America.

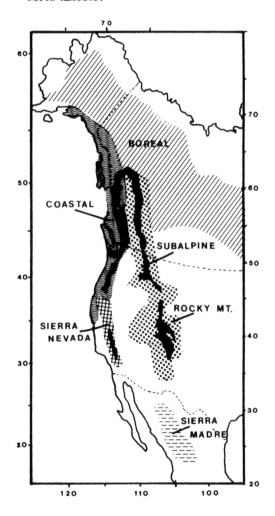

standard deviation of 0.16, 0.57, and 0.27, respectively, yet they exhibit strong correlations with climate variables. Such differences in chronology statistics are to be expected in moderate environments of the Pacific Northwest and temperature-limited environments of the Arctic, since the year-to-year variability of the limiting climatic factors is relatively low (Bryson & Hare, 1974). The collection of large samples is especially important in such areas because individual tree-growth variations can be relatively large. Increasing the sample size to more than 15 or 20 trees per site should increase the ability of chronologies to record growth variations which are common to all trees.

FUTURE PROSPECTS

Despite extensive sampling in western North America, the data set available for dendroclimatic reconstructions can be improved by additional collections. Such efforts should have at least four goals:

  a) to fill in geographical gaps, especially those near important macroclimatic boundaries;
  b) to sample micro-site types which are currently under-represented;
  c) to obtain sequences from long-lived species; and
  d) to update collections from high quality sites which have already been sampled.

The distribution of sites in Figure 4.7 reveals several forested areas which have not been sampled adequately. Especially lacking are samples from the Sierra-Madre Occidentalis and Orientalis, the Cascades and Sierra Nevadas, and Rocky Mountains of Canada. Old trees are still present at high elevations in most of these regions because such areas are inaccessible for logging. These forests, however, have excellent potential for producing high quality tree-ring chronologies since climatic conditions are near the physiological limits for tree growth. Further sampling is also needed across the boundary between the Pacific and Gulf of Mexico airstreams in Mexico and across the Pacific-Arctic airstream boundaries in Alaska and Canada. Such collections should provide important information about the past intensity of these airstreams and their associated climates.

Future sampling must also include trees from a wider variety of sites, since microsite conditions can modify tree-growth responses to climate and thereby add to the diversity of climate information in tree-ring sequences of a given geographic area (LaMarche, 1974b; Fritts, 1976; Brubaker, 1980). Since upper treelines are nearly as ubiquitous as lower treelines in the mountains of western North America, upper treeline sites should be added to the current data set to improve the balance and range of climatic sensitivities. The collection of sites in relatively mesic, continuous forests should also be intensified because such sites can provide significant climate information and because such forests cover areas that are poorly represented in current collections. Coastal forest zones, for example, may yield better quality ring-width index chronologies than traditionally expected. In addition to large sample sizes, careful selection of the most stressful microsites is extremely important in such forests.

Numerous species in western North America are exceptionally long-lived (Table 4.6). It is well-known

Figure 4.7b: Locations of dendroclimatic chronologies in contiguous United States and Mexico as desribed in the text.

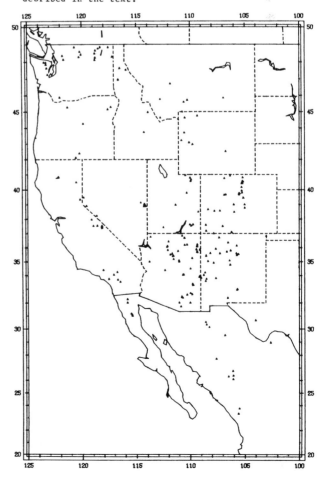

Figure 4.7a: Locations of dendroclimatic chronologies in Alaska and Canada as described in the text.

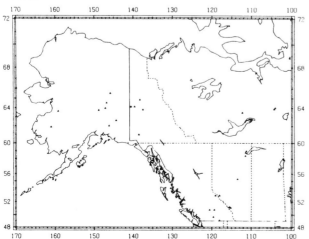

that bristlecone pine species, <u>Pinus</u> <u>aristata</u>, and <u>Pinus</u> <u>longaeva</u>, often yield 1,000-year long chronologies (Schulman, 1958; Ferguson, 1970; LaMarche, 1973, 1974a). However, several other species, for example, Alaska yellow cedar, <u>Chamaecyparis</u> <u>nootkatensis</u>, western redcedar, coastal redwood, giant sequoia, <u>Sequoiadendron</u> <u>giganteum</u>, and Douglas-fir, also have life spans greater than 1000 years. With the exception of Douglas-fir, these species have not been exploited, probably because they occur in mesic environments and/or display basal ring-width asymmetries. These species should be carefully re-evaluated for their potential in dendroclimatic research. Samples should be collected from extreme microsites and from sites at the limits of species distributions. For Cupressaceous species, cores should be taken from above fluted bases to avoid portions of the stem having extreme growth asymmetries.

The network currently used to reconstruct climate by multivariate techniques (Fritts et al., 1979) must be updated because many of the chronologies end in 1963. Tree-ring data since 1963 would increase the calibration period for reconstructions by nearly 30%. The inclusion of these data should increase the ability of calibration equations to describe relationships between regional climate and growth patterns and consequently produce more accurate records of past climate.

DENDROCLIMATOLOGY

Review of research. Dendroclimatic research in western North America began with the founder of dendrochronology, A.F.Douglass of the University of Arizona (Robinson, 1976). He was initially interested in using tree-ring sequences as climate records to invest-igate relationships between sunspot activity and climate. In 1914, he demonstrated a correlation between winter precipitation and ring widths of Arizona trees (Douglass, 1914); a finding which has since been verified by rigorous statistical analyses (Fritts, 1976). During his long and active career he developed important research techniques such as cross-dating (Douglass, 1921), chronology building (Douglass, 1937), and simple ring-width standardisation (Douglass, 1919). In 1936, Douglass established the Laboratory of Tree-Ring Research at the University of Arizona. Scientists at this laboratory have added new procedures and analytic techniques for the evaluation of climate information in tree-ring sequences and the reconstruction of past climates (Fritts, 1976).

The modern era of dendroclimatic research began in the 1950s with publications by Schulman (1951, 1956). Schulman (1958) also initiated important research on bristlecone pine which was later carried on by Ferguson (1968, 1970), Fritts (1969), LaMarche (1968, 1973, 1974a,b) and others (LaMarche & Harlan, 1973; LaMarche & Mooney, 1967, 1972; LaMarche & Stockton, 1974).

In the following decade, H.C.Fritts at the Laboratory of Tree-Ring Research and his co-workers investigated diverse dendroclimatological problems. Their work includes the examination of biological mechanisms that link climate and radial growth (Fritts, 1966, 1969; Fritts et al., 1965b), the development of procedures for standardising ring-width sequences (Fritts et al., 1969) and characterising statistical properties for ring-width index chronologies (Fritts et al., 1965a), and the description of regional patterns in ring-growth variations in western North America (Fritts, 1965). V.C.LaMarche Jr. established the first network (49 stations) of high quality sites for dendroclimatic research. This network has been used extensively in the development of multivariate techniques for reconstructing past climate. One of the most important reasons for rapid research advances during the 1960s was the availability of high-speed computers to carry out routine data processing and complicated statistical analyses. The impact of computers was even greater in the next decade.

During the 1970s, several multivariate statistical techniques were modified for dendroclimatic research

Table 4.7: Summary of dendroclimatic collections in western North America as described in the text.

SPECIES	% OF SITES	AVERAGE CHRONOLOGY LENGTH	LATITUDE	% OF SITES	ELEVATION (M)	% OF SITES
Libocedrus decurrens	0.6	395	20–25°N	0.9	0–500	4.5
Abies concolor	0.6	308	25–30°N	3.7	500–1,000	5.4
A. lasiocarpa	0.3	189	30–35°N	17.9	1,000–1,500	14.9
Picea engelmannii	0.6	453	35–40°N	43.0	1,500–2,000	19.3
P. glauca	7.7	270	40–45°N	9.7	2,000–2,500	38.7
Pinus edulis	15.4	392	45–50°N	11.1	2,500–3,000	9.8
P. flexilis	3.4	484	50–55°N	4.3	3,000–3,500	6.8
P. jeffreyi	2.3	458	55–60°N	1.4	3,500–4,000	0.6
P. leiophylla	6.0	401	60–65°N	7.7		
P. monophylla	0.6	351	65–66°30′N	0.3		
P. quadrifolia	0.6	271				
P. ponderosa	21.7	347				
P. sp.	2.6	407				
Pseudotsuga macrocarpa	2.0	397				
P.menziesii	35.6	380				
Tsuga mertensiana	0.3	646				

(Fritts, 1976). LaMarche & Fritts (1971b) described the usefulness of principal component (or eigenvector) techniques for identifying major variance patterns in spatial arrays of tree-ring data and pointed out the similarity of those patterns to principal component (or eigenvector) patterns of climatic variation. The methodology of response function analysis and canonical transfer functions was developed by Fritts and co-workers at the University of Wisconsin (Fritts et al., 1971). The first quantitative reconstructions of past climates were produced during the mid-1970s (Blasing 1975; Blasing & Fritts, 1975, 1976). Multiple regression techniques have been applied in areas where relatively few tree-ring chronologies exist, while canonical techniques have been used to produce records of climate for large areas with numerous tree-ring chronologies (Fritts et al., 1979). Important techniques were also developed to verify reconstructions (Fritts, 1976). Numerous evaluations of reconstructions have shown that verification tests are the best method for judging the ability of transfer functions to accurately estimate past climate variables (Fritts, 1976; Fritts et al., 1979).

Research accelerated in many other areas during the 1970s. LaMarche inferred long-term palaeoclimate (LaMarche, 1974a) and treeline fluctuations (LaMarche, 1973) from bristlecone pine chronologies in east central California. These chronologies, and those from Utah, New Mexico and Nevada, suggest that bristlecone pine sequences may record evidence of world-wide climatic variations as far back as the middle Holocene (LaMarche & Stockton, 1974). Stockton examined the potential of chronologies from Wyoming (Stockton, 1973), developed a conditional

probability approach to reconstruction (Stockton & Fritts, 1971b), and applied multivariate techniques to reconstruct riverflows (Stockton, 1971, 1975, 1976) in various parts of western North America.

Parker and others (Parker & Meleskie, 1970; Parker & Henoch, 1971) and LaMarche developed X-ray densitometric techniques which permit the study of climatic variations from annual variations in wood density. Since climate-related variations in wood density can occur even when radial growth is uniform (Parker & Henoch, 1971), these techniques provide additional hope that reliable dendroclimatic records can be obtained from wide-ranging, mesic forest zones of western North America. Other new approaches to dendroclimatology, including the analysis of $^{12}C/^{13}C$ (Stuiver, 1978) and $^{18}O/^{16}O$ ratios (Gray & Thompson, 1976, 1977; Burk, 1979), were tested with encouraging results. Finally, many new sites, especially in the Pacific Northwest and Mexico, were collected during the 1970s to increase the coverage of ring-width index chronologies.

Two of the most notable current features of dendroclimatology in western North America are the widespread identification of climate-growth relationships, and the successful estimation of past climatic variations. Climate-growth relationships have been documented from response functions of more than 200 sites from Canada, Mexico, and the United States. Based on 127 response function analyses (Figure 4.8a), Fritts (1974) concluded that precipitation is more important than temperature in limiting growth in semi-arid sites of western North America. He found that at low elevation and southern sites, precipitation tends to be directly and temperature

Figure 4.8a: Mean response functions representing the most important differences among the 127 response functions analysed by Fritts (1974). 41 is typical of arid-site trees of southwestern United States. 39 is of diverse origins, many sites are on steep west-facing slopes. 40 is typical of mountain sites. 45 is typical of very high elevation bristlecone pine sites.

Figure 4.8b Summary of response functions from sites west and east of the Washington Cascade crest (Brubaker, 1980). For each area, the percentage of sites showing a positive response to a climate variable is plotted according to month (14 months, beginning in June prior to the growth year and ending in July of the current growth year).

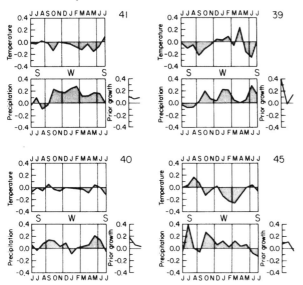

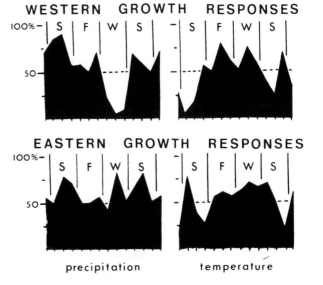

inversely related to growth throughout the year. In contrast, at high elevations, high latitudes, and north-facing slopes, precipitation is often inversely related to growth especially in winter, and temperature is directly related to growth during summer. While species can differ in their response to climate, site differences (such as slope, latitude, and altitude) appear to be more important in causing variations in response functions. Brubaker (1980) analysed 38 response functions from Washington, Oregon, and Idaho (Figure 4.8b). Late spring and summer precipitation and winter temperature were directly related to growth at sites throughout this area. However, winter precipitation and summer temperature were inversely related to growth west of the Cascades, and directly related to growth east of the Cascades. Overall, these studies indicate that a diversity of growth responses to climate are recorded in western North American tree-ring sequences. Transfer functions based on such diverse tree-ring sequences can produce dendroclimatic records with a broad coverage of past climate variables. Since response functions usually display trends which span several months, these studies also suggest that a season (as opposed to a shorter time interval) is the appropriate time interval for dendroclimatic reconstructions.

Achievements in the area of climatic reconstruction are impressive in western North America. This is the only area in the world, for example, where tree-ring sequences have yielded continuous records of climate back to the middle Holocene. According to LaMarche's (1974a) study of high-elevation bristlecone pine, relatively warm summers prevailed in southwest United States during 5500 to 3300 BP, 2200 to 1700 BP, 800 to 600 BP, and 100 BP to the present. Cool summers were typical during 3300 to 2000 BP, 1700 to 800 BP, 600 to 1000 BP. This interpretation is supported by evidence of glacier and treeline fluctuations during the same period in western North America, and historical temperature records from England.

Western North America is also unique as the only area in which canonical techniques have been used to derive quantitative estimates of climatic variations. Fritts et al. (1979) have reconstructed seasonal temperature and precipitation variables (1602 to 1963) over large portions of North America, while Blasing (1975) and Blasing & Fritts (1975, 1976) have reconstructed seasonal atmospheric pressure patterns (1700 to 1963), inferring temperature and precipitation trends from those patterns. These studies indicate that the patterns of climatic variation were complex, and rarely hemispheric in scale. According to Fritts et al. (1979), prior to 1900 in the western United States, temperatures were somewhat cooler during all seasons, winter and spring precipitation was lower in southern but higher in northern regions, summer precipitation was higher in western but lower in eastern areas, and autumn precipitation was greater in the

Pacific Northwest and extreme south-central areas, but less everywhere else. Although the transfer function models of Blasing (1975) and Blasing & Fritts (1975, 1976) differed from those of Fritts et al. (1979), the results of these approaches are generally consistent.

OPPORTUNITIES

Opportunities for future research in western North America are great, and previous research allows the clear identification of important problems. New chronologies are needed to update the existing network of sites, to provide new data from under-represented geographic regions and microhabitats, and to extend records back in time using tree-ring sequences from long-lived species. The analysis of wood-density properties and stable-isotope ratios will provide new climate information to supplement records derived from ring-width indices. Attempts to verify reconstructions with instrumental weather data must be continued. In addition, comparisons of reconstructions derived independently by different researchers can provide an important means of evaluating the quality of climate reconstructions. Increased cooperation between American, Canadian, and Mexican workers will improve the spatial coverage of the existing reconstructions. Perhaps most important, however, dendroclimatologists throughout North America should cooperate to build a reconstruction of recent climate change over the entire North American continent.

EASTERN NORTH AMERICA
E.R.Cook
INTRODUCTION

Eastern North America can be considered as an ecological entity: the eastern deciduous forest biome. The boreal forests of eastern Canada are not discussed since they are part of the continuous forest formation that girdles the subpolar regions of the Northern Hemisphere (see Jacoby, this volume).

The eastern deciduous forest biome is situated in the region of North America where there is an annual surplus of moisture. As one travels westward from the Atlantic coast, the surplus first increases as the highlands of the Appalachian and Adirondack Mountains are crossed and then systematically decreases to zero roughly along the 96°W meridian (Barry & Chorley, 1976). This meridian approximates the westerly limit of the deciduous forest biome remarkably well although some chiefly riparian communities extend beyond it. The moisture sources during the year are storms emanating from the Gulf of Mexico and, to a lesser extent, from the subtropical Atlantic Ocean. Precipitation is evenly distributed throughout the year although some areas have a slightly warm-season dominated regime. The areal distribution of seasonal and annual precipitation is influenced in part by topography and the pattern of the prevailing winds. On the

westerly flanks of higher mountains, a considerable amount
of orographically produced precipitation falls. The
orographic effect causes the formation of moderate, local
rain shadows on the leeward slopes such as in the
Asheville Basin of North Carolina. The zonal westerlies of
the upper atmosphere affect the areal distribution of
precipitation and temperature anomalies. The position of
the semi-permanent western Atlantic pressure trough that
routinely resides over eastern North America, influences
the location of the main storm tracks by steering cyclones
northward along the downstream side of the trough (Barry &
Chorley, 1976).

Lamb & Johnson (1959) have shown that the position
of the western Atlantic pressure trough has varied
significantly during the past 200 years, often in parallel
with changes in the strength of the zonal westerlies.
Longer-term variations have been suggested by the work of
Nichols et al. (1978). Since changes in the flow of the
zonal westerlies have a decided influence on the climate
of eastern North America, a precise understanding of
long-term upper air behaviour is extremely important. By
establishing a network of tree-ring chronologies throughout this region, we may be able to reconstruct not only
temperature and precipitation histories, but also past
pressure patterns.

VEGETATION OF THE EASTERN DECIDUOUS FOREST BIOME

The eastern deciduous forest biome of North America
is a complex matrix of vegetation types that is difficult
to portray in general terms. Forest ecologists have sought
to subdivide it into regions that are distinct, based upon
the composition of the dominant tree species. Perhaps the
best effort in this regard is by Braun (1950). Although
this treatise is in some respects dated, I know of no
other reference on the eastern deciduous forest biome that
is as beautifully descriptive and comprehensive as her
book. I will rely exclusively on Braun (1950) in the
following descriptions of the biome unless otherwise
noted. These descriptions are of the original "virgin"
forest communities. The remnants of these communities are
the primary sources of specimens that can extend back for
several centuries.

The eastern deciduous forest biome is a
closed-canopy, deciduous forest formation extending
westward in North America to roughly 98°W. In limited
areas, the geology and topography will help produce open
canopy and, in high elevations, open alpine environments.
Coniferous species dominate locally over deciduous species
especially in the northern parts of the biome, in higher
elevations, and on less favourable sites.

The forest biome can be divided roughly into nine
major regions which in themselves show considerable
ecotonal variation (Figure 4.9). More recent vegetation
maps (for example, Little, 1971) divide the forest into
ten regions that show exhaustively detailed mosaics of

forest associations within each region; those interested
in such detail or in the distribution of specific tree
species should consult Fowells (1965) or Little (1971).
Figure 4.9 illustrates the tremendous diversity of species
and forest communities that are available for dendroclimatic studies. Over 40 species have been noted as
dominant trees in various forest regions. Although most
forests in eastern North America are second- or
third-growth communities, parcels of virgin growth exist
in all of the forest regions. This is particularly true in
the higher elevations of the Appalachian and Adirondack
Mountains where it was not feasible to log trees. Many of
the virgin primary forest communities that Braun (1950)
described still exist in state and national parks and
forests.

Undoubtedly, there are numerous other virgin areas
that are known to exist only locally or are yet to be
discovered. I have had as much success finding old-growth
trees by wandering through the mountains as I have had by
talking to park rangers and foresters. A dendrochronologist does not look for large trees as much as for trees
with growth characteristics that suggest great age
regardless of size. In many instances, these trees are
growing on very marginal xeric sites and are of little
value to foresters in terms of available board feet of
lumber and future wood production. Pitch pine, _Pinus
rigida_ Mill., chestnut oak, _Quercus prinus_ L., and eastern
red cedar, _Juniperus virginiana_ L., are species that
frequently inhabit this kind of site. However, eastern
hemlock, _Tsuga canadensis_ Carr., may be just as
climatically sensitive and long-lived growing on hummocks
in ravine habitats. This species has a very shallow root
system that may penetrate to the depth of only
250 to 350mm leaving it quite susceptible to drought.

DENDROCHRONOLOGY IN EASTERN NORTH AMERICA

Background and considerations. Dendrochronology in
eastern North America is not new. Several highly important
pioneering studies by Lyon (1935, 1936, 1939, 1949, 1953),
Diller (1935), Hawley (1941), Meyer (1941), and Lutz
(1944) firmly established that dendrochronological
techniques could be applied successfully to several
species of trees growing in this region. Lyon (1936) also
demonstrated the feasibility of inferring past climatic
events from tree-ring sequences of hemlocks growing in New
England. However, except for research by Fritts (1958,
1959, 1960, 1962), Phipps (1961, 1967), and Estes (1969),
the science experienced little progress in the 1950s and
1960s. Since 1970, progress in eastern dendrochronology
has increased rapidly with the availability of high-speed
computers and the development of statistical techniques
for handling and analysing tree-ring data (Fritts, 1976).
This activity led to the development and publication of 39
high quality tree-ring chronologies from eastern North
America (DeWitt & Ames, 1978).

In discussing the status of eastern dendrochronology, I am concentrating my attention on the progress of
the past 10 years and to those chronologies that have been
developed from the prerequisite conditions of precise
dating using accepted cross-dating techniques and
chronology replication. Additionally, for dendroclimatic
studies a minimum chronology length must be considered
that will "significantly" lengthen the available climate
data base. Here we encounter the problem of defining
"significant" since it can vary depending upon which
variable we care to reconstruct. In eastern North America,
temperature and precipitation data are relatively abundant
back to 1830, see, for example, Wahl (1968). Surface
pressure and hydrologic series are generally shorter and
less common in the 19th century. Thus, a rainfall
reconstruction back to 1800 may only marginally extend the
data base, but pressure or streamflow reconstructed back
to 1800 could easily double or triple the instrumental
series' length.

In the context of establishing a global tree-ring
network, another consideration comes to mind. In order to
reconstruct variations in regional and global climate that
will hopefully encompass the full range of behaviour of
the global circulation system in the recent past, the
common period should be long enough to include the two
notable opposing climatic epochs: the Little Ice Age of
the 17th and early 18th centuries and the 20th century
warming period. For this reason, 1700 or earlier is highly
desirable as a common starting year in a global tree-ring
network. The following discussion of currently available
tree-ring chronologies will consider those series that
satisfy this criterion.

The present tree-ring network. Figure 4.10 is a map of the
site locations of all of the tree-ring chronologies known
to this author that begin in or before 1700. The total
stands at 49. The coverage is best along the northeast-
southwest axis of the map. Notable gaps in the coverage
exist in the states bordering the Great Lakes, over
virtually all of southern Canada, and in the states
bordering the Gulf of Mexico. Six coniferous and three
deciduous species are represented eastern hemlock, eastern
white pine, Pinus strobus L., red pine, Pinus resinosa
Ait., pitch pine, shortleaf pine, Pinus echinata Mill.,

Figure 4.9: Forest regions of the eastern deciduous forest biome of North America.

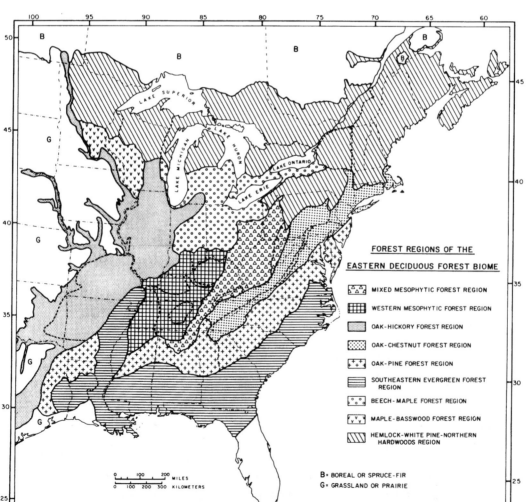

red spruce, _Picea rubens_ Sarg., white oak, _Quercus alba_ L., chestnut oak and post oak, _Quercus stellata_ Wangenh. Table 4.9 lists the species and the chronology length of each site. Caution should be exercised with reference to the several chronologies from Arkansas, Missouri, and Tennessee that were derived from beams in historical structures or represent earlier collections that must be updated with chronologies from nearby living trees.

The network map is best seen as reflecting both the need and the potential for developing a dense grid of tree-ring sites in eastern North America. Since no organised effort has yet been made to establish a network analogous to the ones developed in western North America and the Southern Hemisphere, a large-scale, intensive effort will greatly improve the spatial coverage. Table 4.9 also lists one tree-ring site in Maine that has ring-width and density data. The development and application of X-ray densitometry in dendroclimatology is a revolutionary advance that can increase by several times the useful information derived from tree rings (Jacoby, 1980a; Schweingruber, this volume; Conkey, this volume). This technique should be used to its fullest potential in

the future development of the tree-ring network.

Additional species with dendroclimatic potential. The species in Table 4.9 represent a group with proven dendrochronological potential that should be exploited fully in future collection efforts. However, owing to the tremendous diversity of vegetation types in eastern North America, there are many other species of trees that may be useful for our purposes. Some likely candidates are listed in Table 4.9.

Chronologies for some of the species already exist (DeWitt & Ames, 1978; Sherwood, 1978) but they do not extend back to 1700. The criterion I used for selection was simply one of recorded longevity, using information from Nichols (1913), Hough & Forbes (1943), Fowells (1965), Harlow & Harrar (1969), and Stahle (1978), and my own observations. Only those species with recorded age maxima of at least 250 years were included. In some cases, these maxima may be conservative. For example, Fowells (1965) reports a maximum age for pitch pine as 200 years, but Cook (1976) reported developing a 352-year pitch pine chronology in New York. No effort was made to exclude

Figure 4.10: The tree-ring network in eastern North America. Refer to Table 4.8 for information regarding each site.

so-called swamp species since they may prove to be very useful for dendrochronology. My experience over the past few years indicates that no simple set of site selection criteria exists for the eastern deciduous forests that is analogous to that used in western North America (Cook, this volume). The only prerequisite that must be satisfied is cross-dating between trees of the same species on a site.

This list is undoubtedly incomplete. Some species with unknown age potential that occupy very restricted habitats may also be useful. Two examples are table mountain pine, Pinus pungens Lamb, and Carolina hemlock,

Tsuga caroliniana Engelm. Table mountain pine, in particular, grows on rocky, well-drained sites where growth rates can be extremely slow. Longevity under such circumstances has been found in other species such as pitch pine. Three species already noted deserve additional comment. Baldcypress, eastern redcedar, and Atlantic white-cedar offer the potential for extending the modern chronologies to earlier time periods using subfossil wood accumulated in swamps and from archaeological sites. Harshberger (1970) notes that perfectly preserved white-cedar logs were mined from bogs and swamps in the New Jersey pine barrens for house shingles and other uses.

Table 4.8: Key to the tree-ring sites in Figure 4.10. Where the starting year is incomplete, the site collection is being developed. Those sites with the same map number represent updates of the earlier collections that must be merged. TRW = total ring-width; EWW = earlywood width; LWW = latewood width; MED = minimum earlywood density; MLD = maximum latewood density. All dates are A.D.

MAP	SITE	SPECIES	YEARS	TYPE
1	Dixon Lake, Ont.	red pine	1623-1973	TRW
2	Elephant Mt., ME	red spruce	1667-1976	TRW
3	Elephant Mt., ME	red spruce	1667-1976	EWW
4	Elephant Mt., ME	red spruce	1667-1976	LWW
5	Elephant Mt., ME	red spruce	1667-1976	MED
6	Elephant Mt., ME	red spruce	1667-1976	MLD
7	Nancy Brook, NH	red spruce	1561-1972	TRW
8	Mt. Washington, NH	red spruce	1678-1976	TRW
9	Camel's Hump, VT	red spruce	1635-1971	TRW
10	Winch Pond, NY	white pine	168?-1978	TRW
11	Roaring Brook, NY	hemlock	1599-1978	TRW
12	Roaring Brook, NY	red spruce	1618-1978	TRW
13	Pack Forest, NY	hemlock	1595-1976	TRW
14	Livingstone, MA	red spruce	1697-1971	TRW
15	Mohonk Lake, NY	hemlock	1636-1973	TRW
16	Mohonk Lake, NY	white pine	1626-1973	TRW
17	Mohonk Lake, NY	pitch pine	1622-1973	TRW
18	Mohonk Lake, NY	chestnut oak	1690-1973	TRW
19	Dark Hollow Trail, NY	white oak	1648-1977	TRW
20	Tionesta, PA	hemlock	142?-1978	TRW
21	Cook Forest, PA	white oak	167?-1976	TRW
22	Gaudineer, WV	red spruce	1652-1976	TRW
23	Linville Gorge, NC	white oak	1616-1977	TRW
24	Clemson Forest, SC	shortleaf pine	1684-1973	TRW
25	Clemson Forest, SC	shortleaf pine	1684-1973	EWW
26	Clemson Forest, SC	shortleaf pine	1684-1973	LWW
27	Newfound Gap, NC	red spruce	1686-1972	TRW
28	Norris Watershed, TN	shortleaf pine	1681-1972	TRW
29	Steiner's Woods, TN	white oak	1625-1972	TRW
30	Savage Gulf, TN	shortleaf pine	1700-1972	TRW
31	Warren Country, TN	white oak	1669-1940	TRW
	Fall Creek Falls, TN	white oak	1767-1972	TRW
32	Mammoth Cave, KY	white oak	1648-1966	TRW
33	Ferne Clyffe, IL	white oak	1669-1972	TRW
34	Carter County, MO	white oak	1642-1936	TRW
35	Shannon County, MO	white oak	1558-1936	TRW
	Shannon County, MO	white oak	1725-1972	TRW
36	Ridge House, AK	white oak	1674-1838	TRW
37	Dutch Mills Ruins, AK	white oak	1624-1842	TRW
38	Wolf House, AK	shortleaf pine	1672-1827	TRW
39	Magness barns, AK	white oak	1598-1858	TRW
40	Jackson Cabin, AK	white oak	1654-1849	TRW
41	Pope County, AK	white oak	1642-1939	TRW
	Russellville, AK	white oak	1713-1972	TRW
42	Montgomery County, AK	shortleaf pine	1666-1939	TRW
	Big Brushy Mt., AK	shortleaf pine	1760-1972	TRW
43	Polk County, AK	white oak	1677-1939	TRW
	Brush Heap Mt., AK	white oak	1720-1972	TRW
44	Oak Park, TX	post oak	1699-1974	TRW
45	Pammel State Park, IA	white oak	1641-1977	TRW
46	Ledges State Park, IA	white oak	1688-1976	TRW
47	Itasca State Park, MN	red pine	1672-1971	TRW
48	Saganaga Lake, MN	red pine	1620-1972	TRW
49	Seagull Lake, MN	red pine	1625-1971	TRW

Bald cypress is likewise highly resistant to decay (Harlow
& Harrar, 1969; Stahle, 1978). Stahle (1978) also notes
that well preserved red cedar logs were often found in
Indian burial mounds during archaeological excavations.
Unfortunately, bald cypress and eastern red cedar are two
of the most difficult species to cross-date owing to the
prevalence of intra-annual latewood bands (false rings)
and poor circuit uniformity of the annual rings. Their
eventual dendroclimatic value is uncertain because of
these problems.

Some thoughts on sampling strategy. In a recent comparison
of tree-ring chronologies from eastern and western North
America, DeWitt & Ames (1978) showed that the average
signal-to-noise ratio of eastern chronologies, as inferred
from analysis of variance statistics, was three or more
times lower than that of western tree-ring series for an
equal sample size of cores. They suggested that the number
of sampled trees per site in eastern North America be
anywhere from 26 to 40 (two cores per tree) depending on
how limiting the site is to growth. Alternatively, they
suggested that an equivalent signal-to-noise ratio could
be obtained by increasing the tree-ring network density
three or four times if the sample size per site was not
increased. There are good reasons why the second approach
is much more desirable than the first when additional
species and sites are available within a common geographical region.

The most important aspect of eastern dendrochronology that DeWitt & Ames (1978) do not consider fully is
the closed-canopy nature of the sites with attendant stand
competition and disturbance problems. Polynomial and
spline curve-fitting techniques may remove some of the
consequences of these non-climatic perturbations when they
differ from tree to tree in timing and effect. But if a
disturbance is common to the site (fire or insect damage,
for example), then its manifestation will remain to some
degree in the core ring-width series and in the final site
chronology. Without a priori knowledge of the stand

Table 4.9: Some additional species likely to be useful for
dendroclimatic studies in the eastern deciduous forests of
North America.

SPECIES	REPORTED MAX. AGES IN YEARS
Loblolly pine	300
Longleaf pine	300
Bur oak	200-300
Swamp white oak	300
Northern red oak	300
Sugar maple	300-400
American beech	300-350
Black birch	265-350
Yellow birch	300
Northern white-cedar	400
Atlantic white-cedar	200-1,000
Eastern redcedar	300-500
Baldcypress	400-1,400
Eastern larch	335

history, we cannot determine if an unusual interval of
below-average growth was caused by climate, disturbance,
or a combination of both. This is the most vexing problem
confronting closed-canopy forest dendrochronology.
Increased sampling within the stand is not likely to
resolve this dilemma. Given this situation, I suggest that
the greatest emphasis in sampling be placed on developing
additional chronologies of the same species on
non-contiguous sites and for sampling different species in
the region. With this approach, species-specific and
stand-specific non-climatic variance may be factored out
in developing transfer functions.

The above sampling strategy has an additional
benefit. Since multiple species are involved, we can take
advantage of the differences in each species' growth
response to climate for developing more complete climate
reconstructions.

This discussion leads to the question of tree-ring
network density. How many species and sites will we need
for adequate coverage of eastern North America? As far as
species go, I would say "the more, the better" for the
above stated reasons. The question of overall site density
is much harder to answer. As a direct consequence of high
species diversity in the eastern deciduous forest, I can
easily see the potential for 200 to 300 chronologies.
This estimate is very conservative if we include density
data. Whether or not we need that many series depends on
the quality of the tree-ring data. The diagram relating
the signal-to-noise ratio to sample size in DeWitt & Ames
(1978) suggests that two to four times as many eastern
chronologies may be needed to equal the quality of the
western tree-ring network. This places the minimum at
around 150 to 200 chronologies given adequate geographic
coverage.

DENDROCLIMATOLOGY

Several recent studies have demonstrated conclusively the feasibility of reconstructing past climate
from trees growing in mesic forest regions (Cleaveland,
1975; Cook & Jacoby, 1977; Conkey, 1979a; Blasing et al.,
1981). This sudden burst is due to the development of
highly flexible multivariate response and transfer
functions (Fritts et al., 1971; Stockton, 1975; Guiot et
al., this volume; Lofgren & Hunt, this volume) for
modelling and reconstructing climate from tree rings.

Thus far, the best results have come from reconstructing some measure of growing season moisture supply
such as rainfall (Cleaveland, 1975; Blasing et al., 1981)
or the Palmer Drought Severity Index (Cook & Jacoby,
1977). This is not surprising since most of the tree
species thus far studied show a dominant, direct growth
response to precipitation during the May-August period
coincident with radial growth. This direct response is
often coupled with an inverse temperature response during
the same months. Response functions of four different

species showing this type of relationship are shown in Figure 4.11. However, red spruce, also in Figure 4.11, shows a considerably different growth response that emphasises a direct relationship between radial growth and temperature. Precipitation is far less important for optimum annual ring development. Other response functions of red spruce based on both ring-width and ring density data (Conkey, 1979a,b, this volume) show similar relationships. Thus, this species should be an excellent source of past temperature information.

CONCLUDING REMARKS

I have tried to provide an overview of the potential and progress of dendroclimatology in eastern North America that can serve as an impetus for the future development of a tree-ring network there. Eventually, a dense network will be established in this region that will serve as a unique data set for addressing problems relating to physical, biological, and cultural systems. Whether it is developed in a sluggish, piecemeal fashion (as is the case now) or in a well-organised, expeditious effort depends greatly on two things. One is the willingness of dendro-chronologists to engage in a cooperative collection effort, and the other is the availability of research funds to support such an effort. I do not foresee any serious lack of collaboration between scientists in this type of endeavour. Thus, there is no reason why an excellent tree-ring network cannot be established for

Figure 4.11: Response functions of five tree species growing in eastern North America. The shaded areas represent tree growth responses to climate that are statistically significant at the 95% level.

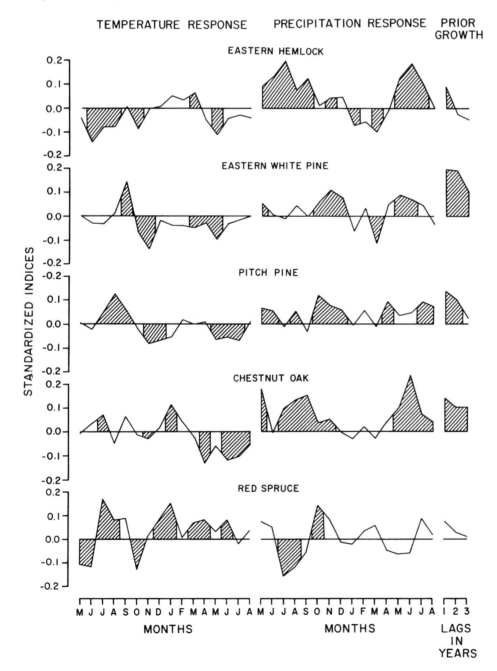

eastern North America from a sustained four to five year collection and development effort, given the necessary support and organisation.

COMMENT

L.E.Conkey

I support Cook's emphasis on increasing the number of collections in eastern North America instead of only increasing the number of cores per site. Many of the oldest trees are in relatively small stands, often not allowing for fulfillment of DeWitt & Ames' (1978) recomm- ended 26 to 40 trees per eastern site. By developing a large grid of chronologies from smaller stands, such sites will still be of great value to eastern dendroclimatology.

Cook suggests that red spruce, Picea rubens Sarg., appears to be a stronger indicator of temperature than other eastern species. This is supported by work I have done with both ring widths and densities of red spruce. Reconstructions of November through March temperature derived from three ring-width chronologies of red spruce and four of other species are more reliable than precipitation reconstructions; one such reconstruction is based on a model which explained 39% of the variance and passed 50% of verification tests (Conkey, 1979a, this volume). Figure 4.12, A, is a low-pass filtered plot of this reconstruction, averaged over several New England stations. Plots B and C in Figure 4.12 show precipitation reconstructions which did not verify as successfully. At

least in this case, the predominance of red spruce chronologies in the predictor set may indeed produce the stronger temperature response.

I am currently developing density and ring-width chronologies from three red spruce sites in Maine, one of which appears in Cook's list. All three sites have five chronologies each, for total ring width, earlywood and latewood widths, minimum earlywood density, and maximum latewood density. Two of the sites are more than 200km apart, and yet the plots of averaged maximum density at the three sites (Figure 4.13) show a great deal of similarity, more than is usually found for ring widths. Correlations among the averaged maximum densities are compared to those among averaged total ring widths (Table 4.10).

At each site, the difference between the ring-width and the density correlation coefficients is highly significant (0.99 level, Spiegel, 1961, p.247). Response functions also show more variance due to climatic factors with maximum density than with ring widths at the same site (Conkey, 1979b). The response, as might be expected, is strongest during the summer when wood of highest density is laid down. The summer response is not to temperature alone, but also to precipitation; high maximum densities are correlated with warm, dry summers, which may produce conditions of water stress in the tree. Warm, moist springs also produce high maximum densities. It thus appears that density of red spruce may allow the

Figure 4.12: Averaged reconstructions of New Eng- land, U.S.A., winter temperature (A), winter precipitation (B), and spring precipitation (C). Line plots are reconstructions, dots are values of actual temperature or precipitation. Each reconst- ructed and actual series is treated with a low-pass filter, and the values are plotted as departures from the calibration period mean (from Conkey,

1979a). A. November-March temperature from six stations: New Haven, Amherst, Blue Hill, Concord, Albany, and New York City. B. November-March precipitation from four stations: Blue Hill, Mohonk Lake, Philadelphia, and St. Johnsbury. C. April-June precipitation from five stations: Amherst, Concord, Hanover, Albany, and St. Johnsbury.

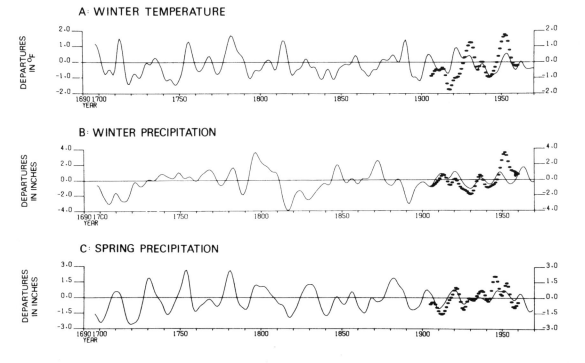

THE NORTHERN HEMISPHERE 134

reconstruction of a wide spectrum of seasonal climatic events. A first exploration of this possibility is described by Conkey (this volume). In addition, maximum density values do not show the effects of suppression and release that are common in mesic-area ring-width series and which may interfere with the climate signal. Autocorrelation is also much lower for density than for ring widths at the same site (Conkey, 1979b). Signal-to-noise ratios (Graybill, this volume) are higher for density, and it appears that fewer trees per site are required for adequate replication, producing a stronger climate signal than from a ring-width chronology with the same number of cores. Thus, it seems likely that developing density chronologies for eastern North America will ameliorate some of the difficult statistical and sampling problems faced by mesic-area dendroclimatology.

COMMENT

R.L.Phipps

While the signal-to-noise ratio may be improved by increasing the network density in eastern North America, as Cook suggests, it is also true that we do not, as yet, have the experience needed to select species or sites with the best climate signal.

Site selection. A generalisation with regard to selecting sites in which environmental factors are especially limiting to growth is to seek sites of water divergence, such as ridges and upper slopes that are convex upward. We have also had some success with collections from areas of convergence and emergence to and including swamps and wetlands (Phipps et al., 1979). Wet-site trees, physiologically attuned to wet conditions, appear quite sensitive to drought, while dry-site trees may be most

sensitive to wet conditions. For wet-site trees, a severe drought may cause enough of a physiological shock that regardless of climatic variation during the next few years, very little ring-to-ring variability will be noted.

Species and tree selection. Evidence suggests that, the more shade tolerant a species, the less it is influenced by its immediate surrounding neighbours. When sampling in closed-canopy stands on flat topography, we feel it is more important to emphasise relative crown radius length than compass direction. Thus, we sample opposing radii beneath the least crowded (greatest radius length) opposing crown radii. Sampling "high stress" sites where density is low is preferable, but the more shade-tolerant a species the less important crown density seems to be. We often find that the variance in common between samples (as determined from analysis of variance) is greater in samples of shade-tolerant beech, *Fagus*, or hemlock, *Tsuga*, from mesic sites than from ridgetop stands of low canopy-density oaks, *Quercus*. An obvious exception to these generalities are very shade-intolerant species (such as many of our *Pinus* species) which have never been part of the understory and which may be relatively climatically sensitive.

Standardisation. A nemesis of working with collections from closed-canopy sites is the difficulty of removal of non-climatic trends, prerequisite to merging the data into a mean chronology. The classical age trend of wide-to-narrow rings may be superimposed upon numerous smaller shifts or perturbations. These perturbations may be classified as either crown size trends or physiological sensitivity trends. Crown size trends are the result of growth-rate changes consequent to factors such as

Figure 4.13: Yearly maximum density at three sites in Maine, U.S.A. Each series is the average maximum density in g/cm³ (from Conkey (1979b). The Elephant Mt. series is correlated with the Sugarloaf Mt. series at r = 0.83 and with Traveler Mt. at r = 0.65. The Sugarloaf Mt. and Traveler Mt. series are correlated at r = 0.84. All correlations are significant at the 0.9995 level.

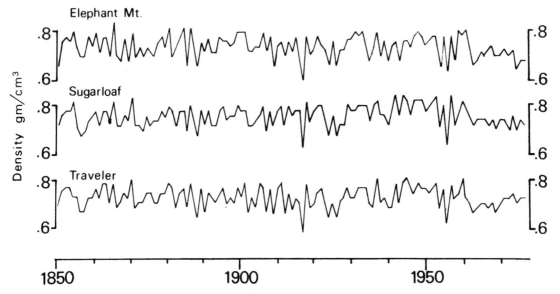

suppression and release, which result in changes in crown size. Crown size trends which do not involve physiological sensitivity trends may or may not be associated with age. Sensitivity is a sought-after characteristic, but what is not wanted is changes or shifts in sensitivity. Except for extreme cases, physiological sensitivity trends are typically more subtle than age and crown size trends. They are generally not removed by curve-fitting procedures, and are still present before data of individual samples are merged into a mean chronology. Merging, however, may have the effect of averaging out or obscuring physiological sensitivity trends of individual samples and thereby lend some degree of time-stability to the mean chronology. On the other hand, it appears that some information from the most sensitive samples of a collection is lost during the process of merging into a mean chronology.

Cook indicates that the typically short time span covered by chronologies of eastern North America may be most valuable in extending streamflow records. To this I should like to add that these chronologies may be useful in examining other climate-related variables such as air pollution. Results of a preliminary investigation suggest that as a consequence of acid rainfall, environmental conditions may now be more limiting to tree growth of some species in areas most severely affected (Puckett, personal communication, 22.10 1979). Cook also indicates the value of a dense network of collection sites. To this I should like to suggest that, in the interest of minimising physiological sensitivity trends of individual samples, collections should contain as many samples as is practicable.

Table 4.10: Correlation coefficients among averaged maximum densities and averaged ring widths at three Maine sites: ELE = Elephant Mt., SUG = Sugarloaf Mt., TRAV = Traveler Mt. All are significant at the 0.9995 level (n = 127).

	MAXIMUM DENSITIES		
	ELE	SUG	TRAV
ELE	1.00	0.83	0.65
SUG		1.00	0.84
TRAV			1.00

	RING WIDTHS		
	ELE	SUG	TRAV
ELE	1.00	0.49	0.32
SUG		1.00	0.63
TRAV			1.00

ALPINE EUROPE

O.U.Bräker

INTRODUCTION

In geographical terms, Alpine Europe can be limited to the Alpine mountains, extending arc-like from the western part in France across 1,100km to the eastern Alps in Austria. These mountains cover an area of approximately 175,000km², of which 3,500km² are glaciers. In geological terms, Alpine Europe summarises all European mountains formed in younger Mesozoic and Tertiary times: the Prealps, the Jura, the Pyrenees, the Apennines, the Dinarids, the Carpathians, and the Balkan Alps have to be included (Figure 4.14). Alpine Europe, in vegetational terms and in its broadest sense including subalpine and alpine forms, covers in addition to the above-mentioned regions parts of the Black Forest and the central German heights lying approximately 1,200 to 2,400m above sea level. The following remarks will be restricted to the geographically defined Alps, although, in appropriate ways, they could be extended to the other areas mentioned too.

The Alps are topographically well subdivided according to altitude, relief (valleys, ridges, plateaux), and exposure. The Alps are influenced by three climate types or regimes (Figure 4.15): Oceanic, Mediterranean, and Continental. Some parts of the Alps channel these influences, others act as barriers. Alpine soils develop on siliceous as well as on calcareous parent rock materials. The various soil types range from acid (pH 3) to slightly above neutral (pH 8), from highly permeable, coarse textured soils in between boulders and rocks (ranker, podzols) to heavy, waterlogged, anaerobic soils (gley), from shallow, nutritionally deficient to deep soils with a high nutrient availability. This variety in topography, climate, and soil produces a small-scale pattern of different vegetational types with various types of plant associations, changing within short distances. Therefore, site characteristics become important for dendrochronological work.

DENDROCHRONOLOGICAL SAMPLING

Large areas of the Alps are covered by forests. Some forest regions are mixed with or are in close contact with pasture or agricultural land. These conditions might have caused, especially in ancient times, some disturbance in tree growth or forest development by livestock activity. The population density in the alpine region nevertheless has been low, especially in ancient times; disturbances in tree growth caused by human activity are slight.

The important living dendrochronological species are conifers. Some of them can be found over large areas, in similar or comparable site conditions. The oldest living trees range from 300 to 500 years in age. Ring widths decrease the nearer the sampling is done to the upper forest border or on sites influenced by specific stress factors. Missing rings may occur in every species, but especially in larch, where they can be caused by periodical damage by the larch bud moth, known since Roman times. In modern times, stressed growth is, in some regions, also caused by air pollution.

Owing to the traditions of the alpine population, coniferous wood of the most common species dating from ancient times can be easily found as timbers in buildings (posts, walls, roof). Owing to glacial advances, old wood is often well preserved in ice or morainic material over thousands of years. During glacial retreat, this wood is liberated and so becomes available for dendrochronological research. Chronologies of alpine species over 500 years long have to be built up by cross-dating processes (Becker & Giertz-Siebenlist, 1970). The most commonly used species in the Alps, with references to some dendrochronologists and publications, are given in Table 4.11. For detailed information see the literature or the International Tree-Ring Data Bank lists.

Figure 4.14: Tertiary fold belts, indicating areas of Alps, Pyrenees (P), Apennines (A), Dinarids (D), Carpathians (C), and Balkan Alps (B). From Bär (1977).

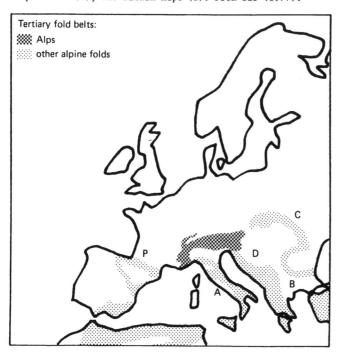

Figure 4.15: Climate regimes significant for Alpine Europe. Diagrams for each shown climate regime indicate temperature curve per annum (T in degrees Celsius) as well as precipitation distribution per annum (N in centimetres). From Bär (1977).

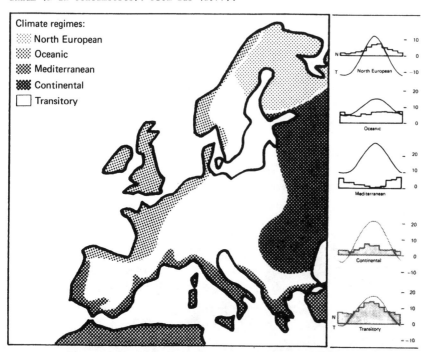

METHODS AND AIMS

Most workers started with ring-width measurements, building up chronologies in order to date samples. The validity of chronologies over large areas has been proven (for example, fir and oak by different authors). Ring width acts during the vegetational period as an integrating or accumulating parameter of exogeneous factors. Subalpine and alpine sites offer a wide spectrum of mixed ecological influences on tree-ring growth. More reliable results for specifying one exogeneous influence on growth can be achieved by selecting special site conditions where one site factor limits growth:

 a) on the northern alpine forest border, where prec-
    ipitation is sufficient, low temperatures normally
    act as the minimising factor; and

 b) in central alpine valleys within the orographic rain
    shadow, warmth and lack of precipitation minimise
    growth.

Another aim is to split ring-width measurements into parameters, which integrate less during the growth period. Several scientists did so in the 1960s using latewood width and found better correlations to exogeneous factors (for example, temperature). Since 1970, remarkable advances in the interpretation of exogeneous influences on growth patterns have been achieved by considering wood-density distributions within the annual rings (densitometry). Comparison of density and width parameters gives some insight into interdependencies and the physiological processes during the growth period. Observations in special cases show that cell wall thickening (density) of the final zone in the year might be subject to some alterations until the start of the earlywood growth in the following year (confirmed by F.Serre, personal communication, 8.7.1980).

With appropriate site selection, modelling and/or reconstruction of climatic influences are possible (Schweingruber et al., 1978b). All of the alpine areas mentioned above might be valuable for these kinds of dendroclimatic research. Smoothed annual sensitivity values over large periods might indicate changes between cold/rough/continental and warm/moderate/oceanic phases. In order to place such regional studies on a sound basis, age-trend elimination processes have to be standardised. Methods differ between width and density parameters. While the age trend in widths follows, in normal conditions, a negative exponential curve, densities show patterns of polynomial type (cubic). Work currently in progress in our laboratory suggests that variance within age of trees changes for widths while it stays nearly stable for densities. It may well be that site comparisons (or compilation of chronologies by cross-dating with short samples of unknown site origin) need the introduction of standardised growth levels at definite cambial ages (for example, the cambial year 50).

TATRA MOUNTAINS

Z.Bednarz

GEOGRAPHY AND CLIMATE

The Tatra Mountains are the only mountain massif in the Carpathian range of a typical Alpine character. Their area is about 950km^2. The highest peak is the Gerlach reaching an altitude of 2,663m. Nature in the Tatra Mountains is protected in National Parks (there are both Polish and Czechoslovakian Tatra National Parks). The Tatra Mountains are divided into three parts: the eastern and western lower limestone areas and the central granitic

Table 4.11: Commonly used species in the Alps.

ALPS	WESTERN PART					EASTERN PART
COUNTRY	FRANCE	SWITZERLAND	ITALY	GERMANY	AUSTRIA	POLAND
SPECIES	larch spruce pine*	spruce pine** larch fir	pine fir	larch pine spruce fir	larch spruce	pine** spruce fir larch
OLDEST PIECES	before 1000 AD	older than 10,000 BP, floating	recent	10,000 BP, floating	before 1000 AD	?
PUBLICATIONS	Serre et al. (1964)*** Thiercelin (1970)*** Huber (1976)*** Serre (1978a) Tessier (1978)	LaMarche & Fritts (1971a) Schweingruber et al. (1978b)*** Rothlisberger (1976)*** Kaiser (1979)		Brehme (1951) v.Jazewitsch (1961)*** Klemmer (1969) Becker & Giertz-Siebenlist (1970) v.d.Kall (1978)***		Bednarz (1978) Lamprecht (1978)***
		Corona (1975)				

  * Pinus silvestris
 ** Pinus silvestris and Pinus cembra
*** Use of other parameters than ring width, e.g. latewood width or densities

area.

The climate of the Tatra Mountains shows many high-mountain features characteristic of the Alps, among others it is characterised by frequent winter temperature inversions and winds of föhn type (Orlicz, 1962). In contrast to the Alps, no glaciers occur in the Tatra Mountains; there are only patches of permanent snow. The meteorological station at Zakopane provides the longest records of meteorological observations in the northern Polish Tatra Mountains, which have been carried out since 1911. In the southern part of the Tatra Mountains, the stations at Stary Smokowiec and Liptovsky Hradek possess data since 1911 and 1881, respectively.

FOREST COMMUNITIES

Within the whole territory of the Carpathian Mountains, the Tatra Mountains are characterised by the most distinctly zoned system of forest vegetation (Fabijanowski, 1962) which, going upwards, comprise:

   a) the lower forest zone (900 to 1,200m above sea level);

   b) the upper forest zone (1,200 to 1,550m above sea level); and

   c) the dwarf pine zone (1,550 to 1,850m above sea level).

The lower forest zone is covered with communities of Carpathian beech, Fagetum carpaticum, forest, montane acid beech, Luzulo-Fagetum, forest, and montane fir, Abietetum, forest. There appears also a Carpathian alder, Alnetum incane, wood, and relict Scots pine, Vario-Pinetum, wood. The upper forest zone is formed by the Carpathian spruce, Piceetum tatricum, and at the timber-line, by a relict stone pine-spruce, Cembro-Piceetum, wood. It is analogous to the Alpine stone pine-European larch, Larici-Cembretum, forest. In these two zones, montane sycamore maple, Phyllitido-Aceretum, patches can also be encountered. The dwarf pine is represented by dwarf pine, Mughetum carpaticum, groves.

MAIN TREE SPECIES FOR DENDROCHRONOLOGICAL INVESTIGATIONS

Stone pine. Stone pine, Pinus cembra L., occupies in the Tatra Mountains, as in the Alps and other European mountains, extremely continental localities in the zone of the timber-line. The oldest stone pine growing in the Tatra Mountains does not seem to exceed an age of 400 years. The stone pine is a species of great value both for Tatra and European dendroclimatology because of its longevity, resistance to pests and fungi, rare seed years, regular rhythm of wood formation, and low genetic variability. Apart from the localities in the Alps and in the Tatra Mountains, its isolated localities in the East Carpathian Mountains (Figure 4.16) should also be taken into consideration in future dendroclimatological investigations. A chronology of stone pine from the Tatra Mountains extending back to 1740 AD (Bednarz, 1976) has been handed over to the International Tree-Ring Data Bank.

Spruce. Spruce, Picea excelsa Lam. Lk., in the Tatra Mountains forms its own separate upper forest zone. It is the most abundant tree in the Tatra Mountains. The tree-ring chronology, developed for that species, for the years from 1765 to 1965 (Feliksik, 1972), is in the International Tree-Ring Data Bank. In future dendro-chronological investigations on spruce, use may be made of the spruce wood artefacts frequently encountered in the Tatra Mountains.

Dwarf pine. Dwarf pine, Pinus montana Mill., prevails above the upper forest zone. Three tree-ring chronologies covering the years 1870 to 1976, 1874 to 1976, and 1871 to 1976 (Table 4.12) have been completed.

A comparison of tree-ring chronologies of dwarf pine, spruce, and stone pine shows that there is a great similarity, especially for signature years. There is thus a possibility of cross-dating wood between these three species.

European larch. The native European larch, Larix decidua Mill., occurs most frequently on the southern side of the Tatra Mountains in the upper forest zone. On the northern side of the massif, in the Polish Tatra Mountains, it

Figure 4.16: Distribution of stone pine in Europe. 1. Alps, 2. Tatra Mts., 3. East Carpathian Mts.

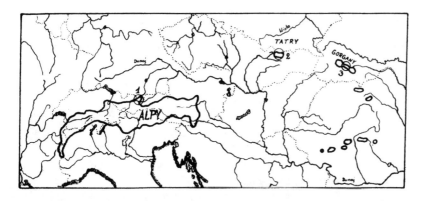

occurs very rarely. Tree-ring chronologies of the native Tatra European larch have recently been developed and include the years 1820 to 1969, 1840 to 1977, and 1806 to 1977 (Table 4.12). They differ markedly from the chronology of European larch of foreign origin, introduced into the tree stands of the Carpathian beech forest in the Strazyska Valley. Sacred objects built of European larch timber, which are numerous in the sub-Tatra region, give hope for interesting dendrochronological studies on that species.

Scotch pine. In the process of the historical development of forests in the Tatra Mountains, the pine tree, _Pinus sylvestris_ L., characterised by high light requirements, had to retreat before spruce and other tree species. The tree-ring chronology developed for this species from the relict locality on Skałka above Łysa Polana includes the years from 1820 to 1975.

Fir. A tree-ring chronology of fir, _Abies alba_ Mill., in the Tatra Mountains and in other areas of Poland is being developed. At present, material has been collected and measured mostly from nature reserves, and from artificially formed tree stands beyond the natural range of the species (Feliksik, personal communication, 10.5.1980).

European beech. The Tatra tree-ring chronology of this species includes the years 1820 to 1977. Since the European beech, _Fagus silvatica_ L., as a sub-Atlantic tree, is a typical autoecological antitype of the stone pine, interesting comparisons of these species can be expected.

Sycamore. The longest chronology of sycamore, _Acer pseudoplatanus_ L., reaching back to 1820 is for sycamore from the lower forest zone in the valley Dolina Bialego. This chronology shows great similarity with the mean growth curve of sycamore from the upper forest zone from a site in the valley Dolina Roztoki.

Carpathian birch. A tree-ring chronology for the short-lived, light-requiring Carpathian birch, _Betula carpatica_ W.K., was prepared for the years from 1890 to 1977 for the upper forest zone in the Slovakian Tatra Mountains.

DENDROCHRONOLOGICAL INVESTIGATIONS

Present dendrochronological investigations in the Tatra Mountains have two aims. First, the preparation of a tree-ring chronology dating back as far as possible in time, based on the material collected from growing trees and, second, the determination of the geographical range

Table 4.12: Tree chronologies from the Tatra Mountains.

SPECIES	LOCALITY	AGE	AUTHOR
	Polish Tatra Mts.		
Fagus silvatica L.	Dolina Bialego	1820–1977	Bednarz, unpubl.
Acer pseudoplatanus L.	Dolina Strazyska	1841–1977	Bednarz, unpubl.
	Dolina Roztoki	1870–1969	Bednarz, unpubl.
Pinus cembra L.	Dolina Sucha Kasprowa	1740–1965	Bednarz (1976)
	Zabie n.Morskim Okiem	1745–1969	Bednarz (1976)
	Szczoty Woloszynskie	1750–1969	Bednarz (1976)
Pinus montana Mil.	Dolina Suche Kasprowa	1871–1976	Bednarz, unpubl.
	Hala Gasienicowa	1874–1975	Bednarz, unpubl.
	Slovakian Tatra Mts.		
Pinus montana Mill.	Lomnicki	1870–1976	Bednarz, unpubl.
	Polish Tatra Mts.		
Larix decidua Mill.	Dolina Strazyska	1825–1977	Bednarz, unpubl.
	Dolina Roztoki	1820–1969	Bednarz, unpubl.
	Slovakian Tatra Mts.		
Larix decidua Mill.	Dolina Valicka	1840–1977	Bednarz, unpubl.
	Slavkovsky	1806–1977	Bednarz, unpubl.
	Polish Tatra Mts.		
Pinus silvestris L.	Skalka nad Lysa Polana	1820–1975	Bednarz, unpubl.
	Slovakian Tatra Mts.		
Betula carpatica W.K.	Slavkovsky	1890–1977	Bednarz, unpubl.
	Polish Tatra Mts.		
Picea excelsa (Lam.) Lk.	Hala Gasienicowa	1765–1965	Feliksik (1972)

of similarity of tree-growth curves in Poland and outside its borders. It should be expected that in consequence of the geographical situation of Poland in the central part of Europe, the tree growth curves will refer both to east and west European chronologies; this has found confirmation in investigations of stone pine (Bednarz, 1975). This fact emphasises their importance from the point of view of possible attempts at development of all-European chronologies of tree rings.

DENDROCLIMATOLOGY

The present state of dendroclimatological invest-igations in Poland, with special focus on the trees of the Tatra Mountains, was reported at the International Botanical Congress in Leningrad (Feliksik, 1975). Among the investigations carried out in recent years, the work on stone pine deserves special attention. It showed a decisive influence of air temperature in the summer months on the width of the annual increment. The relation between the width of the rings and mean air temperature of June and July in the period from 1911 to 1965 is characterised by an 80% coefficient of agreement and by a correlation coefficient of 0.77 (+0.05) (Figures 4.17 and 4.18).

Making use of the relation between the mean air temperature of the months June and July, x, and the width of annual increments of the stone pine rings, y, deter-mined by the linear regression equation, x = 8.13 + 0.20y, the approximate course of temperature variability of June and July in the Polish Tatra Mountains was reconstructed for the years 1740 to 1910 (Figure 4.19). The shape of the

curve presenting this variability proves the existence of marked thermal depressions which occurred, among others, in the years 1800 to 1850. This phenomenon, known as The Little Ice Age, concerned not only the Tatra Mountains but was of a much greater extent (LaMarche & Fritts, 1971a; Lamb, 1972, 1977; Serre, 1978a).

Figure 4.18: Relation between growth rings of stone pine, y, from Sucha Kasprowa Valley in the Tatra Mts. and mean temperatures of air from June to July, x, expressed by a linear regression for the period 1911-65.

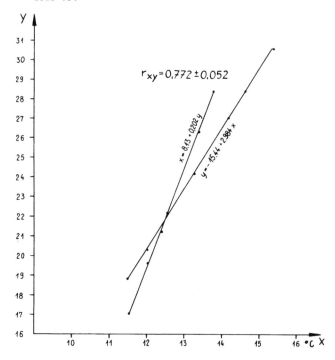

Figure 4.17: Comparison of mean curve of annual rings of stone pine, A, with curves for mean temperature of air in the months June-July, B, mean maximum temperature of air in the months June-July, C, and sums of precipitation from May to August, D.

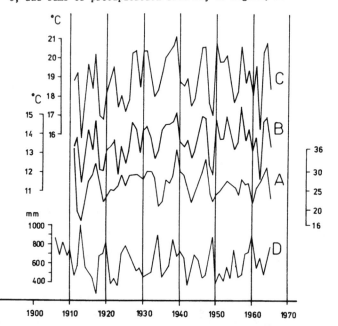

Figure 4.19: Hypothetical course of variability of the mean air temperature of June-July, t, for Zakopane, 1740-1910, reconstructed on the basis of the analysis of annual rings of stone pine from the Sucha Kasprowa Valley in the Tatra Mts.

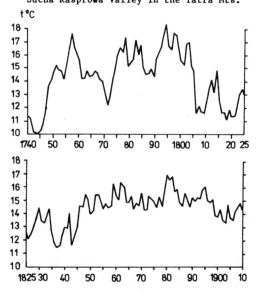

The Little Ice Age has been associated with increased levels of explosive volcanic activity (Lamb, 1970). Great explosive volcanic eruptions emit great quantities of dust and aerosol into the atmosphere which cause a decrease in the strength of direct solar radiation by as much as 20%, as observed in 1912 after the eruption of the volcano Katmai in Alaska (Budyko, 1971; Lamb, 1970). It is believed that, as the solar radiation is reduced, air temperature is considerably lowered. Since in high mountain localities air temperature is the limiting factor to annual growth, a relation may be expected between the periods of increased volcanic activity and the width of rings in trees. In this context, apart from the period 1800 to 1850, attention is drawn to especially deep growth depressions in the year 1913, as well as to marked increment decreases in the years 1902 to 1903, both in the stone pine in the Tatra Mountains and the Alps as well as in the eastern Carpathian Mountains (Bednarz, 1975). The large geographical range of this phenomenon and the fact of its occurrence also in other species of Tatra trees, for example in spruce and sycamore (Figure 4.20), may be, for example, associated with a rapid decrease in the strength of direct solar radiation in consequence of great volcanic eruptions in the year 1902 (Mount Pelee, Santa Maria, Soufriere) and in 1912 (Katmai) This phenomenon deserves special attention with regard to the possibility of making use of tree rings of high-montane trees in studying changes in the strength of solar radiation in the past and their relationship with volcanic activity.

Figure 4.20: The great volcanic eruptions, average monthly values of strength of the direct solar radiation and tree ring chronologies. (a) average monthly values of the strength of the direct solar radiation derived from observations at mountain observatories between 30 and 60°N in America, Europe, Africa, and India, as percentage of the overall mean, from Lamb (1972). Tree-ring chronologies: (b) Pinus cembra L., Tatra Mts.; (c) Pinus cembra L., Tatra Mts.; (d) Pinus cembra L., Eastern Carpathian Mts.; (e) Pinus cembra L., Bavarian Alps; (f) Picea excelsa Lam.Lk., Tatra Mts.; (g) Acer pseudoplatnaus L., Tatra Mts.

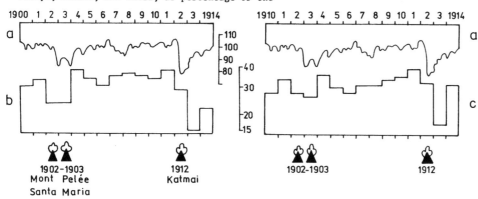

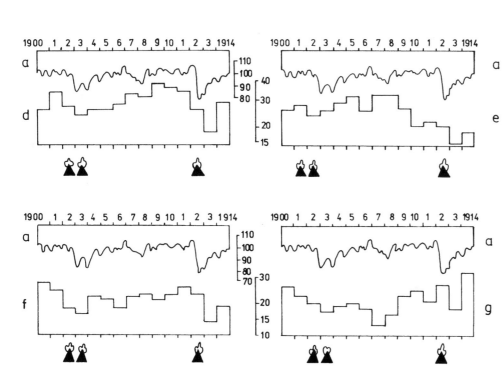

EUROPE

D.Eckstein

INTRODUCTION

Dendrochronological work in Europe started somewhat sporadically in the early decades of this century, mainly in Scandinavia, Russia, and Germany. In the early 1940s, B.Huber in Germany introduced the dendrochronological method on a broad basis and in a systematic way and thus gave an essential impact to its diffusion throughout Europe (Liese, 1978). On the map of Europe (Figure 4.21), the area referred to in the text is delineated. In addition, the current dendrochronological laboratories are shown. It is also recorded for which regions tree-ring chronologies have been established. The temporal dimensions of the chronologies as well as the tree species are listed in Tables 4.13 and 4.14 together with the source of the information. Although it is intended to review the existing geographical and temporal extent of appropriate tree-ring collections systematically, not all studies on the subject in question are mentioned. This is due in part to a necessarily subjective selection and partly due to lack of information in some cases.

Until recently the application of dendrochronology in Europe has been concentrated on the dating of cultural objects and geomorphological processes. Therefore, both the tree species and the regions for the construction of long-term chronologies have been determined mainly by non-dendrochronological criteria. In the foreground of such work there was always the consideration of sufficient availability of wood of an appropriate tree species from the present time continuously back to historic and prehistoric periods. At all times and places, this wood belonged to that species which had its natural range in the region in question over the millennia and provided people with high quality building timber. In the following, the main dendrochronological tree species are considered in more detail.

OAK CHRONOLOGIES

In the climate of central and western Europe, which is not exceptionally extreme either in winter or in summer, forests characterised by summer-green deciduous trees have formed naturally. Their northern and eastern limits of distribution are determined by the increasing coldness of winter and spring and through decreasing length of the vegetative period.

In the whole of the European deciduous forest zone, dendrochronology is based on oak, which occurs here as two botanically distinct species: Quercus petraea Liebl. (sessile oak) and Quercus robur L. (pedunculate oak). The two species cannot be clearly distinguished from each other by means of their wood anatomy. According to Bonnemann & Rohrig (1971), the natural range of the pedunculate oak extends from Scotland to south Scandinavia through central Russia nearly as far as the Urals. From there the eastern limit skips the Russian steppes and comes to the Caucasus Mountains and to Asia Minor. The southern limit takes its course via the Balkans to Italy and northern Spain, from where the western limit runs along the coast of the Atlantic Ocean including the British Isles. The natural range of the sessile oak is somewhat smaller. In the northeast it reaches neither Finland nor Russia. Both tree species avoid high mountains. Today typical areas of sessile oak are, for example, central France and some south German mountains of medium altitude like the Spessart. Pedunculate oaks typically occur along central European streams. Therefore, all dendrochronological laboratories within the range of distribution of the European oak have developed primarily oak tree-ring chronologies.

The first long oak chronology in Europe was initiated for southern Germany by Huber and co-workers. The ring sequence starts with exceptionally old living trees from the Spessart Mountains with more than 600 growth rings and contains further ring series from building timbers. They are from numerous objects from a relatively narrow area of provenances, including the medium altitude mountains of the Spessart, Odenwald, and Vogelsberg, that is, an area north, east, and south of Frankfurt. The chronology goes back to 832 AD (Huber & Giertz-Siebenlist, 1969). This South-German master chronology has been replicated by the West-German master chronology and extended to the 8th century BC by Hollstein (1980). In this chronology, numerous tree-ring series are incorporated from archaeological finds and art-historical remains of a large number of localities from both sides of the Rhine river from north Switzerland, including the southwestern part of Germany, eastern Belgium, Luxemburg and eastern France up to the eastern Netherlands. The validity of both chronologies as a basis for dendrochronological dating has been proven for the southern German area and also for the total range between Czechoslovakia in the east and Normandy and even southeast England in the west. They also serve as a tool for the dating of oak wood in northern Switzerland and even Italy.

Both chronologies have been the starting point for the work of Becker & Delorme (1978), who are trying to construct a continuous tree-ring chronology for the whole Holocene period based on subfossil oaks embedded in gravel terraces of the rivers Danube, Rhine, Main, Weser, and some of their tributaries. These oaks are excavated during dredging. At the present time, this chronology exists in the form of several floating chronologies some of which exceed 2,000 years in length (Becker, 1979). For the Bronze and Neolithic Age, Becker has been successful in establishing a 2,342-year floating master chronology from subfossil oaks, dated by radiocarbon to the period from 1720 to 4050 BC. Furthermore, there are at least three floating postglacial oak chronologies to be mentioned, containing 707, 917, and 399 tree rings, respectively,

altogether covering the period between 6400 and 8550 BP.
In summary, the Central-European oak chronology covers the
past 8,500 years, with several gaps still existing above
all in the range of 700 to 800 BC and some more in earlier
times. This chronology is based on wood samples from
subfossil oaks as well as from archaeological, architectural, and art-historical objects from a wide regional
area of provenances and has been developed by B.Becker
(Botanisches Institut, Universität Stuttgart-Hohenheim)
and A.Delorme (Institut für Forstbenutzung, Universität
Göttingen). They also incorporated already existing
tree-ring series by B.Huber and co-workers (Institut für
Forstbotanik, Universität München), E.Hollstein
(Rheinisches Landesmuseum, Trier), G.-N.Lambert, H.Egger,
and C.Orcel (Musée Cantonal d'Archéologie, Neuchatel), and
B.Schmidt (Institut für Ur- und Frühgeschichte,
Universität Köln).

Since the ring widths of a tree species do not as a
rule show parallel variations all over the range of its
natural distribution, it is necessary to build several
chronologies for the same species for different regions.
An oak chronology of comparable length to that central
Europe is under construction in the north of Ireland
(Pilcher et al., 1977), where oak timber is continuously
preserved in bogs and occasionally also in river gravels
and lake beds. The longest part is a 2,990-year long
master chronology made up from more than 10 different site
chronologies from all over north eastern Ireland and
covering the approximate period from 1000 to 4000 BC.
Some further floating chronologies are placed in time
according to radiocarbon dating, including a sequence of
1550 years approximately 4000 to 5500 BC (Pilcher,
personal communication, 8.7.1980). The chronology of this
region was absolutely dated back to 1001 AD (Baillie,
1977a) and has recently been extended to 12 BC (Baillie
1980a,b). In summary, in the north of Ireland,
chronologies for the last 8,000 years of the post-glacial
period exist, with only a few hundred years remaining to
be bridged for completion. Whether their ring patterns are
also valid for a wider area, at least for the British
Isles, has yet to be shown.

In the British Isles, besides the laboratory in
Belfast, further tree-ring research centres have been
established in the last few years leading to a large
collection of tree-ring material, either absolutely dated
or floating. It is possible to mention only some of them.
For example, a 1,030-year oak chronology was built for
southern central Scotland (Baillie, 1977b) which started
with more than 400-year old living trees and shows a good
cross-agreement to other chronologies of the British
Isles. There are Viking and mediaeval chronologies for
Dublin spanning 855 to 1306 and 1357 to 1556 AD (Baillie,
1977c). In addition to these composite dating chronologies, there are living tree chronologies for northern
Wales (Leggett et al., 1978), Winchester (Barefoot, 1975),

and for southern and eastern England (Fletcher, 1977).
There are also eight index chronologies for Ireland
(Pilcher & Baillie, 1980a) and six index chronologies for
Scotland and England (Pilcher & Baillie, 1980b).

Besides the promising efforts in Stuttgart and
Belfast, leading to oak chronologies for the whole
Holocene period, comparable studies have been carried out
in northwest Germany (Schmidt, 1977a). The necessity for
the building of independent oak chronologies in northern
Germany separately from the existing master curves of
other regions was already recognised some years ago (Bauch
et al., 1967). It has been proven that the long-distance
correlation of the tree-ring variations of southern
Germany ends at the transition from the hilly region to
the northern German lowlands. The material in northwest
Germany is provided by subfossil oaks which are excavated
during the extraction of gravel. The main sites are the
Rhine valley between Bonn and Xanten, the middle part of
the river Weser and recently also the coastal region of
the Baltic Sea of Schleswig-Holstein (Aniol, personal
communication, 8.7.1980). The derived floating chronologies cover mainly the period between 2000 and 7000 BC,
but are concentrated on certain periods rather than being
equally distributed and still contain numerous gaps. For
the same areas, a grid of modern chronologies from 186 oak
trees up to 200 years old has been established in order to
perform preliminary dendroclimatological studies
(Eckstein & Schmidt, 1974; Schmidt, 1977b).

From Denmark in the north to the Netherlands in the
west, six long-term chronologies for narrow regions along
the North Sea coast have been constructed (Eckstein,
1978). These exhibit quite good agreement with each other
and contain at least 800 years each. The material is
provided by architectural excavations and archaeological
objects. The regional chronology for Schleswig-Holstein is
the longest and is coherent back to 436 AD. The early
mediaeval part of this has been verified several hundred
times with oak samples from five prehistoric locations.
Further to the south there are chronologies for the area
of Hamburg back to 1080 AD and for the coastal part of
Lower Saxony back to 1062 AD. For the Netherlands, it has
proven necessary to establish two different chronologies
(Eckstein et al., 1975), one mainly made up with building
timber going back now to 1036 AD and another one made up
from the oak panels of paintings covering the period from
1109 to 1637 AD (Bauch, 1978). An extension to present
times will probably not be possible. Tree-ring variations
in this chronology are extremely different from the first
one. For the southern Weser and Leine upland in the south
of Lower Saxony in Germany, a long-term chronology has
been established using more than 500 samples from recent
and historic origin going back to 1004 AD (Delorme, 1973).

The chronology for Denmark starts with modern oak
trees provided by Bartholin (1973) and has been extended
back to 614 AD by incorporating the tree-ring series of
the timbers from late mediaeval buildings and Viking

settlements (Eckstein & Christiansen, unpublished data). Additionally, there is a more than 400-year long floating curve from the Bronze Age built with the ring series of 12 oak coffins (Eckstein & Schmidt, unpublished data). Bartholin (1975) has also worked on the dendrochronology of oak in southern Sweden. Samples came from 400-year old living trees, churches, castles, and other buildings, as well as from some archaeological excavations. Today, this chronology extends back to 578 AD. Considering the North-German, Danish and Swedish oak chronologies together, it seems that, around the western part of the Baltic Sea, a tree-ring pattern may exist with a fairly good uniformity over a long distance.

With close reference to the already mentioned long oak chronologies from central and western Europe, additional local chronologies have been established for dating purposes, for example, for the Zurich region from 1551 to 999 AD (Schweingruber & Ruoff, 1979), for the interior of Lower Saxony from 1546 to 987 AD (Bauch & Eckstein, 1978), and for western Sweden back to 831 AD (Brathen, 1978). For the territory of the German Democratic Republic, however, it seems to be necessary to construct a special oak chronology, or even two, although some dating has been done based on the South-German master curve. At present there are several floating sections back to the early Mediaeval period, whereas the absolutely dated sequence covers the time back to 1424 AD (Jährig, 1976).

In the eastern part of the natural range of oak trees, long-term chronologies have been built by Russian dendrochronologists. However, it appears that they are less important for dating purposes than the conifer chronologies of the same area. The Russian oak chronologies probably contain a suitable amount of climate information because the growing conditions may become increasingly marginal for oak trees towards the eastern boundary of their distribution. Details are given by Bitvinskas (this volume).

CHRONOLOGIES FOR DECIDUOUS TREE SPECIES OTHER THAN OAK

In central and western Europe, other deciduous tree species besides oak have also been studied, for example beech, elm, ash, lime, alder, and maple. For most of these

Figure 4.21: Map of Europe with the region under consideration. The dots indicate tree-ring laboratories; the encircled areas, including the numbers, refer to Table 4.13 and symbolise regional tree-ring chronologies.

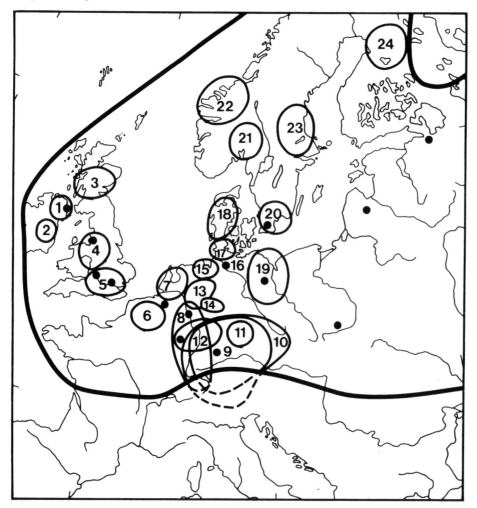

it appears impossible to build a long-term chronology, because they have only been used as a building timber when oak wood was not available in sufficient amounts. Until now it has only been possible to establish a noteworthy tree-ring series for beech, Fagus sylvatica L. Its range of distribution corresponds mostly to that of sessile oak, except that at its western limit beech is more restricted than oak and, for example, does not occur in the coastal region of northwest Germany (Bonnemann & Rohrig, 1971). The beech chronology contains samples from living trees and building timber from objects in southern Germany, mainly around the river Moselle, and goes back to 1320 AD (Hollstein, 1973). Moreover, there is a more than 20-fold

replicated 200-year long floating chronology for the 15th and 16th centuries for southern Germany derived from the tree-ring series of beech wood panels of the paintings of the painter Lucas Cranach (Klein, 1979). Between the South-German chronologies of beech, oak, and even the conifer species fir and spruce, there are remarkable similarities. It may be possible that already existing (or future) floating chronologies of other tree species can be dated absolutely by reference to such chronologies without extending them to the present time. Such dated tree-ring sequences may also be of climatological interest. This may be valid for the chronology of mediaeval elms of London; Brett (1978) has already shown that modern elm trees have

Table 4.13: European tree-ring chronologies: p.t. = present time, oak = Quercus petraea (Matt.) Liebl. + Q. robur L., beech = Fagus sylvatica L., pine = Pinus sylvestris L., spruce = Picea abies Karst., fir = Abies alba Miller.

REGION	SPECIES	REFERENCE NUMBER IN FIG 4.21	PERIOD	SOURCE
NE Ireland	oak	1	p.t. to 12 BC Three parts: 200 BC to 950 BC, 1000 BC to 4000 BC, and 4000 BC to 5500 BC.	Baillie (1977a, 1980a) Pilcher et al. (1977)
Ireland: Dublin region	oak	2	1556 to 1357 1306 to 855	Baillie (1977c)
S Central Scotland	oak	3	p.t. to 946	Baillie (1977b)
N Wales	oak	4	p.t. to 1710	Leggett et al. (1978)
W England	oak		1636 to 1350	Giertz-Siebenlist (pers. comm.)
S & E England	oak	5	1609 to 1120 1193 to 780	Fletcher (1977)
England: Winchester	oak		p.t. to 1635	Barefoot (1975)
France: Normandy	oak	6	1610 to 1280	Giertz-Siebenlist
Netherlands	oak	7	p.t.to 1036 AD 1637 to 1109 AD	Eckstein et al. (1975)
W Germany	oak	8	p.t. to 717 BC	Hollstein (1980)
S Germany	oak	9	p.t. to 832	Huber & Giertz-Siebenlist (1969)
S Germany	oak		p.t. to 690 BC	Becker & Delorme (1978)
			830 BC to 1470 BC 1720 BC to 4050(4850)BC*, three parts from around 6400 BC to 8550 BC	Becker (1979)
S Germany	fir	10	p.t. to 820	Becker & Giertz-Siebenlist (1970)
S Germany	spruce	11	p.t. to 1250	Wieser & Becker (1975)
S Germany	beech	12	p.t. to 1320	Hollstein (1973)
NW Germany	oak	13	various parts between 2000 and 7000 BC	Schmidt (1977a)
Germany: Weser- Leine upland	oak	14	p.t. to 1004 AD	Delorme (1973)
N Germany: Lower Saxony	oak	15	p.t. to 1082 ad	Eckstein et al. (1979)
N Germany: Hamburg region	oak	16	p.t. to 1080	Eckstein et al. (1972)
N Germany: Schleswig-Holstein	oak	17	p.t. to 436 AD	Eckstein (1976)
Denmark	oak	18	p.t. to 614 AD	Bartholin (1973) Eckstein & Christiansen (unpubl.)
Germany (GDR)	oak	19	p.t. to 1424 AD	Jahrig (1976)
S Sweden	oak	20	p.t. to 578 AD	Bartholin (1975)
SE Norway	pine	21	1954 to 1383	Ording (1941) Eidem (1959) Aanstad (1960)
Central Norway	spruce	22	1937 to 1460 AD	Eidem (1959)
Sweden	pine	23	p.t. to 1530 AD	Jonnson (unpubl.)
N Finland	pine	24	1960 to 1181 AD	Siren (1961)

a value as a record of summer rainfall. Similar considerations hold for the maple from the Tatra Mountains in Poland where Bednarz (this volume) has found a good correlation between ring width and summer temperature and radiation. In general, such chronologies will have a primarily local significance.

CHRONOLOGIES FOR CONIFEROUS TREE SPECIES

In the north and east of Europe, coniferous species are able to compete successfully against deciduous trees because of decreasing temperature, as well as decreasing length of the vegetative period. Thus, at the northern and eastern limit of the European hardwood region, a forest type with both conifers and deciduous trees has formed which finally turns into a pure coniferous forest. The same alteration occurs within the central European deciduous tree zone with increasing altitude. Typical examples are the Sudetic Mountains and the Erzgebirge in the east and the Black Forest and the Vosges in the west. In the central and northern European coniferous forests, fir, spruce, and pine are the main species to be considered.

The initial material for the fir chronology for southern Germany was from modern firs, Abies alba Miller,

from the Beskides, the Bavarian Alps, and the Bavarian Forest, the tree-ring series of which go back to 1541 AD. By means of building timber from churches, monasteries, and other architectural objects from more than 20 sites in southern Germany, northern Switzerland, and the Alps, from both north and south of the Brenner Pass, extension has been achieved back to 820 AD (Becker & Giertz-Siebenlist, 1970). The geographical range of this pattern, that is, the distance within which the tree-ring variations run parallel, extends over the whole range of distribution of the fir. Thus, the fir sequence can be used for the dating of timber all over central Europe. The firs in the south Tirol in Italy also exhibit a remarkable similarity in their ring-width pattern to the curve for southern Germany. It even serves sometimes for the dating of northern Italian spruce timber (Corona, 1979), because a suitable spruce chronology does not exist there.

It is difficult to imagine that spruce, Picea abies Karst., in former times lived restricted to relatively small areas with characteristic site conditions: for example, in central Europe, in the Alps, and in some mountains of medium altitude in southern Germany. Spruce has reached its present wide distribution only in the last 200 years, mainly owing to the influence of Man. The

Table 4.14: Chronologies from Europe outside the Alpine and Mediterranean regions collected for, or in a form suitable for, dendroclimatology, prepared from a) International Tree-Ring Data Bank list and b) recent additions. * Ring-width and density chronologies.

SUBMITTOR	SITE NAME	COUNTRY	SOURCE	ELEV.	LAT.	LONG.	SPECIES	YEARS
a) I.T.R.D.B.								
B.Schmidt	Koln Abt.17	FDR	1	131	50°55′N	07°09′E	QURO	197
	Barkhausen	FDR	1	43	52°14′N	08°53′E	QURO	126
	Hessisch-Oldendorf	FDR	1	291	52°11′N	09°17′E	QURO	129
	Hesslingen	FDR	1	210	52°07′N	09°12′E	QURO	132
	Varenholz	FDR	1	95	52°10′N	08°58′E	QURO	186
	Schaumburg	FDR	1	50	52°19′N	09°02′E	QURO	154
B.Becker	Bayerischer Wald	FDR	1	800-1100	48°45′N	13°00′E	FASY	299
J.Pilcher	Rostrevor	Northern Ireland	1	60	55°05′N	06°11′W	QUPE	226
D.Eckstein	Oldenburg	FDR	162	5-15	53°00′N	08°00′E	QURO	124
	Schleswig	FDR	162	20-40	54°30′N	09°30′E	QURO	170
B.Becker	Beskid Mts.	Czechoslovakia	1	600-920	49°50′N	18°20′E	ABAL	243
	Beskid Mts.	Czechoslovakia	1	650	49°50′N	18°20′E	FASY	284
M.Baillie	Raehills	Scotland	1	170-200	55°20′N	03°30′W	QU	152
B.Becker	Rohrbrunn	FDR	1	400-500	49°53′N	09°24′E	FASY	279
	Rothenkirchen	FDR	1	500-600	50°15′N	10°15′E	PCEX	126
	Schussbach	FDR	1	410	49°29′N	10°35′E	PCEX	132
Tree-Ring Group, Liverpool Polytechnic	Maentwrog	Wales	1 & 5	50-100	52°56′N	03°54′W	QUPE	265
R.Morgan	Towy Valley, Dyfed	Wales	1	30	51°52′N	04°10′W	QURO	162
	Castle Howard, York	England	1	65	54°07′N	00°55′W	QUPE QURO	265
	Padley Wood Sheffield	England	1	300	53°19′N	01°37′W	QUPE	103
b) Recent additions								
J. Pilcher & M. Baillie	Ardara	Ireland	1	30	54°45′N	08°23′W	QUPE	129
	Killarney	Ireland	1	30	52°00′N	09°33′W	QUPE	170

generation of a spruce chronology in central Europe has
been somewhat delayed, mainly because of insufficient
existing or known historical material. It was only a few
years ago that Wieser & Becker (1975) succeeded in
producing such a master curve made up of the timber of the
famous timber-framed houses in Mittelfranken (Franconia)
in southern Germany, which goes back to 1250 AD.
Surprisingly this sequence reveals a high cross-agreement
with the South-German oak chronology. Furthermore, there
is a spruce chronology for central Norway back to 1460 AD
(Eidem, 1959), showing a good agreement with north German
spruce trees. According to some dendroclimatological
analyses with spruce from the Tatra Mountains (Feliksik,
1972), this tree species can be assumed to be well suited
for palaeoclimatological research.

Of all European tree species, pine, Pinus sylvestris
L., covers the largest area and has, at the same time, the
greatest altitudinal extent. In western Europe from
Denmark to France, however, pine is missing in the
lowlands. Whereas, in central Europe, no pine chronologies
are known, there have been noteworthy pine master series
in Scandinavia for some decades, above all in Norway.
There, in the 1940s and 1950s, different chronologies of
various lengths were generated by several authors for the

southeast and west and have already been used for dating
purposes. However, at present, no dendrochronological
laboratory is known to exist in Norway. The longest
chronology for pine has been published by Ording (1941),
Eidem (1959) and Aanstadt (1960) for the southeast and
goes back to 1383 AD. In Sweden, a pine chronology has
been generated by B.Jonnson (personal communication, 1964)
back to 1530 AD. The present situation of dendrochronology
in Finland is similar to that in Norway and Sweden. After
Siren's (1961) pine chronology back to 1181 AD for the
northern treeline, no further attempts are known to have
been made to establish regional chronologies. However, in
the aforementioned countries, comprehensive
dendroclimatological analyses in the widest sense are
performed (for example, Mikola, 1971; Jonnson, 1976;
Karenlampi, 1972). Recently ring-width and density
chronologies have been established in the Scottish
Highlands by Schweingruber et al. (1978b). Two of these
exceed 200 years in length. Further collection and
analysis in the British Isles is being carried out jointly
by M.K.Hughes (Department of Biology, Liverpool
Polytechnic) and F.H.Schweingruber (Eidgenossische
Forstliche Versuchsanstalt, Birmensdorf). Bitvinskas (this
volume) reports on the conifer chronologies of the
European U.S.S.R.

Table 4.14: Continued.

SUBMITTOR	SITE NAME	COUNTRY	SOURCE	ELEV.(m)	LAT.	LONG.	SPECIES	YEARS
	Lough Doon	Ireland	1	30	52°50′N	08°40′W	QUPE	129
	Eniscorthy	Ireland	1	30	52°50′N	06°32′W	QUPE	168
	Glen of Downs	Ireland	1	100	53°08′N	06°05′W	QUPE	170
	Cappoquin	Ireland	1	150	52°08′N	07°54′W	QUPE	166
	Glenluce	Scotland	1	20	54°53′N	04°50′W	QU	181
	Lockwood	Scotland	1	155	55°14′N	03°28′W	QU	405
	Scorton	England	1	50	53°56′N	02°45′W	QU	166
	Oxford	England	1	70	51°48′N	01°07′W	QU	198
	Blickling	England	1	40	52°49′N	01°13′E	QU	263
	Bath	England	1	45	51°22′N	02°19′W	QU	266
	Ludlow	England	1	185	52°21′N	02°44′W	QU	150
J.Pilcher	Chinon	France	1	110	47°14′N	00°22′E	QUPE	176
	Fontainebleau	France	1	136	48°27′N	02°41′E	QU	449
	Fontainebleau	France(2)	1	136	48°27′N	02°41′E	QU	146
	Guines	France	1	140	50°50′N	01°51′E	QU	152
	Chambord	France	1	100	47°34′N	01°30′E	QURO	248
	Halatte	France	1	140	49°14′N	02°34′E	QU	261
M.Hughes	Cannock	England	1 & 5	150	52°46′N	02°01′W	QU	335
P.Leggett	Peckforton	England	1	150	53°07′N	02°47′W	QU	216
M.Hughes	Coed Crafnant	Wales	1	150	52°50′N	04°04′W	QUPE	156
D.Cartwright & M.Hughes	Rassal	Scotland	1	60	57°25′N	05°34′W	FREX	157
	Benglog	Wales	1	300	52°49′N	03°46′W	FREX	159
M.Hughes & S.Milsom	Maentwrog	Wales	1	50-100	52°56′N	03°54′W	QUPE	261
F.Schweingruber	Cadnam Density	England	1*	100	50°52′N	01°41′W	PISY	134
	Drimmie	Scotland	1*	200	56°37′N	03°20′W	PISY	149
	Inverey	Scotland	1*	500	57°00′N	03°30′W	PISY	272
	Mar Lodge	Scotland	1*	400	56°59′N	03°29′W	LADE	134
	Glen Affric	Scotland	1*	300	57°17′N	05°00′W	PISY	243
F.Schweingruber, D.Cartwright & M.Hughes	Ballochbuie	Scotland	1*	380	56°59′N	03°19′W	PISY	267
	Glen Derry	Scotland	1*	460	57°01′N	03°34′W	PISY	206
	Shieldaig	Scotland	1*	10	57°30′N	05°37′W	PISY	132
	Loch Marree	Scotland	1*	100	57°38N	05°20′W	PISY	223
	Coulin	Scotland	1*	250	57°32′N	05°21′W	PISY	308

Species Key: PISY - Pinus sylvestris, FASY - Fagus sylvatica, ABAL - Abies alba, FREX - Fraxinus excelsior,
QU - Quercus sp., PCEX - Picea excelsa, QURO - Quercus robur, LADE - Larix decidua, QUPE - Quercus
petraea.

CONCLUSIONS

Dendroclimatology in Europe is determined mainly by four facts. The area under consideration is poor in tree species. The trees mostly are short-lived (200 to 300 years). The forests have been influenced by Man since prehistoric times. Only a small part of the area is covered by forests. I have tried to review systematically the long-term chronologies of various tree species existing in central, western, and northern Europe. In comparison to an earlier report (Eckstein, 1972), most of the progress in the collection of tree-ring data has been achieved in the British Isles, in southern Sweden, and Germany. The map in Figure 4.13 shows the already existing dense network of tree-ring chronologies. They go back at least to the late Mediaeval time. Recently, dendro-chronological activities have become known for Czechoslovakia (Kyncl, 1976; Velimski, 1976). Blanks on the map exist mostly in southeast Europe, where no noteworthy dendrochronological attempts have been made, not even using the existing chronologies of neighbouring regions. In such cases, it is desirable to stimulate and support dendrochronological studies.

The same is true for central and northern France, despite the available possibilities there to achieve dendrochronological results using the South- and West-German master chronologies. Five living-tree oak chronologies have recently been completed by the Belfast Laboratory (Pilcher, personal communication, 7.5.1981) for north and central France, including one of 449 years. Historic building material has also been collected. Further independent chronologies should be constructed for this area. Such work can be based on the already existing material of V.Giertz-Siebenlist covering the period from 1610 to 1280 AD and the recent collections mentioned above. Moreover, it is desirable that the old dendro-chronological tradition in Scandinavia be revitalised and continued in extending the conifer chronologies by incorporating material of historic origin. In summary, however, the area under consideration may be covered sufficiently for pure dating purposes. Of course, all chronologies are open towards the past and their temporal range may sometimes not give enough satisfaction in the opinion of the prehistorians who make use of them. But since, on the other hand, in several areas (for example, northern Germany), archaeologists have been concentrating mainly on mediaeval times, an extension of the chronologies will only be possible when older periods are included in the sphere of archaeological interest.

The fact that it was possible to build the chrono-logies shows that they contain climate information. The question is only whether the proportion of the climate in the variation of the ring widths is large enough to be extractable. In some cases, this could be proven explicitly (Feliksik, 1972; Bitvinskas, 1974; Eckstein & Schmidt, 1974; Jonnson, 1976; Schmidt, 1977b; Brett, 1978;

Hughes et al., 1978b; Schweingruber et al., 1978b). The question that now arises is how the available tree-ring data should be prepared or processed in order to be used for dendroclimatological work. It will certainly not be possible to take the existing chronologies as they are.

In addition, we have to differentiate between two possible main aims: first, the large-scale or global reconstruction of the past climate, which is mainly of climatological interest and needs a high concentration of all tree-ring data available in some few research centres, and, second, the reconstruction of local or small regional climate in the framework of the modelling of the environ-ment of prehistoric man which is asked for mainly by prehistorians. This may be done by regional tree-ring centres. These two aims do not compete with each other, but may require a different approach.

COMMENT

J.R.Pilcher

A great deal of tree-ring work has been carried out in Europe in the last 20 years, but it is clear that most of this work, at least in its present form, is unsuitable for climate studies. Almost all the existing chronologies have been constructed for dating purposes. That is to say, the criteria for inclusion or for exclusion of samples from a chronology are purely those of convenience in producing a useful dating chronology. It could be argued that as these criteria are based, albeit subjectively, on cross-dating quality, they should be the same criteria as would apply in the construction of a climate chronology. In practice this is not so for two reasons. First, it is often expedient to add to a dating chronology trees that clearly "date" but show a poor statistical correlation. They may be helpful in the early stages of chronology building, but would need to be removed if the chronology were to be used for climate studies. Second, most dating chronologies cover too large a geographic area.

All the long chronologies from Europe depend more or less on the use of timbers from buildings or other Man-made structures. The origin of these timbers may be uncertain and there may be no modern analogue of the forest from which the timbers were cut. Although the chronology may be presented as a continuum running up to the present day, can we be sure that the earlier portions represent trees growing on similar soils and at similar altitudes to the modern portion?

However severe these problems appear to be, they must be faced. Europe has the best climate data base in the world. If we are to make a useful contribution we need to be able to extend this data base to a significant extent. To further compound the problem, the living trees in Europe are, in general, not of great age. We have 600-year old oaks in the Spessart Mountains (Eckstein, this volume), 400-year old oaks in central France (Pilcher, unpublished data), southern Scotland (Baillie,

1977c), and in southern Sweden (Bartholin, 1975). Few other sites with suitable long-lived trees are known and hardly provide an adequate grid of sites for any spatial study of past climate. On the other hand there seems to be a reasonable chance of producing a grid of small-area historic chronologies with a spacing of about 800km or five degrees latitude and three to four degrees longitude Each of these area chronologies could be extended back to at least 1350 AD, and in many cases to before 1000 AD. If tight enough control could be maintained on chronology quality and on timber provenance, a grid of this type could be of real value.

If, on the other hand, one restricts the study to living trees, the period covered would be much shorter. We were hopeful of obtaining numerous sites in the British Isles extending to 1750 AD, but of the 19 modern chronologies of oak so far produced in the U.K., only nine extend significantly before 1800 AD. Four site chronologies extend back to 1750 or earlier (Hughes, personal communication, 1.4.1981). In fact a number do not extend before 1830 AD with adequate replication for climate work. A similar limitation has been found in some areas of France (Pilcher, unpublished data). However, it would probably be reasonable to envisage a grid of sites at 80 to 160km spacing (1 to 2° latitude by 1° longitude) that extends from the present to 1800 AD. Clearly this is a shorter span than that of many instrumental weather recording stations. It is, however, considerably longer than the detailed grid of instrumental records at a similar spacing to the tree-ring sites. Thus some useful extension of a detailed climate grid could be obtained if a suitable transfer function can be developed. For a project of this type very detailed verification of the reconstruction produced would be possible from the available long instrumental records.

I now look to the future and try to envisage what tree-ring data the area is capable of producing that will be of value in climate reconstruction. This is best considered on three timescales.

The short timescale (approximately 1750 to 1970 AD). As mentioned above, a close grid of modern oak tree-ring chronologies is possible with a spacing of 80 to 160km, each from a single wood or forest stand and with good replication. Existing chronologies of this type may need to be checked for climate information content and some will need to be remade with additional trees. Major gaps in this grid exist in Spain, Italy, and France, with smaller gaps in all areas. The only limitation on this work is people and money. It could be completed in as little as three years with adequate funding; a small tree-ring laboratory can produce 10 to 15 chronologies a year. Provided, at the same time, the problems of providing a suitable transfer function can be solved, this grid could yield some very detailed mapping of the recent past climate of Europe.

The medium timescale (1350 to 1970 AD, perhaps extended to 1000 AD). The limit of 1350 AD for these historical chronologies is based on evidence that a major tree-depletion phase ending at about this date with depopulation caused by plague and other factors (Baillie, 1980b). Provided care is taken in the selection of timbers for inclusion, a grid of historic building chronologies with an 1300km spacing should be possible. Extension to 1000 AD or before may be possible in some areas, but probably not for all areas. Most of these chronologies can be based on existing work, but will have to be remade with the removal of trees from outside the defined area and the addition of new trees to bring the level of replication up to the required standard. This work could be most effectively carried out by workers in their own areas sampling to general guidelines such as discussed at the Second International Workshop on Global Dendroclimatology (Hughes et al., 1980). Completion of the chronologies of this type is not as predictable as that of living-tree chronologies, being dependent on the cooperation of historians, archaeologists, and the owners of buildings. Given adequate funding, this work could probably be completed in five years. Calibration of such a grid will pose the most severe problems because, as already mentioned, most of the previously extensive European forests have disappeared. The small remnants of existing forest may not form a good analogue for calibration, thus great care will have to be taken in verification of any reconstructions made.

The long timescale (5000 BC to the present). At the present time, there exist the backbones of two chronologies of about 7,000 years length, one from Germany (Becker & Frenzel, 1977) and the other from the north of Ireland (Pilcher et al., 1977). Both cover almost the whole of the 7,000 years, but with several short gaps. The greater part of the chronologies are thus floating. Sections of the floating sequences are being used at present for radiocarbon calibration and recent work has suggested a link between past radiocarbon variations and climate (see Wigley, this volume). The long sequences of dated timber are also available for any other isotope studies that might yield palaeoclimate data. Any attempts to extract climate data from the ring widths will necessarily be limited and the results of a very general nature. If the existing floating sequences from the two areas can be cross-dated some information might be derived on the relative continentality of the European climate in past periods. If both sequences can be made absolute by closing the gaps then the great length and calendrical accuracy of the records suggest that strenuous efforts should be made to derive some climate data from them.

There is probably the potential in Europe for about three more such long tree-ring sequences derived from natural subfossil timbers. However, it is not clear at present that the potential climate information would justify the vast amount of work involved. With advances in isotope techniques (Long, this volume), this position might change.

EUROPEAN RUSSIA

T.T.Bitvinskas

INTRODUCTION

The long-term study of biological processes is a complex and time-consuming problem. In a study of this nature, the short length of existing climate records and the lack of information on climatic variations of low frequency (for example, of periods 100 to 600 years) prevent the determination of the constancy and dynamics of any shorter climatic rhythms (for example, of periods 11 to 22 years). On the basis of research work carried out in the Soviet Union (Bitvinskas, 1974), it is possible to arrive at the following conclusion: dendroclimatological investigations based on natural changes in the annual rings of trees can serve as a valuable basis for forecasting the probable course of future environmental conditions.

At the Institute of Botany of the Lithuanian Academy of Sciences (in the Dendroclimatochronological Laboratory), a research programme on the annual rings of wood has been in progress since 1968 (Bitvinskas, 1968). In the Republic, this research started in 1953, when the author (then a student) carried out the first dendroclimatological investigation of black alder thickets of the Birzai virgin forest (Bitvinskas, 1961). Since then extensive investigations have been carried out in the Republic, of which the following deserve mention: the dynamics of growth of pine stands of the Lithuanian SSR and possibility of its prediction (Bitvinskas, 1964); the dynamics of oak stand increments in the Lithuanian SSR, 1970 to 1975 (Bitvinskas & Kairaitis, 1975); correlation of width of fir annual rings and climatic factors in Lithuania (Cerskiene, 1972).

Since then, dendrochronological methods have been applied when determining fluctuations of water level of lakes in eastern Lithuania (Pakalnis, 1972) and the variability of radial increments of pine and their relationship to environmental conditions (Karpavicius, 1976). During this time, new programs for computer processing of annual ring information were developed, and a start was made on the use of an automated system for the measurement and preparation of annual ring data for the computer (Maleckas, 1972; Maleckas et al., 1975). Radiocarbon measurements obtained in the laboratory are used for the verification (synchronisation) of separate series of annual rings, that is, for the relative dating of samples of unknown age.

Particular attention is being paid to the construction of very long tree-ring chronologies. In the Uzpelkiu Tyrelis peat bog in northwest Lithuania, the first thorough investigation of past environmental conditions was carried out by means of the dendroclimatological method. Stumps and trunks preserved in the surface strata of the peat bog were used for this study (Bitvinskas et al., 1972). It was shown that it is possible to construct millennia-long tree-ring chronologies from continuously overlapped sequences. This has been made possible by a complex method of investigation including dendroclimatological, radiocarbon, pollen, and geobotanical analysis of the investigated samples of wood and peat. The chronology of Uzpelkiu Tyrselis is 2,200 years long extending back from the present and, at present, appears to be the longest in eastern Europe. The construction of chronologies of up to 6,000-7,000 years long using timber from marshes of the northwestern U.S.S.R. should be possible.

LIVING-TREE CHRONOLOGIES

The age structure of the forests of the Soviet Union permits an investigation of the variability of tree rings on a large scale throughout the forest and forest-steppe zones of European U.S.S.R., Siberia, the Caucasus, and other mountainous regions. The Laboratory of the Lithuanian Academy of Sciences does not confine itself to the small territory of Lithuania, but carries out investigations in the territories of other republics of the Soviet Union. The material on a transect through the Murmansk region, Karelia, Leningrad, and Novgorod-Pskov has already been published, as has work on the Latvian SSR, eastern Lithuania, western Byelorussia, and western Ukraine, including the Transcarpathian region (Bitvinskas & Kairaitis, 1978). Another dendrochronological transect under construction at present runs approximately along the 56 to 54 parallels N of the U.S.S.R. and runs from Lithuania to the Far East (Bitvinskas, 1978).

Enormous regions of the Soviet Union have been, and are being, investigated by dendrochronologists. Nevertheless, it is clear from these investigations that in the more inhabited areas and in the zone of intense forest exploitation, few trees exceed 250 to 350 years of age. The species examined were Pinus silvestris L., Picea excelsa L., and Quercus robur L. Longer chronologies are either represented by smaller quantities of samples or have to be constructed by means of cross-dating (that is, they are composite chronologies).

RESEARCH RESULTS

About 250 chronologies of Pinus silvestris L., Quercus robur L., Picea abies L., Larix sibirica Lebed., and other forest stands have been constructed. Detailed information on climate changes in the Lithuanian SSR and other regions of the Soviet Union was collected. The interdependence between the amplitudes of 22-year cycles of solar activity and the stand increments were examined. Climate factors which have an influence on the stand

increment were determined as well. A series of complex
hydrothermic indices reflecting the variability of the
dynamics of annual rings was also drawn up. Regularities
in the dispersion of trees in stands of which changes in
the width of annual layers and the sensitivity of trees to
environmental conditions were examined.

The possibility of constructing extremely long
chronologies from timber taken from peat bogs and sand and
gravel pits was proved (Bitvinskas et al., 1976), as well
as the effectiveness of investigations into the yearly
information on radiocarbon content in annual rings of
trees. Convincing correlation between radiocarbon and
solar activity was obtained on the basis of tree-ring data
accumulated by the laboratory. Regularities of the
distribution of climate trends in different stages of
solar activity were examined statistically (Bitvinskas,
1974). The profile method was applied to the examination
of the variability of radial increments and of their
relationship with solar-physical components. The oppos-
ition of prevailing trends in five stages of solar
activity from 22-year cycles was found in northern
latitudes and southern regions of the Murmansk-Carpathians
transect.

Investigation of the occurrence of stumps in peat
deposits in Lithuania showed that the complex examination
of such deposits by radiocarbon, pollen, and dendrochronological methods yields excellent information on past
environmental conditions. Information on climatic cycles
of average duration (11 and 22 years) and on secular
environmental changes are reflected in the width of pine
annual rings, in changes in pollen and botanical
composition, and in the extent of decomposition of the
peat (Bitvinskas et al., 1976). The results obtained in
several of our laboratories showed that radiocarbon
analysis of annual rings of the recent century helps to
trace the extent of anthropogenic effects on the biosphere
(Dergachev & Sanadze, 1974).

CONCLUSIONS

Dendroclimatological investigations have been
carried out in the Lithuanian SSR as well as in areas
to the north and south of the Republic and in other areas
of the Soviet Union. This research demonstrates the
success of the methods employed and also shows the large
unexplored resources of dendroclimatic information
available in the Soviet Union. The obvious value of
cooperation between scientists of different disciplines
both in the Soviet Union and throughout the world is
clearly demonstrated.

THE MEDITERRANEAN AREA

A.V.Munaut

INTRODUCTION

From a dendrochronological point of view, the common
climate characteristic of the regions comprising this area
is the simultaneity of the warmest and the dryest season.
Such a climate strongly affects tree growth since evapo-
transpiration reaches its maximum when the water supply is
lowest. Water stress is the major limiting factor although
in the mountainous regions, low temperatures play an
important role. Roughly, the Mediterranean area extends
from the Atlantic Ocean to the Himalaya, between the high
pressure subtropical belt and the temperate regions. To
the east, it is in contact with countries influenced by
the monsoon. The sunny, hot and dry summers are dependent
on the seasonal shift of the desert regime, while winter
rains fall by displacement of northerly cyclonal activity.
The Mediterranean area is thus a very sensitive area,
favourable to the detection of annual or long-term
variations of climate affecting large parts of Europe,
North Africa, and Asia.

VEGETATION

Although adapted to summer dryness, the Mediterranean flora is not uniform. Indeed, two climate gradients
affect the vegetation: increased warmth to the south, and
continentality to the east. Other phenomena can modify
these general trends. Location, orientation, and the
altitude of mountain masses are important factors of
differentiation. Also the migration of floras during the
climate vicissitudes of the Quaternary must be taken into
consideration. Finally, topographic and edaphic
peculiarities contribute to the diversification of the
local composition of the vegetation. Nevertheless, in most
conditions, the forest is the climax. This is a favourable
feature for the dendrochronologist who can choose trees
growing in various environments. The diversity of suitable
species giving various response functions also extends the
possibilities for climate reconstructions.

THE INVESTIGATED SPECIES

According to literature and information given by
several dendrochronologists, many species have been
sampled and the analyses are still in progress. The
species sampled are summarised under each laboratory.

Laboratoire de Botanique Historique et Palynologie de
Marseille-St-Jerome (F.Serre-Bachet, M.L.Tessier and
A.Pons):

France: <u>Abies alba</u>, 5 sites; <u>Larix decidua</u>, 3 sites;
<u>Picea excelsa</u>, 1 site; <u>Pinus halepensis</u>, 3
sites; <u>Pinus laricio</u>, 2 sites; <u>Pinus</u>
<u>sylvestris</u>, 5 sites; <u>Fagus sylvatica</u>, 3 sites;
<u>Quercus pubescens</u>, 3 sites.

Italy: <u>Pinus leucodermis</u>, 1 site.

Tunisia: <u>Pinus pinaster</u>, 1 site; <u>Tetraclinis</u>
<u>articulata</u>, 1 site; <u>Quercus faginea</u>, 2 sites;
<u>Quercus suber</u>, abandoned.

Syria: <u>Quercus cerris</u>, 2 sites.

All this material is sampled from living trees. For
further details, see Serre et al. (1964), Borel & Serre
(1967), Serre (1969, 1973, 1977, 1978a,b).

Dendrochronology Laboratory, Tel Aviv University
(Y.Waisel, N.Liphschitz and M.S.Lev-Yadim):

Israel: <u>Cupressus sempervirens</u>; <u>Pinus halepensis</u>;
<u>Pistacia atlantica</u>.

Sinai: <u>Juniperus phoenicea</u>; <u>Pistacia khinjuk</u>.

Turkey: <u>Pinus halepensis</u>; <u>Pinus nigra</u>.

Cyprus: <u>Cedrus brevifolia</u>; <u>Juniperus foetidissima</u>;
<u>Pinus brutia</u>; <u>Pinus nigra</u>.

Crete: <u>Cupressus sempervirens</u>; <u>Pinus brutia</u>.

Iran: <u>Juniperus palycarpos</u>; <u>Pistacia atlantica</u>;
<u>Quercus persica</u>.

In addition to the living trees, 56 specimens of <u>Cedrus</u>
<u>libani</u> (between 1 and 1900 AD), 48 of <u>Cupressus semper-</u>
<u>virens</u> (between 1200 and 1900 AD), 37 of <u>Pinus nigra</u>
(between 1600 and 1900 AD), and 34 of <u>Cupressus</u>
<u>sempervirens</u>, have been sampled in 38 archaeological sites
located in Israel and eight in the Sinai.

Cornell University, New York (P.Kuniholm, H.Levin,
B.Miller):

Turkey: From information sent by P.Kuniholm, it
appears that most of his work has to be done
in connection with archaeological problems
(for example, Kuniholm & Striker, 1977). The
absolute chronology is 670 years long and has
been established with <u>Pinus nigra</u> growing on
the Anatolian Plateau. A 3,000-year floating
chronology has been established for the
Anatolian Plateau and the potential exists for
a continuous master chronology extending from
the present to 9000 BP. Amongst the material
used is: <u>Cedrus libani</u>, <u>Abies</u> sp., <u>Pinus</u>
<u>nigra</u>, <u>Quercus</u> sp. and <u>Juniper sp.</u>

Greece: <u>Quercus mibra</u>, 1 site; <u>Quercus coccifera</u>, 1
site; <u>Quercus</u> sp., 1 site; <u>Picea abies</u>, 1 site;
<u>Pinus nigra</u>, 2 sites; <u>Pinus leukodermis</u>, 3
sites; <u>Abies cephallonica</u>, 1 site; <u>Abies</u> sp.,
1 site; <u>Juniperus</u> spp., 1 site.

The Belgian Group (A.Berger, A.Munaut, J.Guiot (Louvain-
la-Neuve), L.Mathieu, L.Fraipont (Fac.Sc.Agrono. de
l'Etat, Gembloux)):

Spain: <u>Abies pinsapo</u>, 2 sites, 40 samples.

Morocco: <u>Abies pinsapo</u>, 5 sites, 158 samples; <u>Cedrus</u>
<u>atlantica</u>, 75 sites, 2,392 samples; <u>Cupressus</u>
<u>atlantica</u>, 6 sites, 62 samples.

On the whole, 11 genera and 27 species dispersed from
Spain to Iran are under investigation in four laboratories. For further details of the research of the Belgian
group, see Miessen (1975), Helleputte (1976), Guiot et al.
(1978), Matagne (1978), Munaut et al. (1978), Portois
(1978), Berger et al. (1979), De Corte (1979), Guiot
(1980a,b), Lefébure (1980), and Rome (1980).

Other research in the Mediterranean area is
discussed by Gassner & Christiansen (1943), Devaux et al.
(1975), and Metro & Destremau (1968/1969).

TECHNIQUES AND METHODS

Standard techniques are used to provide total ring
widths. In the Belgian group, Fraipont (Gembloux) has
developed X-ray densitometry and the first cedar samples
are under study. Programs RWLST and INDXA (Graybill, this
volume), provided by the Laboratory of Tree-Ring Research,
are in use at the Cornell University, Marseille, and
Louvain, so there is the possibility that the format of
the results will be standardised.

New models to provide both response and transfer
functions have been developed at Louvain (Guiot et al.,
this volume). These models are based on digital filtering,
principal component, cross spectral, discriminant and
cluster analyses, and on generalised least squares and
spectral multiple regressions. Specific statistical
methods are also used in order to validate the climate
reconstructions. The experience of the group in this kind
of research dates back from 1974 and excellent results
have already been obtained. At Marseille, such programs
are not yet operational and response functions have been
calculated elsewhere.

SOME SPECIFIC PROBLEMS

Human activities. For many thousands of years, human
activities have deeply perturbed the natural Mediterranean
forest. The most destructive action is, of course,
clearing for cultivation, although in some circumstances
Man is able to maintain in the same area an agro-sylvo-
pastoral equilibrium. The "forests" of <u>Argania spinosa</u> in
Morocco are perfect examples of such a subtle treatment.
In Islamic countries, cemeteries and holy places
constitute reservations, usually small, where large trees
are protected.

Cutting trees for fuel and charcoal does not
completely destroy the forest but damages it to produce
various stages of degradation. In these more or less open
woodlands, the trees are very often small, twisted, and
too young for dendrochronological use. However, the most
disturbing factor remains overgrazing, especially by
goats. When the herd is passing through the forest, the
goats eat not only herbs and small bushes but also foliage
of hardwoods and conifers. When the woodland is over-
grazed, herders do not hesitate to climb on large trees to
cut living branches. We have seen an old cedar forest

(more than 300 years old) almost completely trimmed and some large trees standing dead. For these reasons, the dendroclimatologist must be extremely cautious when sampling in open woodland. Almost all the trees have suffered such maltreatment and stress is so heavy that the effects of the climatic factors are probably concealed. In closed forests, the risk is not so important. Nevertheless, each tree must be checked before sampling to eliminate those showing more or less obvious anthropogenic deteriorations.

The acquisition of meteorological data. In some parts of the Mediterranean area, the meteorological network is sparse, record length is short, and the series are sometimes incomplete. Therefore, each regional investigation must be associated with a climatological study, in order to select the best and most viable series. This problem is very acute in the mountainous regions, where conditions prevailing in the settlements, and therefore in the meteorological posts, are not necessarily representative of the climate affecting the forest. It is sometimes convenient to seek correlations between distant stations exposed to the same general influence. This choice is made easier when using field observations. This can be illustrated by a study of Cedrus atlantica in Morocco.

Emberger (1939) concluded that human activity restricted the cedar forests to only some parts of their potential area. More recent research based on intensive field work (Peyre, 1979), has shown, however, that, especially in the marginal parts of the Cedrus area, their location is strictly dependent on local conditions, such as snow cover or cloud dynamics. Ecological observations in the field are absolutely necessary for the interpretation of dendroclimatological results (or the absence of results). The sciences of both ecology and climatology must be coupled in all dendrochronological work.

Hypersensitivity and anomalies of growth. Although interesting from a climatological point of view, some sites must be abandoned for technical reasons. In severe conditions (both climatic and anthropogenic) annual growth is strongly disturbed. The rings are asymmetric, partially or totally missing, their boundaries are indistinct, and their annual nature open to question. Growth irregularities are more frequent in hardwoods, Argania, Olea, Pistacia, and sometimes Quercus, while multiple or missing rings especially affect conifers. For instance, hypersensitive Cedrus may show a tenth of its rings missing.

PROSPECTS

The prospects for dendroclimatological work in the Mediterranean area are good (Bertrand, 1979), but an international effort is needed to sample and process the

abundant material that is available and to solve some of the complex problems that the region poses. For that reason, an international multidisciplinary working group was founded in 1979 in order to gather not only dendrochronologists already working in this area, but also people interested in ecological or climatological problems.

COMMENT
F.Serre-Bachet

Problems with climate data and tree-ring sites. In the Mediterranean region, as elsewhere, whenever a dendroclimatological study is considered, the problem is raised of the nature, length, quality, and reliability of the available meteorological data. It is quite evident that, as Munaut states, "each regional investigation must be associated with a climatological study".

In France, most of the meteorological records over relatively long periods (more than 40 to 50 years) in the Mediterranean region concern only rainfall (Garnier, 1974). Long temperature sequences are rare because the existing pluvio-thermic stations have only been working with some regularity since about 1950. In North African countries, such as Morocco and Tunisia, there exist fairly long precipitation records: about 50 years in Morocco (Guiot et al., 1979), 75 years in Tunisia (Aloui, 1978), but the corresponding records of temperatures are intermittent. In Syria, Chalabi (1980), working in the northwestern region, had at his disposal for two dendroclimatologically interesting stations only 15- and 18-year records provided by the meteorological department of the Defence Ministry. In any country, periods corresponding to political instability or simply to holidays have repercussions on the record length.

Three aspects of the reliability of records coming from a meteorological station have to be considered: the precision of the observations recorded; the homogeneity of the observations; and the agreement of the meteorological station with the meteorological characteristics of the tree-ring site studied. On the last point, one can question the significance of response functions which are calculated with data from meteorological stations situated in surroundings differing from the site(s) where the samples were taken. This problem is particularly acute with regard to mountainous sites where the influence of snow cover can never be studied directly and where daily summer storms sometimes bring quantities of rainfall that can hardly be compared with those recorded in lower sites.

A second crucial point in this region concerns the difficulties that are encountered (as emphasised by Munaut) in the reconstruction of past climate owing to Man's activities. The first traces of this action are found in France as early as 7000 BP (Triat-Laval, 1979). In fact, in most cases, at least in France, reconstruction is made difficult because of the absence of old enough

living trees to start long chronologies. The unrestricted
exploitation of forests or their devastation by fire
few hours, are accountable for this. There remains the
possibility of studying mountainous regions that are
generally less affected, but also less representative of
the prevalent Mediterranean climate.

Possible species. As regards species that can be studied
in dendroclimatology, there seems to be a certain
limitation in broad-leaved trees in relation to the
persistence of foliage. We have noted in Marseilles that
the difficulties in analysing rings increase as a function
of this persistence. Thus, if no major problem is faced as
far as deciduous Quercus pubescens or Quercus cerris are
concerned, difficulties appear with half deciduous Quercus
faginea and it is quite impossible to get over them in the
case of Quercus suber and, to a lesser extent, Quercus
ilex, both sclerophyllous species. Now, sclerophyllous
oaks have been, at least since the regular manifestation
of Man's action (Triat-Laval, 1979), the most widely
distributed trees in the Mediterranean region. Among
conifers, Cupressaceae, Cupressus, Juniperus, Tetraclinis,
are also widespread and there are also great difficulties
in defining their rings.

Dendroclimatological results. Some results of studies made
in the Mediterranean region by the dendroclimatologists of
Louvain-la-Neuve and Marseilles are summarised below, and
in Table 4.15. They are not complete but are only intended
to give an idea of the dendroclimatological potentialities
of the species analysed in the two laboratories. Some
general points can be made. Except for some genera, Abies,
Larix, Cedrus, a small number of years only are covered in
each site. Mean sensitivities range between 0.12 and 0.38
with a mean value of 0.21, approximating the value 0.20
found by B.Huber (Munaut, 1966) for species from central
Europe. Percentages of variance ascribable to annual
factors vary from 2.2% to 52% with a mean value of 29%.
Serial correlation coefficients also vary and are
generally quite high in Morocco.

Considering the postglacial history of this region,
which is now quite well known, at least in western
Mediterranean regions (Beaulieu, 1977; Reille, 1975, 1977;
Triat-Laval, 1979), some of the available species may well
be heterogenous - from the genetic point of view. This
could lead within the same species to variable responses
to the same type of climate. Thus, it is necessary to
multiply the sampling sites so as to be able to
differentiate the genetic from the ecological effects on
tree growth.

Table 4.15: Summary of dendroclimatological results for the Mediterranean from: 1a) Serre (1978a,b); 1b) Serre
(unpubl. data); 2) Tessier (unpubl. data); 3) Guiot et al. (1979); 4) Aloui (1978); 5) Frezet et al. (1979); 6) Chalabi
(1980).

SPECIES	COUNTRY	STATION		PERIOD	NO OF YEARS	MS	SERIAL r	ANOVA PERIOD	Y	MEAN RING-WIDTH	NO OF TREES
Larix decidua	France	Les Merveilles	a)	988-1974	987	0.239					31
1a			b)	1187-1974	788	0.283	0.200	1887-1974	33.4%		7
Abies alba	France	Mt-Ventoux	a)	1653-1975	323	0.174	0.574	1950-1975	21.5%	101.46	18
1b			b)	1664-1975	312	0.205	0.576	1901-1975	36.9%	91.71	7
Pinus silvestris	France	Drome	a)	1884-1978	95	0.208	0.512	1950-1977	25.0%	86.27	15
			b)	1898-1978	80	0.155	0.537	1930-1977	42.2%	116.84	10
								1950-1977	42.4%		10
			c)	1895-1978	84	0.160	0.273	1928-1977	19.0%	82.80	10
								1950-1977	21.5%		12
		Drome 2	a)	1909-1978	70	0.188	0.364	1950-1977	25.6%	148.89	13
			b)	1904-1978	75	0.182	0.354	1950-1977	8.0%	89.76	11
		Drome 3		1864-1978	115	0.227	0.267	1935-1977	28.6%	109.83	11
2								1950-1977	29.8%		11
Pinus halepensis	France	Marseille 1		1807-1973	167	0.376	0.471	1890-1940	42.2%	127.53	7
1b		2		1913-1978	66	0.311				221.40	6
Cedrus Atlantica	Morocco	301		1790-1975	186	0.18	0.28	1871-1970	25%	210.00	10
		302		1754-1975	222	0.15	0.83	1901-1970	52%	140.00	11
		304		1700-1975	276	0.19	0.60	1871-1970	41%	55.00	11
		307		1537-1976	440	0.21	0.69	1870-1970	31%	106.00	12
		310		1604-1977	374	0.17	0.61	1871-1970	23%	124.00	18
3		321		1840-1977	138	0.15	0.63	1900-1977	18%	303.00	20
Pinus pinaster 4	Tunisia	Tabarka		1928-1977	50	0.150				235.92	10
Quercus pubescens	France	Drome 1		1886-1976	91	0.259	0.226	1933-1976	23.6%	56.08	13
		Drome 2		1902-1976	75	0.302	0.501	1933-1976	44.2%	98.28	10
		Drome 3		1850-1976	127	0.204	0.639	1851-1976	46.8%	70.35	7
		Drome 4		1934-1976	43	0.125	0.363	1944-1976	14.1%	184.51	14
		Drome 5		1928-1976	49	0.290	0.077	1954-1976	26.7%	209.77	4
		Drome 6		1850-1976	127	0.224	0.706	1940-1976	2.2%	134.92	6
5		Drome 7		1931-1976	46	0.269	0.483	1954-1976	26.9%	286.35	6
Quercus faginea 4	Tunisia	Tabarka		1888-1977	90	0.199				184.98	14
4		Qued Dalma		1911-1977	67	0.263				169.11	10
Quercus cerris 6	Syria	Baer		1900-1975	76	0.170	0.130	1960-1975	26%	111.31	20
		Slenfe		1917-1975	59	0.144	0.440	1960-1975	36%	157.47	20

Response functions accumulated up to now in Marseilles indicate that the percentage of the total variation of ring width accounted for by predictor variables is higher when these variables are either rainfall and maximum temperatures or rainfall and minimum temperatures than when they are rainfall and mean temperatures. This has been noted particularly in the case of Pinus sylvestris and Quercus pubescens stations in the Drome. Though it seems that one could be satisfied with rainfall records alone in the calculation of response functions of trees from low altitude sites, the use of both rainfall and temperature data proves to be absolutely necessary for high altitude sites (Chalabi, 1980). There does not seem to be any notable distinction between coniferous and broad-leaved species as regards significant climatic factors revealed by response functions, at least when functions are calculated with meteorological data of the 12 months from October to September. The most important months influencing growth owing to their rainfall and temperature are generally October and December of the previous year and April to June and September of the current year. Their direct or inverse relation to ring width is, for a particular species, either the same in all the stations studied or peculiar to some of them.

Prospects and cooperation. It is imperative for scientists to work together in the Mediterranean zone, and the working group created in 1979 on A.Munaut's instigation is but the expression of this need. This cooperation should be strengthened, in particular with regard to data processing techniques, so as to make possible comparisons between results obtained from diverse sites and to achieve the necessary syntheses at an earlier date. In certain countries bordering the Mediterranean, dendroclimatology should be able to develop thanks to the participation of indigenous research workers. In Marseilles, for example, we have the possibility of receiving Moroccan, Algerian, Tunisian, or Syrian students, most of them forest scientists, who are much interested in this original type of work. Their knowledge of the flora, vegetation, and topography of their country are great advantages and should not be neglected.

ASIA
Zheng Sizhong, Wu Xiangding & Lin Zhenyao
STATUS OF DENDROCLIMATOLOGY
The alternation of wide and narrow rings in trees was known long ago by woodcutters who also noticed relationships between climate and ring widths. Dendroclimatic work in Asia was initiated first in Japan. Hirano (1920) studied a single Cryptomeria japonica of about 250-years age in Miyazaki, Central Japan. He found that the variation of the ring-width series showed a 33-year cycle.

Dendroclimatology in China began in the fourth decade of the 20th century. The late Professor Zhu Kezhen (Chu Ko-chen) recognised that the method of dendroclimatology could be used to reconstruct past climate and he advised his students to make an attempt at this approach. Cheng (1935) took about 70 samples of trees in Beijing (Peking) for tree-ring research. Figure 4.22 shows the location of places referred to in this section. These collections were made up mainly of three species of trees, Pinus bungeana, Juniperus chinensis, and Thuja orientalis. Simple linear regression was used to calibrate the ring widths of Pinus bungeana with annual precipitation. The period of analysis was 1879 to 1933 and the correlation coefficient was +0.44. The author concluded that tree growth in Beijing was controlled mostly by moisture conditions. Although dendrochron- ological and dendroclimatological studies in Soviet Asia started more recently, Soviet scientists have made great efforts in tree-ring research in Central Asia and the Far East in recent decades. De Boer (1951, 1952) published papers on tree-ring measurements in Java (Indonesia) from 1514 AD and their relation to the sunspot cycle of weather fluctuations.

Dendroclimatological work was done somewhat sporadically in Japan. Yamazawa (1929), Shida (1935), Enmoto (1937), Hoyanagi (1940), Yamamoto (1949), and Saito (1950) worked on this field independently. None of them did systematic research in dendrochronology, but the regions where they studied were not limited to Japan. They included Taiwan and Nei Monggol (Inner Mongolia). The species that have been utilised for research in Japan are Cryptomeria japonica, Castanea crenata, Chamaecyparis obtusa, and Picea ajanensis. There are about eight tree-ring charts and tables showing annual ring-widths in papers on dendroclimatology in Japanese from these authors. All the series are based on single trees. The length of these series ranges from about 200 to 1,000 years. The analyses for ring-width series showed some long- and short-term cycles, which were contrasted and verified with historical data in Japan. A short survey of Japanese tree-ring analyses has been made by Arakawa (1960). According to his view, the reasons for the slow development of tree-ring analyses in Japan were almost the same as in Europe. There are no very old trees and the climatic factors are very complex, so tree-ring analysis was not immediately successful.

Since dendroclimatological work began in China in the early 1930s, climatologists, botanists, and geographers have become successively involved. The regions for tree-ring exploration extend over all China, but are found mainly in the arid and semi-arid regions of northwestern China. Teng (1948), a botanist, presented a paper discussing the relationships between ring width and climate in Gansu (Kansu). Three samples were taken from Qilian (Kilien) Shan, where the annual precipitation is about 375mm, and one from Bailong Jiang (Peilungkiang), where the annual precipitation is approximately 625mm. The three samples from Qilian Shan all belong to one species, _Picea asperata_. The trees from which the samples were taken were located about 32km apart. The sample from Bailong Jiang belongs to _Abies cheniensis_. No attempt was made to cross-date rigorously, but the rings in the different trees seemed to cross-date naturally. From an analysis of ring variations in terms of cycles, there seemed to be a fairly close correlation between the sunspot period and cycles shown by ring width. The sunspot maxima correspond quite well with narrow rings in the trees. Further, the tree-ring minima coincide with times of drought known in the history of Gansu. Such correlations are rather striking in the samples from Qilian Shan, but less definite in the sample from Bailong Jiang.

In order to utilise the water resources of the Huang He (Yellow River), it is necessary to know the variation of riverflow in the past. In spring 1955, cross-sections of 30 trees were collected from five sites in the river basin. These specimens include five tree species, the

sampling being of an exploratory nature. One finding of this preliminary work was that ring widths of elms in the suburbs of Lanzhou (Lanchow) seem to be closely correlated with rainfall, particularly in the growing season.

During the 1970s, there has been some advance in dendroclimatological work in China thanks to the use of electronic computers. The climatologists at the Geographical Institute of China have focussed their dendroclimatological efforts in two major areas of China. First, they are examining tree-ring evidence and proxy or natural climatic indicators from the Xizang (Tibet) Plateau, because of the importance of this area to the climate in all of eastern Asia (Wu & Lin, 1978). Second, climatic change in northeastern China is receiving attention because the lowest winter temperatures occur in this area.

In Soviet Central Asia, Mukhamedshin & Sartbaev (1972) and Mukhamedshin (1974) give results of dendro-chronological studies on 1,580 specimens of junipers. From 18 trees aged 452 to 1,214 years, felled at an altitude of 2,900 to 3,500m on a north slope of Altai Range, a 1,214-year chronology was established. It is

Figure 4.22: Geographical orientation.

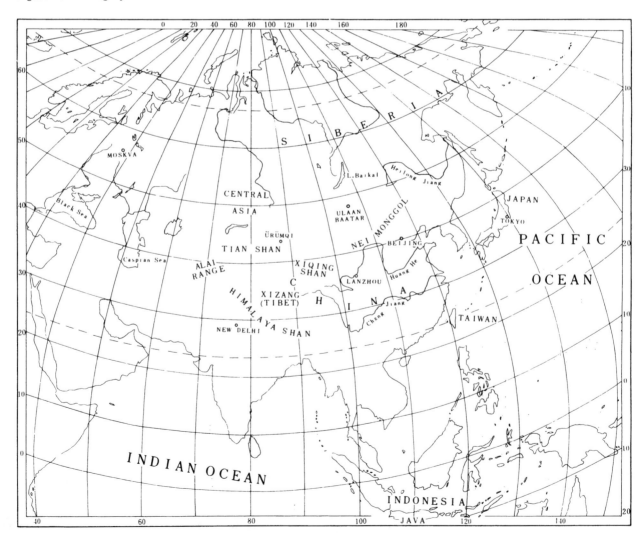

found that in the arid lower regions, the growth of
Juniper seravschanica and Juniper semiglobosa is
determined mainly by moisture, and that high temperatures
in the growing season (expecially in July) have an adverse
effect. In the colder and moister higher regions, the
increment of Juniper turkestanica is determined mainly by
temperatures in June and July. In addition, Molchanov
(1976) made some dendroclimatological studies in this
region. He presented a tree-ring chart, indicating the
decade means of the ring-width index from about 1970 back
to 1330. The objective of the chart is to show the
long-term cycle of drought in Central Asia. In the years
of drought, which occur frequently, the ring widths
decrease.

In Siberia, a detailed study of tree-ring variations
in relation to fluctuations in the water level of Lake
Baikal was presented by Galazii (1972). He concluded that
the cyclic fluctuation of water level was caused by the
cyclic variation of moisture conditions around the lake
area. The cycles vary in length from 50 to 52 years to 80
to 84 years and include some short-term cycles, such as
7-11, 24-25 and 39-40 years. An extensive survey of
dendroclimatology in the Soviet Far East was made by
Tarankov (1973) with 138 references. He reviewed the main
methods of dendrochronological and dendroclimatological
analyses, and presented dendrochronological series for the
last 100 to 400 years for Taxus (cuspidata), Pinus
koraiensis, Picea jezoensis, Abies holophylla, Abies
nephrolepis, and Larix gmelinii, growing in the Soviet Far
East. Cycles in radial growth of the tree were determined,
the most frequent cycles observed being 5-6, 10-11, 20-25,
30-40, and 80-90 years long.

THE AVAILABILITY OF SUITABLE TREE-RING DATA

Dendroclimatic research in Asia is likely to have a
broad potential, judging from the work that has been done.
Trees of many genera seem to reach considerable age and
have well-defined rings showing evidence of climatic
sensitivity. In the arid regions of Central Asia, the
younger trees are 100 to 300 years old and the most
long-lived trees are 600 to 700 years old. Further, there
are trees reaching more than 1,000 years and Juniperus
turkestanica having reported ages of as much as 2,000
years (Mukhamedshin, 1974). There are Chamaecyparis obtusa
and Cryptomeria japonica with known ages over 800 years in
Japan (Enmoto, 1937). China has a number of tree species
attaining great ages, the most famous long-lived tree
species being Ginkgo biloba reaching more than 3,000
years. It is likely that careful site selection in the
arid and semi-arid regions of the country could yield long
ring-width records representative of past precipitation
fluctuations. Trees near the upper treeline in the high
mountains could potentially be used as indicators of past
temperature fluctuations. For example, it was reported
that a Sabina przewalskii at an altitude of 3,670m in

Qilian Shan reached more than 900 years, and that the
tree-ring series indicates past temperature fluctuations.

Northwestern China offers the greatest immediate
potential for the application of tree-ring data to
palaeoclimatic problems. There are at least six genera
with at least one species that is suitable for
dendroclimatological work. These genera are Abies, Larix,
Picea, Pinus, Sabina, and Ulmus. The experience of
dendroclimatological work in regions of Qilian Shan and
Tian (Tien) Shan indicates that trees near the upper
treeline in high mountains are temperature sensitive and
trees near the lower treeline are drought sensitive. The
growth of trees in northeastern China is often limited by
low temperatures occuring during the summer months. Trees
of at least three species are sensitive to temperature.
There are Fraxinus mandshurica, Pinus sibirica and
Larix sibirica. In a recent paper by Gong et al. (1979),
simple linear regression was used to calibrate the two
series of ring-width indices of Larix sibirica at Gen He
(Gen Ho) in Heilong Jiang (Heilungkiang) province with
temperatures from May to September. The period of analysis
was 1957 to 1975, and the correlation coefficients were
+0.82 and +0.69, (P < 0.001). There are many possible
species and promising sites in northeastern China that
have not yet been examined. In the Xizang Plateau region,
at an altitude of 4,000m near the upper treeline, the ring
widths of Sabina recurva reflect primarily regional
temperature fluctuations (Wu & Lin, 1978).

The primary task facing the Asian dendroclimatologist is the extension of the tree-ring record backward
in time, the utilisation of new species, and the sampling
of trees in new regions of the continent.

COMMENT

M.K.Hughes

This region differs in several respects from most
areas where work on dendroclimatology has been done. Of
particular significance is its location as the eastern
seven-eighths of a very large landmass reaching from high
northern latitudes to equatorial regions and spanning over
140 longitude. Asia contains a great range of climatic
conditions, from monsoon lands in the south to the Arctic
in the north. The hot and cold deserts of the centre of
the continent, along with the highest mountains on earth,
play an important part in the partitioning of the
continent on climatic, phytogeographical and ecological
grounds. Zheng et al. have indicated the great potential
for dendroclimatic research in Asia, emphasising the need
for extensive new sampling.

It would seem reasonable to accord highest priority
to those regions where other proxy climate records, such
as historical accounts, are fewest. It may be that
combined grids of tree-ring chronologies and other
annually dated proxy records will be the most appropriate
in Asia. If this approach is to be used, problems of

calibration and verification of reconstructions will
arise, since the areas without historical records will
often be those most remote from modern instrumental
stations. In planning new sampling in Asia, this problem
should be borne in mind.

As the largest landmass in the Northern Hemisphere,
Asia is of great significance to the development of a
hemispheric tree-ring data base. In addition to the new
data needed, examination is required of existing Soviet
data from Siberia, Soviet Central Asia, and the Soviet Far
East, reported in the Soviet literature. In order to
assess the extent to which these data are suitable for
inclusion in the kind of global or hemispheric data base
discussed in this volume, detailed collaboration will be
required between tree-ring scientists in the Soviet Union
and those based elsewhere. The International Tree-Ring
Data Bank offers an excellent framework for such
collaboration.

INTRODUCTION

The Editors

The aim of dendroclimatology is to extract the climate signal in the annual rings of trees and use it to provide a proxy record of climate for times and places where the instrumental record is absent or inadequate. These proxy records are used to produce reconstructions of one or more climate variables in the time, frequency, and sometimes the spatial domains. Reconstructions from tree rings are of potential value in four main fields. First, they may be used to provide an extended climate data base to be used in the testing of models of climate. Second, they may provide a longer and more representative data base for the calculation of climate and climate-related statistics. Third, they may provide detailed descriptions of climate in distant periods which may be used as analogues of possible future changes in climate. Fourth, they may be used in the verification of other proxy records of climate, including historical (or documentary) data, for pre-instrumental times.

The bases of the methods currently available for the preparation and testing of climate reconstructions from tree rings have been discussed in Chapters 1 and 2 of this book, whilst the existing data base, and the potential for its improvement, has been described in Chapters 3 and 4. In order that the reader may see the potential of dendroclimatology, short accounts of a number of dendroclimatic reconstructions are brought together in this chapter. They differ in the variables reconstructed, the type and size of tree-ring data base used, and the region of the world to which they refer. They also vary in the extent to which it has, so far, been possible to subject them to the process of verification (Gordon, this volume). In each case, the authors have presented interim results which show both the breadth of dendroclimatology's potential and the current state of the art.

All but the two final contributions are examples of a relatively new development in dendroclimatology. This is the successful reconstruction of single-station records from relatively small tree-ring grids of up to 15 chronologies. Fritts and Gordon use a large tree-ring data base to derive averaged results for a wide region. Fritts, to illustrate points made in his overview, describes recent developments in his group's programme of reconstructing climate variables over a wide sector of the Northern Hemisphere using the largest existing tree-ring data base.

In the first contribution, Guiot, Berger and Munaut explore novel approaches to reconstructing temperature in the European Alps and take advantage of the availability of long instrumental records in that region. This has allowed them to test their reconstruction against independent data for the whole of its length. Conkey describes a reconstruction of winter temperature at a group of New England sites using seven ring-width index chronologies. She also reconstructed summer temperature at one station from a stepwise multiple regression procedure into which five ring-width and density variables from a single site were entered; only maximum density was selected as a predictor. Cook shows the potential value of reconstructing composite climate-related variables such as drought indices in his account of a long-term drought sequence reconstructed from tree rings. Holmes, Stockton and LaMarche, and then Campbell, describe the recent reconstruction of riverflow records in Argentina and Tasmania, respectively. Their papers serve to introduce the reader to some of the results now emerging from recent work in the Southern Hemisphere. This group of five papers describing reconstructions based on relatively small tree-ring data bases is completed by LaMarche and Pittock's account of preliminary temperature reconstructions for various stations in Tasmania. It is of interest that these papers cover several of the areas to which the methods of dendroclimatology have but recently been applied, namely Europe, the eastern U.S.A., South America, and Australasia. Recently, two further reconstructions have appeared from such areas, the Pacific Northwest of North America (Garfinkel & Brubaker, 1980). and southern Africa (Dunwiddie & LaMarche, 1980a).

The initial impetus for the development of dendroclimatology came from work using chronologies from the southwestern United States, based at the Laboratory of Tree-Ring Research, University of Arizona. Fritts and Gordon describe the use of a large group of chronologies, carefully selected from their laboratory's prodigious data base, in the reconstruction of annual precipitation in California. Their paper is a valuable illustration of many of the points on transfer functions and verification made earlier in this book; notably by Lofgren and Hunt and by Gordon. At the Second International Workshop on Global Dendroclimatology, Professor H.C.Fritts (Laboratory of Tree-Ring Research) was asked to take a dendroclimatic overview of the proceedings in the light of his extensive experience in and contribution to the field. The paper at the end of this chapter is based on that overview, illustrated by recent work by Fritts' group. As well as providing a valuable account of this latest work, Fritts emphasises points of procedure which, in his experience, are of great importance in arriving at the best possible reconstructions of past climate from tree rings.

During the 1970s, dendroclimatology has been largely concerned with the development of techniques to handle the data arising from sampling in regions outside the cradle of dendroclimatology in the semi-arid southwest of the United States, building on the pioneering work of the group at the Laboratory of Tree-Ring Research and a handful of researchers elsewhere. It is likely that the present decade will see the fruits of this work: the reconstruction of a range of climate and climate-related variables for many geographical areas.

Several general points concerning the development of global reconstructions emerged during the Second International Workshop on Global Dendroclimatology (Hughes et al., 1980). Even in the most intensively worked areas, there are still gaps and deficiencies in the tree-ring records that need to be filled. A greater diversity of species and of sites would be of value in all areas. The natural climatic regions of the globe cut across political and cultural regions, demanding a greater degree of international cooperation than is usual in most branches of science. The International Tree-Ring Data Bank should form a starting point and nucleus for a positive approach to cooperation. In Europe especially, the political units are too small for climatic reconstructions on a national scale. The newly-initiated climate programme of the European Economic Community provides an obvious framework for dendroclimatic cooperation throughout a large part of Europe.

It is clear that, in most areas, the major obstacle to progress lies in the shortage of workers and finance, which in many cases has made it difficult to reach the goal of producing well-verified reconstructions, rather than in material to work on. In all areas where cross-dateable trees have been found, a distinct climate signal can be demonstrated. In most areas, a close grid suitable for at least regional temperature and pressure reconstructions is clearly possible. Progress in the tropics is hampered at present by the lack of suitable techniques for treating the materials that are available. Further research is urgently needed in this area. In the higher latitudes of the Southern Hemisphere, the relatively small land area means that the potential dendroclimatological cover is less than in the Northern Hemisphere. It is likely that dendroclimatology in the Southern Hemisphere will be directed more towards the reconstruction of key climate indices rather than towards the reconstruction of dense grids of climate variables. The latter may, however, be feasible in the Northern Hemisphere. With close cooperation between scientists of all nations, it should be possible to produce reconstructions for the whole Northern Hemisphere for, at the very least, the last 200 years.

AN ILLUSTRATION OF ALTERNATIVE TRANSFER FUNCTION METHODS IN SWITZERLAND

J.Guiot, A.L.Berger & A.V.Munaut

INTRODUCTION

The transfer function methods discussed by Lofgren & Hunt (this volume) permit reconstruction of past climates in areas where many sites are available and where the climate is sufficiently homogeneous. However, the dendroclimatologist often does not have sufficient sites to use such a technique. Fortunately, alternative methods exist for extracting most of the climate information contained in tree-ring data even when the number of sites is limited. Results obtained from this kind of information are not always quantitative, but provide a satisfactory initial approach. Here, we will illustrate such a method using only four sites located in Switzerland. It is based on filtering techniques, principal components analysis, cluster analysis, and discriminant analysis.

DATA PREPARATION AND PRELIMINARY ANALYSIS

Interannual variability in tree-ring widths generally appears to be complex and, because trees often have a response time of more than one year, it can be better to eliminate this kind of short-term variation with a low-pass filter. The techniques of filtering are presented in Box & Jenkins (1976) and Mitchell et al. (1966). In this case, a six-weight, low-pass filter with a cut-off period of six years is used. Tree-ring indices were obtained from four sites located in the Grisons (Switzerland; altitude: 1,800 to 1,900m) over a period extending from 1760 to 1972. In order to analyse the relationship with climate, 24 climate parameters (12 monthly values for both temperature and precipitation) were chosen at the nearest station to the sites, Bivio, 6km away. These series extend from 1901 to 1972 (Schüepp, 1961; Uttinger, 1965). Both the tree-ring and climate data were filtered in the same way.

In order to simplify comparison between long-term variations of (both) climate and tree-ring indices, principal components of the four chronologies were extracted. The first two principal components accounted for 82% of the total tree-ring variance. These two chronologies were correlated with the 24 filtered climate parameters. The first is correlated with too many climate parameters to be interpreted easily, while the second (23%) is mainly correlated with the mean August temperature ($r = -0.87$). Thus, a reconstruction from 1760 to 1972 is possible (Figure 5.1). This reconstruction is a direct transposition of the second principal component to the scale of August temperature.

Often it is impossible to isolate one climate parameter and to reconstruct it in such a straightforward way. Then, the method of discriminant analysis may be of value. This method cannot provide quantitative information on climate, but can answer qualitative questions relating to the occurrence of wet/dry or warm/cold climatic states.

DISCRIMINANT ANALYSIS

This technique applies a classificatory approach to the reconstruction of past climate. In our case, we have a sample of 213 years' tree-ring measurements, climate data being available for the 72 most recent years. These last 72 years were divided into groups on the basis of the tree-ring data alone. It was first necessary to establish criteria for this classification (Step 1). After this, discriminant analysis was used to test if there was any linear combination of the climate data that would effectively separate the groups produced in Step 1. This was Step 2 of the procedure. Such linear combinations are expressed as vectors of weights on the original climate data and are known as <u>discriminant functions</u>. If an effective separation of the groups is achieved for the 72 years, the tree-ring classification of the 141 years not used in the discriminant analysis may be interpreted climatically in terms of the discriminant functions. This is Step 3.

Step 1. The classification of years on the basis of tree-ring indices was made objectively using a cluster analysis technique (Berthet et al., 1976). The sample of 72 observations was divided into three groups with the following ranges of coordinates for the first two principal components (PC1 and PC2) of the four tree-ring chronologies:

Group 1: the amplitudes PC1 and PC2, highly negative;
Group 2: the amplitudes PC1 and PC2, weak (in absolute value); and
Group 3: the amplitudes PC1 and PC2, positive.

Step 2. Techniques of discriminant analysis are described in detail by Anderson (1958) and Romeder (1973). The technique used here - factorial discriminant analysis - is one of data analysis because no statistical hypothesis has to be tested (Romeder, 1973). The analysis computes a set of orthogonal axes for which the groups are best separated and most compact. A stepwise procedure is used. Discriminant function 1 uses August temperature and succeeds in classifying 60% of the 72 years in the correct group, as assigned in Step 1. Discriminant function 2, which also includes August precipitation, improves the proportion correctly classified by 2.5%. Further steps did not yield any improvement in the classificatory effectiveness of the set of discriminant functions. Thus, the analysis was stopped. The two discriminant functions produced by this analysis failed to place 37.5% of the 72 years in their correct tree-ring groups.

This classification is optimal for the tree rings, but not for the climate. Since only August temperature and rainfall entered the first two discriminant functions, a second discriminant analysis was carried out. In this case, only August temperature and rainfall were presented to the analysis. Only August temperature entered the discriminant functions at a significant level, 86% of the 72 years being placed in their correct tree-ring groups.

Figure 5.1: Reconstruction of the mean August temperature in Bivio from the second principal component of four tree-ring sites in the Grisons (Switzerland). Solid line: reconstruction of August temperature in Bivio; dashed line: observed August temperature in Bivio (1901-1972); dotted line: observed August temperature in Bever (1869-1900); chained line: observed August temperature in Basel

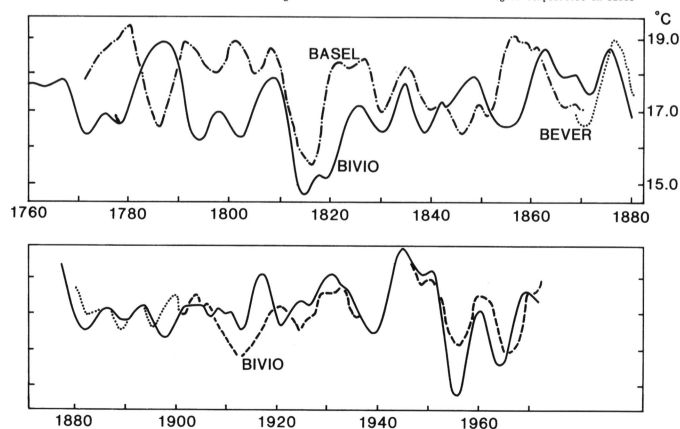

It may be concluded that the three groups of years based
on the classification of their tree-ring properties
(Step 1 above) may be separated fairly successfully on the
basis of their August temperatures alone. The years of the
groups have mean August temperatures (and standard
deviations) as follows:

    Group 1: $10.7^{o}C$ $(0.4^{o}C)$;
    Group 2: $9.9^{o}C$ $(0.4^{o}C)$; and
    Group 3: $9.4^{o}C$ $(0.3^{o}C)$.

The members of these groups will be referred to as years
with warm, mild, and cool Augusts, respectively.

Step 3. The years 1760 to 1900 were classified into Groups
1, 2, and 3 as in Step 1. These three tree-ring groups of
years were characterised as having warm, mild, or cool
Augusts in Step 2. Figure 5.2 shows each year's
classification on this basis. Figure 5.1 shows some maxima
(or minima) more accentuated than others, but, in Figure
5.2, they are all on the same level. Discriminant analysis
allows us to locate the observations in comparison with
the centroid of the three groups. As a consequence, a
particular observation belonging to one particular group
may be located at a point in discriminant space which is
very distant from the group's centroid. This indicates a
climatic state unlike any one during the 72 years of
calibration: these cases have a double line in Figure 5.2.

ANALYSIS OF THE RECONSTRUCTION

Figures 5.1 and 5.2 show a period of relatively warm
Augusts before 1790, followed by colder ones until the
1830s. A warming occurs until the 1870s, followed by the
cold period 1875 to 1920 which corresponds to glacial
advance (Heuberger, 1974). The maximum advance during this
period culminated between 1916 and 1920 (Hoinkes, 1968).

The verification of such a reconstruction requires
comparison with other climate series. For the period 1869
to 1900, we have used a climate series for Bever (Schüepp,
1961) located less than 50km from Bivio. This series has
also been filtered and is represented in Figure 5.1
(dotted lines). It can be seen that the agreement is as
good as for the calibration period. For an earlier period
(1770 to 1868), we have used August temperature at Basel
(Schüepp, 1961), also filtered. Both the Basel and the
Bivio reconstructions are similar between 1805 and 1852.
Some differences exist between 1770 and 1805 and between
1850 and 1860. The discrepancy at the beginning of the
series is due perhaps to the youth of some of the trees at
that time. Nevertheless, it can be seen that the high
temperature of this period is well recorded by the trees.
A similar comparison can be made between Figure 5.2 and
the control series. Figure 5.2 is effectively a schematic
of Figure 5.1. The cold period of 1875 to 1915 is clearly
visible in Figure 5.2; Augusts of this period were cold or
mild, on average. Our reconstruction of past August
temperature in Switzerland also agrees with the results of
Schweingruber et al. (1979) who have successfully

Figure 5.2: Reconstruction of warm, mild, and cool Augusts through discriminant analysis. Double line
indicates a reconstruction which must be taken cautiously (see text).

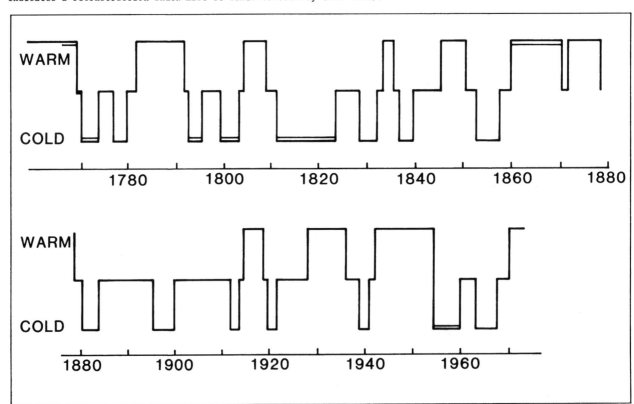

reproduced tne mean summer temperature of several stations using tree-ring data.

A LONG-TERM DROUGHT SEQUENCE FOR THE HUDSON VALLEY, NEW YORK

E.R.Cook

INTRODUCTION

During the 1960s, the northeastern United States experienced a drought of extreme severity that seriously depleted municipal water supplies in major population centres such as New York City and Washington, D.C. For much of that time, the centre of drought severity was located in the Hudson Valley region of New York. By the end of July 1965, the extensive New York City water reservoir system had been depleted to only 45% of capacity compared to a normal 80% for that time of year (Andrews, 1965). Heavy rains in September 1966 broke the drought, and generally adequate precipitation since then has caused this episode to be virtually forgotten leaving important questions about drought variability unanswered. These concern the probability that a drought of similar magnitude will occur in the future and the soundness of expectations of drought variability in the Hudson Valley based upon the last 80 to 100 years of meteorological data.

Although most of the eastern deciduous forests of North America have been cleared at one time or another, scattered remnant stands of centuries-old trees still exist. The information that can be derived from the ring-width parameters of these old trees is a unique and extremely valuable resource.

A DROUGHT RECONSTRUCTION FROM TREE RINGS

Cook & Jacoby (1977) have investigated the feasibility of quantitatively reconstructing July drought in the Hudson Valley using the annual tree-ring series of moisture-responsive trees growing in that region. The measure of drought used was the Palmer Drought Severity Index (PDSI, Palmer, 1965a). Although it is called a drought index, a series of PDSIs reflects both soil moisture deficits and surpluses. Thus, PDSIs are time series of relative wetness and dryness. The computation of PDSIs from month to month includes an autogressive component that makes the value for any one month a weighted average of current and previous moisture

conditions. For this reason, the July index is an effective expression of moisture supply for the full growing season of the trees. Cook & Jacoby (1977) found that July PDSIs could be reliably reconstructed from tree rings, and they produced a July PDSI reconstruction back to 1730. Since that time, additional tree-ring chronologies have been developed from the Hudson Valley and have been incorporated into a new reconstruction back to 1694 (Cook & Jacoby, 1979). This 279-year sequence is shown in Figure 5.3. The reconstruction was verified against independent data (Cook & Jacoby, 1979) using three statistical tests that all passed at the 97.5 to 99.0% significance level.

ANALYSIS OF THE RECONSTRUCTION

Scrutiny of the July PDSI series reveals that in terms of both intensity and duration, the 1960s drought has never been exceeded since 1694. This finding supports Palmer's (1965b) more intuitively based estimate that, because the drought was of such intensity, "we should ordinarily expect it to occur in this region only about once in a couple of centuries". The reconstruction also indicates that wet and dry intervals tended to be more persistent in the past, although the dry anomalies were less severe than the 1960s drought. This tendency is especially evident prior to 1900. Such an increase in the persistence of wet and dry anomalies could have a significant impact on reservoir design and the ability of current reservoir systems to meet demands during longer periods of below-average precipitation.

Another way of analysing the PDSI reconstruction is to study different classes of PDSI variability. The frequency of occurrence of two classes was investigated: moderate-to-extreme drought (PDSI < -2.0) and moderate-to-extreme wetness (PDSI > +2.0). These classes are the most important ones to look at because large departures from the norm have the most impact on society. Figure 5.4 shows the frequency histograms of the PDSI series for its entire length (A), and the 18th century (B), the 19th century (C), and the 20th century (D). The histogram of the Mohonk Lake, New York PDSI data calculated from actual meteorological observations (E) is included to illustrate a bias in the reconstruction. By comparing the histograms of the actual and reconstructed series (D and E), we see that the tree rings underestimate

Figure 5.3: A tree-ring reconstruction of July Palmer Drought Severity Indices for the Hudson Valley, New York, The reconstruction is a series of regression estimates based on six tree-ring chronologies and five tree species (eastern hemlock, eastern white pine, pitch pine, white oak, and chestnut oak).

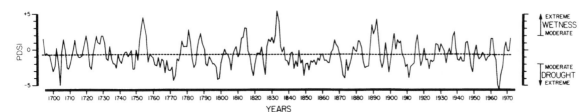

the frequency of moderate-to-extreme wetness by about
10%. This bias was not unexpected since moisture supply
above that which is needed for growth will not be very
well recorded in the annual ring widths. The frequency of
moderate-to-extreme drought in the tree-ring estimates is
very close to that in the actual data, 16% and 18%
respectively. Thus, the reconstruction is a more reliable
indicator of dryness than wetness.

The histograms indicate that the frequency of
moderate-to-extreme drought was 27% in the 18th century,
20% in the 19th century, and only 16% thus far in the 20th
century. The long-term average from 1694 to 1972 is 22%.
These percentages indicate that the frequency of
moderate-to-extreme drought has been significantly below

average in the 20th century, in spite of the serious event
in the 1960s. The histograms also indicate that the
frequency of moderate-to-extreme wetness was somewhat
higher in the past. The previously noted bias suggests
that the frequencies should be corrected up to 15% for the
18th century, 18% for the 19th century, and 13% for the
20th century. The 1694 to 1972 average is 16%. Again, the
20th century has a below-average frequency of wet
anomalies.

The below-average frequency of both wet and dry
years in the 20th century suggests that spring-summer
climatic variability in the Hudson Valley region has been
anomalously low. When the variance of the 1694-1972 period
is compared to that for the 1900-1972 period, July PDSI in

Figure 5.4: Frequency histograms of July PDSI for the entire reconstruction (A), the 18th century (B),
the 19th century (C), the 20th century (D), and for the Mohonk Lake, New York, meteorological station in
the Hudson Valley (E).

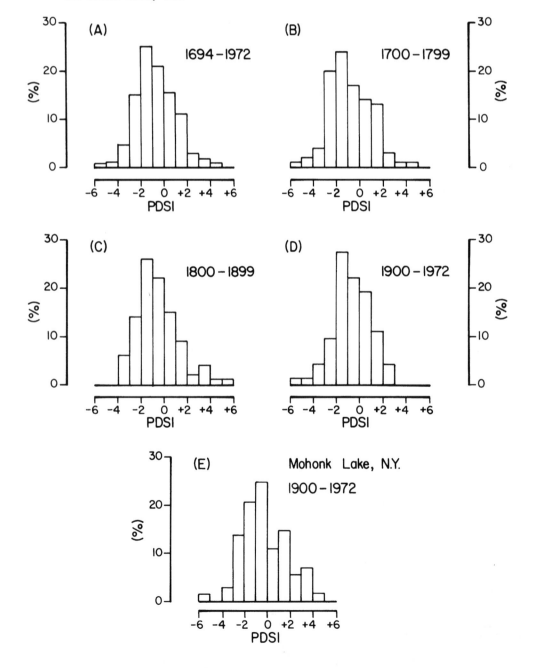

the latter period is 16% less variable. This lower variability is particularly evident during the period 1931 to 1960, once used to define "normal" climate, where the variance is 48% below that of the total period. In terms of variance, that period appears to be one of the most abnormal intervals in the past 279 years.

TEMPERATURE RECONSTRUCTIONS IN THE NORTHEASTERN UNITED STATES

L.E.Conkey

INTRODUCTION

Climate reconstructions have been developed successfully using trees from areas of low precipitation, such as the western United States (Fritts et al., 1979; Meko et al., 1980). Reconstruction of climate from trees in more humid areas such as the eastern United States has only recently begun. Because trees in such environments

Figure 5.5: Location of tree-ring sites and climate stations for temperature reconstructions, New England, U.S.A. Letters are tree-ring sites: A, Camel's Hump, VT; B, Nancy Brook, NH; C, Livingston, MA; D-G, Shawangunk Mts., NY; H, Elephant Mt., ME. Numbers are climatic stations: 1, Eastport, ME; 2, Burlington, VT; 3, Portland, ME; 4, Concord, NH; 5, Albany, NY; 6, Amherst, MA; 7, Blue Hill, MA; 8, Hartford, CT; 9, Mohonk Lake, NY; 10, New Haven, CT; 11, Block Island, RI; 12, New York City, NY; 13, Farmington, ME. Sites A – G and 1 – 12 were used in the analysis of winter temperatures; sites H and 13 were used in the summer temperature analysis.

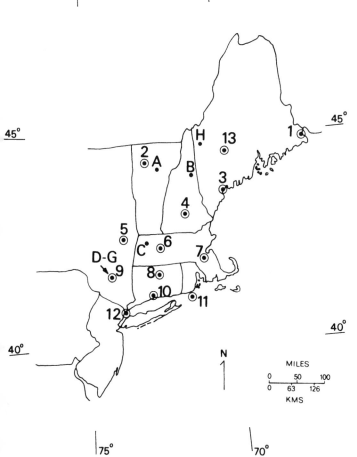

may be subjected to climatic stress less frequently than in areas of extreme cold or aridity, tree growth was previously believed by many to be too closely related to local microclimate and stand dynamics to provide a reliable record of macroclimate (Schneider & Conover, 1964; Phipps, 1970). However, recent studies have demonstrated the feasibility of dendroclimatic reconstructions using eastern trees (Conkey, 1979a; Cook, this volume). Two such reconstructions are presented here. The first one is derived from a spatial array of tree-ring width index chronologies; the second uses wood density data from an X-ray densitometric analysis. The reconstructed records are tested for reliability, and are discussed in terms of how the estimates may be improved by developing further methods in humid-area dendroclimatology.

TEMPERATURE RECONSTRUCTIONS

Winter temperatures. November through March temperatures from 1697 to 1970 were reconstructed, based on a canonical regression analysis using seven ring-width index chronologies (A-G, Figure 5.5) and 12 instrumental records (1-12, Figure 5.5) from the northeastern United States (Conkey, 1979a).

Because ring widths have been shown to be dependent on climatic conditions that prevailed in years previous to the current growing season, and because the width of the ring formed in one year can be related directly or indirectly to the widths of rings formed in following years (Fritts, 1976), tree-ring widths of three years (years t-1, t, and t+1) were used to estimate climate for each year, t. Here the tree-ring data are the independent variables, the predictors, and the instrumental data are the dependent variables, the predictands. Canonical correlations were calculated among all tree-ring and temperature variables, and canonical variate pairs were regressed to derive a predictive equation which best described the climate-tree-growth relationship during the period of calibration, 1906 to 1960 (Glahn, 1968; Clark, 1975; Conkey, 1979a; Fritts et al., 1979).

The canonical regression equation explained 38.5% of the variance of winter temperature at the 12 climate stations during the calibration period. Estimates of temperature at each of the 12 stations were then calculated for the years 1697 to 1970 from the regression equation. These were compared with instrumental records from the same stations for those periods (from 9 to 109 years, according to the length of each station's record) that were not included in the regression. These comparisons are, therefore, independent of the data used to generate the regression equation. The degree of similarity between the instrumental data and the reconstructions during these independent periods was tested according to verification methods described by Fritts (1976), Conkey (1979a), and Gordon (1980, this volume). The total percentage of verification tests passed at P < 0.05 by all stations was 50%, indicating a relatively high degree of similarity between the series. The six individual station reconstructions passing 42% or more of the verification tests were then averaged to create a regional estimate of past New England winter temperatures (Figure 5.6).

Closer examination of the verification results (Conkey, 1979a) indicates that the tests that were passed most frequently were those emphasising year-to-year, or high-frequency, similarities between the instrumental and reconstructed records prior to 1906. In other words, the reconstructed series show a pattern similar to the actual data with respect to changes from one year to the next, but they do not match the observed longer-term trends. Plots of the reconstructed and instrumental series after application of a low-pass digital filter (Figure 5.6, lower plot) show a gradual increase in values for the instrumental series (symbols) from 1900 to 1950, but this low-frequency change is conspicuously absent in the reconstruction for the same period (line plot).

The failure of the reconstruction to consistently reproduce longer-term changes in conditions possibly indicates inaccuracies in the tree-ring or climate data used in calibration. In preparing the tree-ring widths for use in calibration, each individual series was standard-ised, deriving yearly indices by fitting exponential or polynomial curves and then dividing by the value of the

curve. The indices were averaged for the two sampled cores from each tree and for all trees to produce each of the seven chronologies. This procedure removes ring-width variations due to episodes of released (large widths) and suppressed (small widths) growth that are due to individual tree and stand dynamics. Unfortunately, it is thought that use of polynomial curves in standardisation may also remove some of the long-term climate signal (Conkey, 1979b), possibly creating a reconstruction that does not exhibit as much low-frequency variation as do the instrumental data. This point has been discussed elsewhere in this volume.

Instrumental data are another potential source of error in calibration, possibly exaggerating long-term trends compared to actual conditions. There are at least three possible explanations for the existence of trends in an instrumental temperature record (Mitchell, 1953; Conkey, 1979a):

a) changes in observational circumstances, such as changes in exposure, elevation, or location of the thermometers, changes in types of instruments, or changes in observation time or calculation of the daily values;

b) local environmental changes in the vicinity of the thermometers caused by changes in vegetative cover, by growth of cities with addition of buildings and road paving affecting wind currents and surface

Figure 5.6: Averaged reconstruction of annual winter (November-March) temperature from six stations in New England, U.S.A.: Concord, NH (Site 4 in Figure 5.5); Albany, NY (5); Amherst, MA (6); Blue Hill, MA (7); New Haven, CT (10); and New York City, NY (12).

Upper plot is annual values; lower plot is with the application of a low-pass digital filter (Fritts, 1976). Line plots are the reconstructed values from ring-width index data; symbols are averaged instr-umental data.

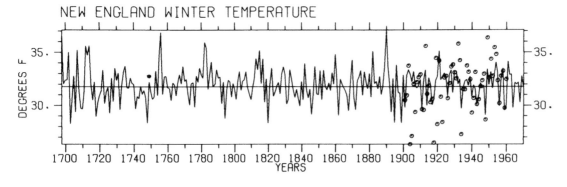

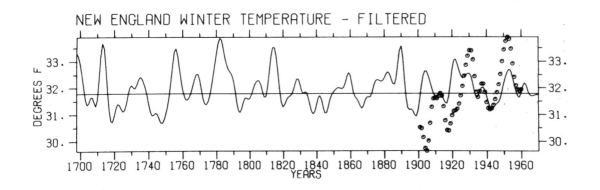

albedo, or by the addition of atmospheric pollutants
affecting solar radiation levels; and/or

c) actual change in macroclimatic conditions.

Any and possibly all three of these conditions may affect
the instrumental records used, such that shifts due to
non-climatic causes may mask or exaggerate any actual
macroclimatic trend.

There is some evidence to suggest that globally
winters did become, on the average, more mild from the
late 19th century on (Willett, 1950; Mitchell, 1961). In
addition, for at least one of the six stations used in the
averages presented in Figure 5.6, New Haven (Connecticut),
other factors of observational technique and local
environmental change apparently amplify the change in mean
winter temperatures in the instrumental record (Mitchell,
1953; Conkey, 1979a), thereby exaggerating the difference
between the observed and reconstructed means at that
station. Both the addition of artificial long-term trend
to the climate data and the removal of low-frequency
climate information from the tree-ring series may
partially explain the discrepancy between the long-term
trends of the instrumental and reconstructed series shown
in Figure 5.6.

Summer temperatures. Mean July through September temperatures from 1670 to 1976 were reconstructed from the
results of a stepwise multiple regression analysis using
ring-width and wood density index chronologies from a site
in the mountains of Maine, U.S.A. (Conkey, 1979b), and an
instrumental record from a nearby station.

Yearly values of minimum earlywood density, maximum
latewood density, earlywood width, latewood width, and
total ring width were obtained from cores of Picea rubens
Sarg. (red spruce) from Elephant Mt., Maine, by the X-ray
densitometric technique (Conkey, 1979b; Schweingruber,
this volume). Recent dendroclimatological studies have
shown that wood density, particularly maximum latewood
density, is closely related to climatic events occurring
in the summer months when the denser latewood is formed.
Parker & Henoch (1971) found high correlations between
maximum density and mean maximum August temperatures and
runoff in Canada. Schweingruber et al. (1978a,b), working
on Swiss alpine trees, found maximum density to be related
to summer temperatures. It was thought that maximum
density, perhaps in combination with other ring
parameters, would, therefore, be a successful indicator of
summer temperatures in New England.

Daily mean temperatures for July, August, and
September at Farmington, Maine (13, Figure 5.5), a town
50km southeast of the tree-ring site, were averaged. A
stepwise multiple linear regression (Nie et al., 1975) was
conducted, using as predictors index chronologies of the
five parameters of tree-ring width and density and as
predictands Farmington summer temperatures, for the years
1920 through 1976. Maximum latewood density alone was

significantly (P < 0.05) selected in the stepwise
procedure, accounting for 45% of the temperature variance.
The resulting regression equation was applied, using the
entire series of maximum density values at Elephant Mt.,
to produce a record of reconstructed summer temperatures
at Farmington from 1670 to 1976. Instrumental data from
Farmington previous to and including the 1920 to 1976
calibration period are plotted with the reconstruction for
comparison (Figure 5.7).

Tests of verification (Gordon, 1980) were applied
comparing the reconstructed values with the independent
Farmington instrumental data from 1900 to 1919. Three of
the six verification tests passed at P < 0.05, including
all tests of similarity at high frequencies and one
indicating similarity at low frequencies. The simple
linear correlation coefficient for the verification period
was 0.38, significant at P < 0.05. The low-pass filtered
series (Figure 5.7, lower plot) do show similar long-term
trends during the period of overlap of the reconstructions
and instrumental data, although the reconstruction shows
less trend and smaller amplitude than the instrumental
record. One characteristic of maximum density time series
is their low autocorrelation or persistence (Conkey,
1979b). This is manifested in the strong agreement of the
two series at high frequencies, and in the smaller
magnitude of low-frequency variations in the reconstructed
series as compared with the instrumental record. It is
likely that the use of some other physiologically
meaningful tree-ring parameter in addition to maximum
density will provide a reconstruction of summer climate
that is more reliable over a greater frequency range.

CONCLUSIONS

At the present time, it is clear that values of
maximum latewood density provide a means to increase the
climate information obtainable over that from ring widths
alone. A linear regression using maximum density alone
from a single site explained a higher percentage of
climate variance than more complex canonical regressions
using ring widths from several sites. High intercorrelations among ring-width and density variables prohibited
the multiple linear regression from adding ring-width
information to an equation including maximum density in
this case. Current studies are attempting to combine
variables in a physiologically meaningful way to avoid
such problems of variable intercorrelation and to increase
both explained variance and the reliability of future
climatic reconstructions.

EXTENSION OF RIVERFLOW RECORDS IN ARGENTINA

R.L.Holmes, C.W.Stockton & V.C.LaMarche Jr.

Based on material previously published in <u>Water Resources Bulletin</u>, <u>15</u>, 1081-1085, 1979).

INTRODUCTION

Knowledge of past variations in runoff or riverflow is of great interest to hydrologists, administrators, and engineers for such projects as dams, irrigation systems, hydroelectric installations, erosion and flood control, public water supply, navigation, and bridge construction. In most cases, gauged riverflow records exist for only a few years, and, in the best of cases, the data extend back only 100 years. Although such relatively short records are conveniently used as a basis for planning and engineering design, this record length appears to be too short to determine with any confidence the true flow frequency or the frequency, severity, and duration of droughts. Detection of long-term trends or possible cycles is impossible over such a short time, but may be feasible using long streamflow series reconstructed from records of tree-ring growth (Stockton, 1975). The Laboratory of Tree-Ring Research of the University of Arizona is undertaking a project for palaeoclimatic reconstruction from long tree-ring records in the Southern Hemisphere, and a 972-year record of estimated precipitation at Santiago, Chile, has already been developed (LaMarche, 1975). The results of a first attempt to reconstruct annual runoff for two important river basins in Argentina are presented here.

DATA BASE

During the austral summer of 1975-76, members of the Laboratory of Tree-Ring Research and of the Argentine Institute of Glaciology and Nivology took some 800 increment core samples from nine species of trees in the Patagonian Andes of Argentina, from northern Neuquén province to southern Tierra del Fuego (latitude 37°48′ to 54°51′ S). The core samples were mounted and surfaced and the annual rings were dated using the technique of cross-dating (Stokes & Smiley, 1968). Then, the width of each ring was measured to the nearest 0.01 mm. The series of measurements from each core was standardised to eliminate the biological age trend, and all of the standardised series were combined within each site to form a chronology. Two coniferous species turned out to have outstanding qualities for tree-ring studies – <u>Araucaria araucana</u> (Mol.) C.Koch and <u>Austrocedrus chilensis</u> (D.Don) Endl. <u>Araucaria</u> is found in the Argentine province of Neuquén and at corresponding latitudes across the Andes in Chile. <u>Austrocedrus</u> ranges as far north as latitude 32°S in Chile and southward in Argentina to 44°S (see Holmes, this volume).

The Río Neuquén and Río Limay in the Patagonian Andes of Argentina drain an area extending from latitude 36°S to about 42°S. They flow, respectively, southeast and northeast and meet at the city of Neuquén from which point the river has the name Río Negro and flows some 550km east-southeastward into the Atlantic Ocean (Figure 5.8).

Figure 5.7: Reconstruction of average summer (July-September) temperature at Farmington, ME, from maximum tree-ring latewood density values from Elephant Mt., ME. Upper plot is annual values; lower plot is after the application of a low-pass digital filter. Line plots are the reconstructed values from maximum density indices; symbols are instrumental data.

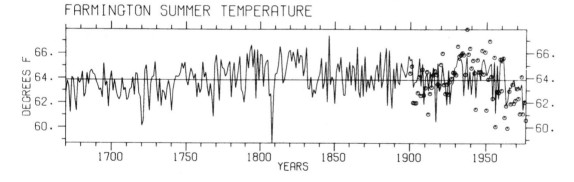

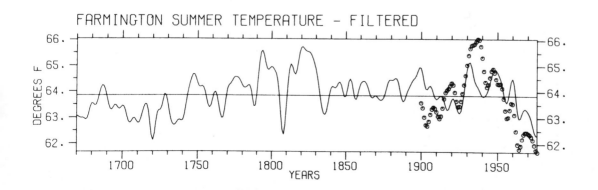

These major rivers have good quality continuous gauging station records beginning in April, 1903. The upper parts of their basins contain the densest network of tree-ring chronologies in Argentina, so these rivers were logical choices for a first attempt at flow reconstruction.

The chronologies used in the riverflow reconstructions were chosen on the basis of two criteria: first, location of the sampled stand of trees within the drainage basin of one of the two rivers; and second, a high correlation among the ring-width index series from trees within the site (a measure of the degree of common response to climate). Seven chronologies were thus selected, five of Araucaria and two of Austrocedrus. All seven were used in each reconstruction.

In the southern Andes, the prevailing air movement is from the west, thus the area drained by the Neuquén and Limay rivers is in the rainshadow of the Andes. Precipitation drops off steeply from over 2,000mm per year at the base of the Andes to under 200mm some 70 to 80km to the east, out into the Patagonian steppe. Most of the precipitation contributing to the runoff recorded at the gauging stations falls in the high mountains and foothills of the Andean chain. The stands of trees from which chronologies were developed are in the foothills or on the edge of the steppe. Thus, the tree-ring sites are close to the areas receiving most of the precipitation, but in areas sufficiently dry that moisture availability is frequently limiting to growth processes in the trees, so that dry years are reflected in narrower than normal rings (Fritts, 1976).

RECONSTRUCTIONS

Annual runoff of the Neuquén and Limay rivers, as gauged at the locations shown in Figure 5.8, was reconstructed using the seven tree-ring data series and the multivariate technique of canonical analysis. Details of the technique of canonical analysis used are outlined in Glahn (1968) and by Lofgren & Hunt (this volume). The technique provides a least squares equation maximising the linear relation between two sets of data. In this case, the predicted variables are the gauged annual riverflow data from the Neuquén and Limay rivers and the predictor data included modified series for each of the seven tree-ring data sites. Data from each tree-ring site were lagged such that the annual runoff value for year t was estimated from tree-ring data for years t+2, t+1, t, and t-1. Such lagging of the predictor variables is in agreement with a general autoregressive moving average (ARIMA) model which has been used by Stockton (1975) to model tree-ring series and by Wallis & O'Connell (1973) to model runoff series. The solution of the equations for the years 1903 to 1966 (n = 63) provided a transfer function from which estimated (reconstructed) values of past flow for the Neuquén and Limay rivers could be obtained from the much longer tree-ring chronologies. The annual runoff series were reconstructed back to the year 1601 although only that portion of the record from 1880 to 1966 is shown in Figure 5.9 for comparative purposes. In both cases, the correlation between the measured and reconstructed series is 0.73. The measured streamflow data appear to show more

Figure 5.8: Map of western Argentina, showing the locations of the Rio Neuguén and Rio Limay stream-gauge stations and the tree chronology sites.

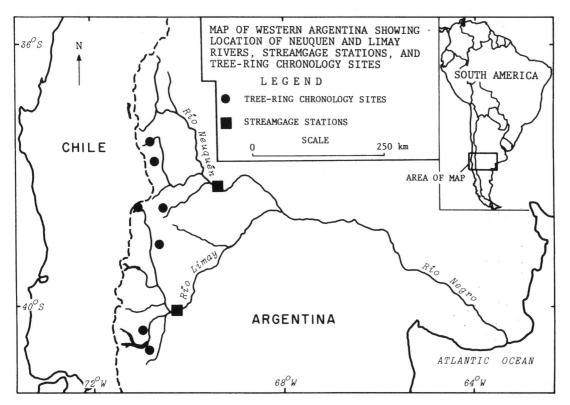

extreme high and low values than the reconstruction, particularly for the years of greatest flow and in the early part of the record.

CONCLUSIONS

This is a first attempt at reconstructing streamflow data from tree rings in South America, carried out without applying some of the statistical refinements that may be used in the future to extract additional and more precise information. Nevertheless, the good correlations suggest that:

a) the tree-ring series used have been precisely dated, confirming the applicability of traditional tree-ring techniques in more exotic locales;

b) the trees selected for sampling respond markedly to climatic variability;

c) the climatic factors influencing tree growth are related to total yearly runoff in these two river basins; and

d) the statistical techniques employed to perform the reconstructions are suitable for extracting much of the climatic and hydrological information contained in the tree-ring chronologies.

By employing the techniques described above, it seems feasible to extend runoff records back as far as the tree-ring chronologies go - in this case back to the year 1140 AD in at least one locality. The same techniques may be applicable to other areas in the Southern Hemisphere for which suitable tree-ring chronologies exist or may be developed. We plan to attempt in the near future to extend hydrological records for central Chile and for the province of Mendoza in Argentina.

PRELIMINARY ESTIMATES OF SUMMER STREAMFLOW FOR TASMANIA

D.A.Campbell

INTRODUCTION

The Australian state of Tasmania has only short instrumental climate and even shorter hydrological records. The confidence of hydrologic planning is severely restricted because the statistics on which it is based must be estimated from available data which are not necessarily representative of long-term conditions. It has been calculated that errors in sample estimates from annual runoff series of 40 years or less may reach 20% in the mean, 60% in the variance, and as much as 200% in the first order serial correlation coefficient (Rodriguez-Iturbe, 1969). Synthetic series generated from short

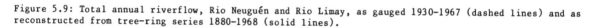

Figure 5.9: Total annual riverflow, Rio Neuguén and Rio Limay, as gauged 1930-1967 (dashed lines) and as reconstructed from tree-ring series 1880-1968 (solid lines).

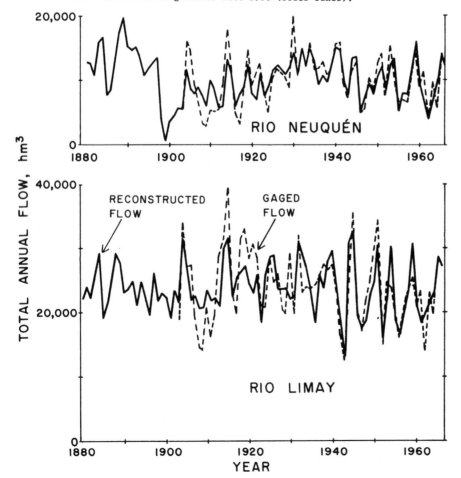

series will therefore be less reliable than those derived
from longer instrumental records.

Where practicable, extension of hydrologic records
using proxy data to increase the accuracy of estimates is
desirable. Extended records from tree rings may be used to
judge how well the recorded hydrological data reflect the
long-term past conditions and, consequently, how much
confidence should be placed in them for predicting future
streamflow. Under certain conditions, more reliable
estimates of the mean, standard deviation, and first order
correlation coefficient may be made from a reconstructed
record (Stockton & Boggess, 1980); that is, probabilities
may more reliably be assigned to events of particular
magnitudes or duration. Until now, rainfall records, the
longest of which commence in the 1880s, have provided the
only data for hydrologic planning in Tasmania. While
reliability of the hydrologic predictions is increased,
the gain in record length using precipitation data to
supplement streamflow data is only a few decades. Tree-
ring chronologies, however, may potentially be used to
extend hydrologic records up to several hundred years
(Stockton & Fritts, 1971c; Stockton, 1975; Stockton
& Boggess, 1980).

The results of a preliminary study using a set of 11
tree-ring chronologies and flow records from eight streams
in Tasmania, are presented. Tree-ring width and summer
streamflow (cumulative November to March) are both
affected by temperature and precipitation. It was, therefore, possible to describe empirical relationships between
the main variance components of the set of tree-ring site
chronologies and those of riverflow using a 16-year
calibration period, 1958 to 1973. The derived transfer
function was then applied to the 199-year tree-ring record
to estimate the five-month November through March
streamflow at eight gauging stations for the period 1776
to 1973.

DATA BASE

Dendrochronological data. The ring-width series
(chronologies) available for this study were collected for
a project to reconstruct past climate (LaMarche et al.,
1979d; LaMarche & Pittock, this volume). The sites were
selected to cover a wide geographical range and to include
several species, so that the data are not necessarily from
optimum locations for comparison with streamflow.
However, within a limited area, actual tree sites were
chosen to maximise the climate signal, a strategy that
would have been followed had the samples been collected
specifically for a dendrohydrologic study. As a general
policy, a minimum sample size of 10 trees was considered
to be desirable, but several of the series consist of data
from fewer trees.

The set of 11 chronologies comprised one chronology
each of Nothofagus gunnii (tanglefoot beech), one of
Athrotaxis selaginoides D. Don (King Billy pine), two of

Athrotaxis cupressoides D. Don (pencil pine), and seven of
Phyllocladus aspleniifolius (Labill.) Hook (celery top
pine). Although the longest chronology, MTF, extends from
1028 AD, the common period from the whole tree-ring data
set is determined by the shortest record, DRR, to begin in
1776. The tree-ring sites (Figure 5.10) are generally in
high rainfall areas and frequently include dense forest
stands, very different from the environments found to be
best suited for dendrochronological research in the
southwestern part of the United States. Yet, these sites
appear to yield chronologies sensitive to both
precipitation and temperature variations. The apparent
paradox is thought to be explained by the soils in which
the sampled trees grow. These soils tend to be shallow and
often are coarse-textured, so that local soil-moisture
deficits may occur despite the high precipitation. In
addition, the sites (except Pine Lake) from which the
chronologies were derived are on slopes and can be
considered to be well-drained. The Pine Lake site occurs
on a relatively level relict block-field with a shallow
water table.

Two Athrotaxis sites, Mount Field (MTF) and Pine
Lake (PNL), are at 1,200m elevation and have coarse, rocky
soils derived from dolerite. Most of the trees sampled at
Mount Field are growing on steep, east-facing slopes while
those at Pine Lake are scattered among large dolerite
boulders on only very slight slopes. The moderate
gradients of Weindorfer Forest (WDF) and the adjacent
Cradle Mountain (CMT) site are west-facing. The soil depth
was not measured but appears to be greater than at the
other sites; the forest floor is covered with organic
matter including logs in various stages of decay. The
seven Phyllocladus chronologies were developed from sites
with a range of parent materials, slopes, and aspects.
All these sites except Holly Range Road (HRR) have fairly
shallow soil; at Holly Range Road, a deeper clayey soil
has formed on schist. The Dove River Road (DRR) stand,
unlike others sampled, is relatively even-aged, and the
cored trees are generally younger than those from which
the other ten site chronologies were derived.

Hydrological data. Eight rivers with natural flow records
of at least 15 years were selected for study (Figure
5.11). Monthly streamflow data were extracted from the
Australian Water Resources Council (A.W.R.C.) Catalogue or
provided by the Hydro-Electric Commission (H.E.C.) of
Tasmania. No other natural flow records from Tasmania,
having at least 16 years' data to 1973, were available
(B.Watson, H.E.C., personal communication, 25.9.1979).
Several stream gauges established after 1958 now have
records complete to the present, but the available
tree-ring data whose final year is 1974 and the inclusion
of a lag of one year in the structure of the joint
analysis precluded the use of these shorter flow records.
All values are expressed as point depth in millimetres for

easier comparison of series from a wide range of catchment areas (254 to 2,539km²). H.E.C. gauging station numbers were used to identify streamflow records. The year in which the season begins has been assigned to flow data, so that, for example, the five-month November to March records are designated by the year of the November.

All eight stations record their minimum monthly flow in summer and their maximum in winter (June, July, or August), a reflection of the winter-dominant precipitation regime in western Tasmania. The seasonality of both precipitation and riverflow is less pronounced in the more eastern lower rainfall drainage systems of the Florentine, Gordon, Huon, and Derwent rivers (Stations 40, 46, 119, and 10087, respectively, Figure 5.11). Several records show a secondary minimum flow in June.

ANALYSIS AND RESULTS

The general approach to the chronology and streamflow data analyses is outlined in LaMarche & Pittock (this volume). The chronology and climate data networks used in this study and in theirs are identical. Response

function analysis and separate principal component analysis of the dendrochronological and hydrological data sets were followed by multiple and canonical regression analysis between the original data sets and the principal components of each. The transfer function developed from the canonical analysis of the principal components of flow and principal components of tree-ring indices was used to generate seasonal streamflow estimates at each gauging station for the period of the tree-ring data. These estimates were verified where possible and compared qualitatively with drought indices developed for selected precipitation stations (Watson & Wylie, 1972).

Response functions. Two series of response functions were calculated, one set between each of the 11 ring-width chronologies and selected climate data, the other set between each of the eight seasonal runoff records and appropriate climate data. The response function results guided the selection of a season when tree-growth variations were most strongly linked through climate anomalies to variations in streamflow.

Figure 5.10: Location of tree-ring sites. Species code: star, _Phyllocladus asplenifolius_; open square, _Athrotaxis selaginoides_; closed square; _Athrotaxis cupressoides_; circle, _Nothofagus gunii_.

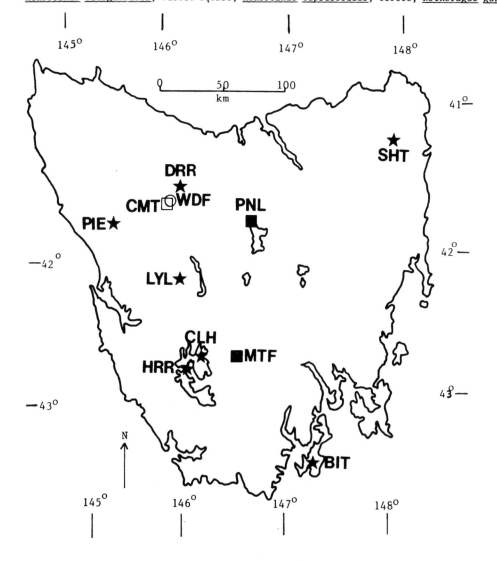

The development of various climate respone functions for the Tasmanian chronologies is described by LaMarche & Pittock (this volume). This study used the 18-month climate interval from the October preceding the summer of ring growth, included three years' prior growth, and terminated the stepwise regression analysis with the step at which the entering variable had less than 10% probability of entering at a higher F-level. The response function analysis for one chronology only reached the fifth regression step; the F-levels of variables entering the final steps ranged from 3.0 to 39.0 in the 11 response function analyses (for further details, see Campbell, 1980). The most frequent response to precipitation over all chronologies implies that wider rings form when either the current or previous summer, or both, have been wetter than average, and that narrower rings form after wetter than average winters. The apparent negative response to winter precipitation may be a positive response to solar radiation, as discussed by LaMarche & Pittock (this volume).

In the streamflow response functions, meteorological data for the 12 months from the beginning of April were compared with the April-March and November-March cumulative streamflow at each station. One year's prior

streamflow was included in the model. The results were predictable, with lower temperatures and higher precipitation being associated with increased total streamflow for both the 12-month and five-month totals. In no case was prior streamflow significant in determining current streamflow. The greatest variability in monthly streamflow occurs in all records in the summer months. This is also the season of tree-ring formation. The climate response functions for both the tree-ring chronologies and the streamflow records show, in general, positive responses to precipitation and negative responses to temperature for the summer months November to March. It seemed, therefore, that streamflow for the summer period was the most likely to be strongly linked to tree growth.

Principal component analysis. The results of the principal component analysis of the set of 11 Tasmanian tree-ring chronologies are discussed by LaMarche & Pittock (this volume). The first three components were used here in regression analyses.

A principal component analysis was performed on five- and 12-month streamflow totals from eight gauging stations, which were used as variables, for the 17-year period, 1958 to 1974. The first three principal components extracted from correlation matrices accounted for 98.1% of the variance in the five-month records and 97% in the 12-month series. Only the results for the November-March season are given, since it is was the amplitudes of these components that were used in reconstructing streamflow.

All eight records show strong common behaviour, as shown by the high (89%) variance accounted for by the first component, and values for this component were almost identical in all series. The amplitudes of the first component thus provide a sound estimate of the mean response of all watersheds involved. The record from the most northern gauge at station 159 exhibits a different secondary mode of variation from the others with a large negative value for the second component. This is possibly attributable to the location of Arthur River watershed in the path of summer northeasterlies. The other stations would be much less influenced by these storm systems. Only 6% of the total variance in the whole set of records is accounted for by the second component, and 11-12% might be expected to be due to chance alone. Despite this, the second component of the summer (November-March) streamflow was found to be significantly correlated with tree-ring width. The third component seems to show a north-south gradient with the higher positive values to the north, negative values to the south, and several stations showing insignificant weights for the component. The third component was made available to the multiple regression analyses, but was excluded from the later canonical analysis with tree rings.

Figure 5.11: Map of western Tasmania showing locations of stream gauging stations. Broken lines indicate associated watersheds.

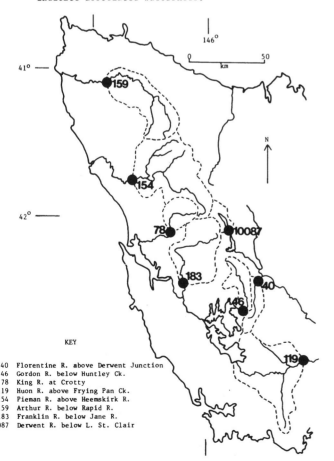

KEY

40	Florentine R. above Derwent Junction
46	Gordon R. below Huntley Ck.
78	King R. at Crotty
119	Huon R. above Frying Pan Ck.
154	Pieman R. above Heemskirk R.
159	Arthur R. below Rapid R.
183	Franklin R. below Jane R.
10087	Derwent R. below L. St. Clair

Data characteristics. Since violations of assumptions of normality and independence of predictor variables may seriously diminish the validity of regression results and the subsequent transfer function, these characteristics of the data sets were noted. In the 16-year calibration period, the tree-ring data, both the individual site chronologies and the amplitudes of the first two components derived from principal component analysis of the set of 11 series, are autocorrelated. Neither the observed November-March streamflow values nor the first two component amplitudes of the set of eight records, are significantly autocorrelated in the same period. Only one of the series in both data sets, was significantly skewed within the 16-year period. This was the first amplitude of November-March streamflow which was significantly negatively skewed and has a more peaked distribution than normal. The number of observations was too small for any other test of normal distribution to be made. In regression analyses, the difference in persistence between the chronology and hydrological data sets was dealt with by lagging independent tree-ring variables.

Multiple regression analysis. Individual stream gauging station data grouped to include the five-month cumulative total monthly streamflow for November through March were used as the dependent variable in a series of multiple linear regression analyses. The independent variables were the first three components of tree growth lagged so that streamflow in the current year, t, was compared with ring width years t-1, t, and t+1. Ring widths for three years were included to account for the autocorrelation existing within the tree-ring chronology amplitudes. If only those tree-ring variables entering at significant (95%) F-levels are considered, the first and second component of year t+1, and the first component of year t consistently entered the regression equation. In several cases, the second component in year t was also included. For the five-month (November-March) total, the first three variables entered the regression for each of the eight gauging stations.

These regression results confirm independently the results of the response functions in that the relationships between tree growth and streamflow were stronger in the summer months. The sign of the correlation coefficient, as each tree growth variable entered, was also consistent with information given by the response functions of the two main genera of the trees whose ring-width indices were included. That is, the amplitude of component 1 of tree growth reflecting _Phyllocladus_ chronology variations was correlated positively with current summer streamflow and negatively with flow in the previous summer. The second-component amplitude reflecting more the _Athrotaxis_ chronologies was positively correlated with streamflow both in the previous and current summer. In no case were the t-1 year tree-ring data significant in

predicting streamflow in year t.

Since the first component of this five-month series (for 1958 to 1974) accounts for 89% of the variance in the recorded streamflow at eight gauging stations, it is reasonably representative of the overall or mean response of the set of stations. It, as well as the second and third components, was used as the dependent variable in a multiple regression analysis with the independent variables being the first three component amplitudes of tree rings for year t and t+1. The results of these regressions confirmed those for individual stations, but the F-levels at which the variables, the first two components of the t+1 and then the current year, entered, were considerably improved. The first two component amplitudes of the five-month period were significantly correlated with tree growth at the 95% level. The correlations between the amplitudes reconstructed from the transfer functions or regression equations and the actual amplitudes derived from the principal component analysis of streamflow were 0.73 for the first amplitude of November-March streamflow and 0.57 for the second amplitude of November-March runoff. Correlations greater than 0.48 are significantly different from zero at the 95% level.

It was originally intended to terminate this project after the initial regressions because of the scarcity and shortness of streamflow records and the relatively sparse, but wide, spatial coverage of tree-ring sites. The results of the regression between streamflow amplitudes and tree-ring amplitudes, however, encouraged an attempt at preliminary streamflow reconstructions.

Canonical analysis. Two equations derived from multiple regression analysis can be used to reconstruct the two streamflow amplitude series separately. Also, the two amplitudes can be optimally combined by performing a canonical correlation analysis. For one matrix, the amplitudes of the first two components of November-March streamflow were used, and for the other, the first two component amplitudes of the tree-ring data lagged to include year t+1 as well as year t. This gave a total of six variables in the analysis.

Following the technique that Stockton et al. (1978) used for Palmer Drought Severity Index (PDSI) reconstructions, canonical equations were derived and solved for the calibration period, 1958 to 1973. The two resultant canonical regression equations accounted for 46.6% of the variance in the two amplitudes of November-March streamflow for the calibration period. The equations derived from the canonical and from the multiple linear regression both emphasise the importance of the components in year t+1, showing that the first component of streamflow has a strong positive correlation with chronology component 1 and a strong negative correlation with chronology component 2.

Solutions to the canonical regression equation were

applied to streamflow amplitudes estimated for the period of tree-ring record, 1776 to 1974. These reconstructed amplitudes were transformed to normalised streamflow values at each of the eight gauging stations in the original streamflow component network, by applying component weights.

DISCUSSION

The implications of the use of a negatively skewed, peaked series in canonical analysis are not fully understood. The violation of the assumption of normality affects the standardisation of the series and the calculation of canonical weights which indicate the degree of involvement of each station record in each of the two patterns (of variation) common to the tree-ring and streamflow data sets. It is thought that the results of the canonical analysis are probably valid, but that they need to be interpreted cautiously.

Good agreement between the reconstructed record and observed values in the calibration period is prerequisite for valid reconstructions (Gordon, this volume). Several tests of association between variables were performed on data in the 16-year calibration period, 1958 to 1973. There were five measures of association to which significance tests were applied, in addition to the calculation of the reduction of error (RE). Verification using independent data outside the calibration period was possible at only three of the gauging stations, a minimum of eight years being considered necessary for verification. Longer records were unavailable at the time of the analysis. The results of verification tests, in the calibration as well as in the independent years, are presented in Table 5.1.

In the calibration period, stations 119, 154, and 10087 were the only three that passed all association

tests, each of the others failing at least one of the sign tests. Stations 40 and 159 showed the poorest correspondence between estimated and observed data with only 36% and 30% variance calibrated, respectively; all the other stations had at least 50% variance calibrated, with the results for station 119 (Huon River above Frying Pan Creek) being the best. This station had 65% variance calibrated, corresponding to a correlation coefficient, r, of 0.83.

In the verification period, station 78 had a relatively low but still statistically significant r and a marginally significant first difference sign test. An examination of the plotted reconstructed time series for this station (Figure 5.12) suggests that the record was quite well duplicated except for the period 1937 to 1942 when the estimated series has very little variance and a lower mean than for most of the record. This period is reconstructed similarly at all eight stations.

The means and ranges of the reconstruction are neither consistently higher or lower than those of the corresponding series. An examination of Figure 5.12 reveals the underestimation of the higher streamflow values, even in the calibration period. Autocorrelation is higher and more persistent in the reconstructed records than in the observed, and this reflects the serial dependence within the tree-ring series used to estimate the longterm streamflow ones. The value of the autocorrelation coefficient of the estimated records remains significant and positive in the first five lags except for station 159. This persistence may reflect more the autocorrelation structure of the tree-ring chronologies and chronology amplitudes than that of the gauged streamflow records. The tendency in the observed series for high November-March streamflow totals to be followed immediately by low ones is absent in the reconstructions.

Table 5.1: Summary of results of association between recorded and reconstructed streamflow.

STATION	PERIOD	N (YRS)	CORRELATIONS		RE+	SIGN TESTS #		t TESTS# ON PROD. MEAN	D.F.	NO. OF PASSES
			r	FIRST DIFF.		FROM MEAN	FIRST DIFF.			
CALIBRATION										
Station 40	1958-1973	16	0.60*	0.61*	0.36	No	No	Yes	9	3
Station 46	1958-1973	16	0.77*	0.75*	0.60	No	No	Yes	10	3
Station 78	1958-1973	16	0.72*	0.70*	0.52	Yes	No	Yes	11	4
Station 119	1958-1973	16	0.83*	0.80*	0.68	Yes	Yes	Yes	8	5
Station 154	1958-1973	16	0.73*	0.72*	0.53	Yes	Yes	Yes	10	5
Station 159	1958-1973	16	0.52*	0.50*	0.30	No	Yes	Yes	9	4
Station 183	1958-1973	16	0.77*	0.75*	0.60	No	No	Yes	10	4
Record 10087	1958-1973	16	0.78*	0.76*	0.61	Yes	No	Yes	11	5
VERIFICATION										
Station 40	1922-1933									
	1951-1957	18	0.70*	0.75*	0.40	Yes	No	No	4	4
Station 78	1924-1957	34	0.30*	0.49*	-0.38	No	No	No	32	3
Station 119	1949-1957	9	0.57*	0.70*	0.11	No	Yes	No	2	2

*   Significant at 95% confidence level
RE+   Counted significant if positive but it is not included in number of passes
Yes (no) indicates that the result is significant (not significant) at the 95% confidence level

Stockton (1975) discusses the conditions under which the estimates of the mean and variance of the reconstructed series will be superior to those from the shorter observed record, and he reviews the work of several others who have addressed the question. One of the conditions necessary for an improved estimate of the mean is that of the absence of autocorrelation in the lengthened series (Stockton, 1975, p.70). The problem of introduction of autocorrelation into the reconstructions from highly autocorrelated tree-ring series needs to be resolved to determine whether the apparent serial correlation in the reconstructed series is climatically or biologically determined. Until this is clarified, statements about the relative qualities of the means and variances calculated from the recorded and from the estimated series cannot be justified.

Drought duration intensity indices for runs in excess of two years (Watson & Wylie, 1972) for Gormanston on the western edge of the catchment between 1895 (the beginning of the rainfall record) and 1970 were compared with the reconstructed series for the King River at Crotty (Station 78). These drought indices (Table 5.2) were calculated by Watson & Wylie (1972, p.7) using the formula of Yevjevich (1967):

$$\text{drought intensity} = [(x-\bar{x})/\text{SD}]/N \qquad (5.1)$$

where x is annual rainfall, $\bar{x}$ is mean annual rainfall, SD is standard deviation of annual rainfall and N is duration of deficiency in years. These drought data can be considered to be independent checks of the validity of the reconstructions. It must be kept in mind, however, that the reconstructions are for only a five-month season spanning parts of two calendar years, so that they do not coincide with the 12-month calendar year on which the drought indices are based. Nevertheless, since the drought data are for periods of two years or more, periods of prolonged low summer streamflow might be expected to be reflected in more intense droughts even though the comparisons are necessarily qualitative. The 1905 to 1908

drought is possibly reflected in the reconstructed record with lower summer streamflow in 1907 to 1909. In the intense drought periods, 1931 to 1933, 1935 to 1938, and 1948 to 1951, the reconstructed values are all below the long-term mean, but the 1917 to 1922 drought is not indicated.

CONCLUSIONS AND RECOMMENDATIONS

This investigation was handicapped by paucity of data. Within the bounds of statistical validity, many apparent links between tree rings, climate, and streamflow could not be explored further without, at least, better replication of tree-ring chronologies. Nevertheless, almost half the variation in a set of eight summer streamflow records could be explained by the main components of the available tree-ring chronologies in Tasmania. The tree-ring sites are widely spaced and are not ideally located to give information about the watersheds with which they are compared. It is possible, therefore, that the relationships between tree rings and streamflow shown in this study could be considerably strengthened with more strategic selection of tree-ring sites.

There is evidence that the various tree species respond differently to climate so that greater representation of all dateable species may give added information about climate and hydrology in particular seasons. While summer streamflow was most strongly related to tree-ring width in this investigation, streamflow for a 12-month period was also correlated with tree growth.

The use of tree rings to reconstruct hydrological records has traditionally been restricted to dry climates. This study presents tentative evidence that tree-ring chronologies developed from temperate rain forest species may also provide proxy hydrological information. The reconstructions of streamflow made here probably have inflated serial correlations and this problem needs to be addressed if future dendrohydrological studies are undertaken. The limited data available for verification

Figure 5.12: Reconstructed and observed November-March streamflow (depth in mm) at station 78. Broken vertical lines indicate the calibration period.

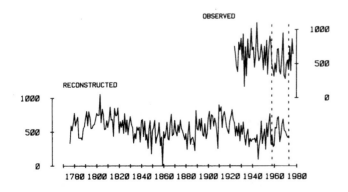

Table 5.2: Drought duration intensity for runs in excess of two years at Gomanston.

PERIOD	DROUGHT INTENSITY
1897–1900	0.99
1901–1903	0.43
1905–1908	0.44
1917–1922	0.67
1931–1933	0.80
1935–1938	0.50
1948–1951	0.97
1958–1961	1.17

suggest that the records estimated from the tree-ring series are reasonable. In Tasmania, dateable trees up to 2,000 years in age are known. The possibility of developing long chronologies from sites located in watersheds for comparison with the existing hydrological records, coupled with evidence presented in this study, offers considerable potential for using tree-ring chronologies to augment streamflow records in Tasmania.

PRELIMINARY TEMPERATURE RECONSTRUCTIONS FOR TASMANIA

V.C.LaMarche Jr. & A.B.Pittock

INTRODUCTION

The analysis of past climatic variations by study of tree-ring based proxy records is just beginning in the Southern Hemisphere (see Chapter 3, this volume). Early results include estimates of total annual precipitation in central Chile (LaMarche, 1978) and of stream runoff in Argentina (Holmes et al., 1979, this volume) and Tasmania (Campbell, 1980, this volume). In this report, we present the first temperature estimates from Southern Hemisphere tree-ring data, representing some of the initial output from a long-term programme of dendroclimatic analysis in south temperate latitudes (LaMarche, 1975; LaMarche et al., 1979a,b,c,d,e).

Dendroclimatic applications are expanding rapidly into new geographic areas and using unfamiliar tree species, many growing in ecological situations far removed from those in which some of the basic concepts and techniques were originally developed and applied (LaMarche, this volume). The Tasmanian work reported here and by Campbell (this volume) demonstrates how modern analytical techniques, coupled with dendrochronological, ecological, and climatological insights, can be used in a systematic way to develop empirical transfer function models yielding climate proxy records for new areas.

Tasmania was chosen for our first concentrated efforts at dendroclimatic reconstruction in the Southern Hemisphere because it provided a relatively dense network of long, high quality meteorological records together with an adequate tree-ring data base. The general geography, forest vegetation, ecological conditions, and broader climatic setting are discussed by Ogden (this volume) and by Salinger (this volume). Dendrochronological background, species characteristics, and some chronology statistics are discussed by Ogden (1978a,b; this volume) and the chronologies and site descriptions are published in LaMarche et al. (1979d).

DATA BASE

Tree-ring data. All the available Tasmanian chronologies were incorporated in the analysis. Eleven ring-width index chronologies from four species were used, including one from _Athrotaxis selaginoides_, two from _Athrotaxis cupressoides_, one from _Nothofagus gunnii_, and seven from _Phyllocladus aspleniifolius_. Three of these

(MTF, LYL, and BIT in Figure 5.13) represent composite records obtained by combining two or more chronologies for the same species growing on nearby sites (Ogden, this volume, Table 3.6). All are based on well-dated material, are well replicated, and were developed using standard indexing and averaging methods (LaMarche et al., 1979d). Because of the approach followed in this initial work (which uses principal components), the beginning date of the shortest record (1776 AD; WDF) determined the starting date of the reconstructed temperature record. It should also be noted here that, by convention, the date assigned to the annual ring is that of the calendar year in which growth begins, normally in the austral spring or early summer months of October through December.

Climatic data. Precipitation and temperature data were used in this study for two distinctly different purposes. The first was to model the climate-growth relationship empirically through calculation of a response function (Guiot et al., this volume) that expresses the effect of departures from long-term mean climatic conditions on current and subsequent ring width. The second was to develop a transfer function (Lofgren & Hunt, this volume) for estimating the value of a given climate parameter as a function of a combination of tree-ring variables, using data for a common calibration period, and to test the validity of the relationship by comparing these estimated values with actual instrumental data from a period different from that used in calibration.

The original data were obtained on magnetic tape from the Australian Bureau of Meteorology and consist of total monthly precipitation and monthly average daily maximum temperature values for individual stations. Maximum temperature was used in this study because it was thought to be more representative of conditions in the free atmosphere than minimum or mean values, and particularly to be freer of local boundary-layer affects such as cold hollows and shallow inversions. Separate grids of 15 temperature (Table 5.3) and 78 precipitation stations were then selected on the basis of their location, length of record, continuity, and homogeneity. Selection criteria and procedures are discussed in detail by Campbell (1980).

Response function calculations were made using monthly precipitation and temperature data estimated for each tree-ring site by interpolation based on principal component analyses of the respective climatological data grids; a procedure also discussed in detail by Campbell (1980). We believe this approach may yield better estimates of the actual climate at a remote tree-ring site than provided by any individual station record or arbitrary combination of records of surrounding stations. The principal components of the data for the 15-station temperature grid were also used for transfer function development, reconstruction, and verification.

METHODS AND RESULTS

Introduction. Different combinations of multivariate analytical techniques, including stepwise multiple linear regression, principal components, and canonical regression, were used in different stages of the analysis. These were chosen and applied in the light of our experience with this particular data set rather than as part of a predetermined scheme. Clearly, completely different approaches or different combinations of the same procedures might have been used to arrive at similar results, and no single approach is appropriate under all circumstances.

The general strategy we used was first to develop response functions for each tree-ring site that seemed to best characterise the tree-ring response to climatic variations. Then, the response functions were examined individually and collectively to identify those variables and those seasons for which the climatic influence on tree growth was greatest, and to identify potentially important lagging relationships. Next, we grouped the climate data into seasons to produce annual time series of values for each observational record, which emphasised those components of the climate signal having the strongest relationships to tree growth. Then, in part to reduce the number of variables needed for subsequent analyses, the grids of seasonal climate data and the tree-ring data grids were separately subjected to principal component analysis. In effect, this procedure should enhance the common signal present in each original data set, while reducing or eliminating unwanted noise. The next operation was to combine subsets of the climate and tree-ring components in a stepwise canonical regression analysis to generate an empirical transfer function in which tree-ring data are the predictor variables and climate data are the predictand variables. Finally, this last relationship was used to estimate annual values of a seasonal climate

Figure 5.13: Locations of tree-ring sites and of stations used for verification of temperature reconstructions.

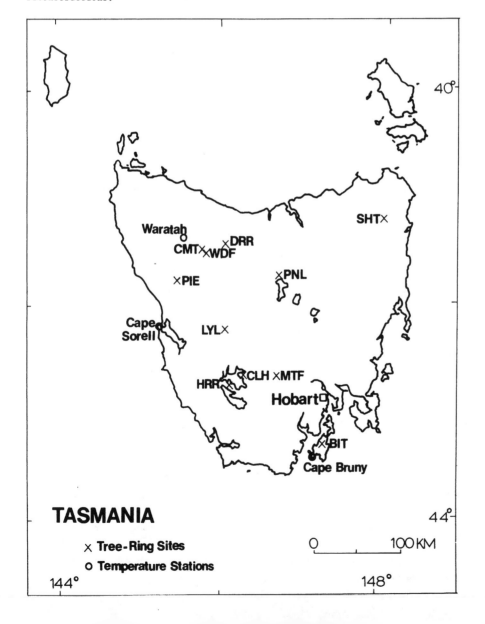

parameter at each original observational station, and the results tested by verification using independent data.

Response functions. We calculated response functions using each of the Tasmanian chronologies following the general procedure described by Fritts et al. (1971), further elaborated by Fritts (1974, 1976), and discussed in Chapter 2 of this volume. Because little was known initially about the phenology of these species, their physiological requirements, ecological relationships, or important growth-limiting factors, we experimented with variations on the standard schemes. Response functions were computed for each chronology using both a 12-month and an 18-month climate interval, and each was computed with and without the inclusion of three years prior growth as variables in the stepwise multiple linear regression (MLR) analysis. Our best results, as measured by the maximum percentage variance in the chronology explained by climate, and by the smallest number and highest relative rank of the climate principal components included in the last MLR step, were consistently obtained using an 18-month interval without prior growth. All analyses were done using interpolated monthly climate data for the period 1941 to 1970. The 18-month climate interval extends from October (austral spring) of the year previous to initiation of current growth and through the intervening summer and winter. It ends in March (austral autumn) of the calendar year following the year in which current growth begins. Thus, it extends over three calendar years and includes two complete growing seasons.

Study of the response functions for different species and sites in Tasmania shows some striking general features. Climate during the year prior to growth is, consistently, at least as important as current climate, and temperature rather than precipitation is the dominant factor related to tree growth. These results are not surprising when we consider that all the coniferous species are evergreen, most of the sites are in mesic, closed-canopy forests, and all receive moderate to high

amounts of precipitation, well distributed throughout the year. Another feature is the grouping of the temperature response patterns by genus, irrespective of the site. The apparent precipitation response is not only much weaker, but is highly variable from site to site, even within the same species.

Figure 5.14 presents the temperature portion of the response functions calculated for Mount Field Athrotaxis cupressoides, and Holly Range Road Phyllocladus asplenifolius, respectively. Both analyses used the 18-month climate interval, and prior growth was not included. Climate principal components cumulatively accounting for 80% of the climate variance were screened in the MLR analysis, and included if they entered the regression equation at an F-level of greater than 1.0. Climate accounted for 54% of the chronology variance in Phyllocladus (HHR). Seven climate components (ranked 1,2,3,4,5,8,10) entered the regression, and the resulting response function has 12 monthly elements significant at the 95% confidence level. For Athrotaxis (MTF), 50% of the variance is climate related, and only five climate components (1,2,3,6,9) entered the regression. The function also has 12 significant elements. Athrotaxis shows a dominant positive response to higher than normal summer temperatures in the preceding and current growing seasons. This is consistent with the high altitude of the site, where low temperature may be expected to have a limiting effect on growth-related processes, especially photosynthesis.

In contrast, the response of Phyllocladus to temperature is generally negative; notably, to the temperature of the previous warm season. The reasons for this pronounced lag in Phyllocladus's response to climate are not understood, but seem to be characteristic for this genus. We observe the same phenomenon in New Zealand. It is very probably related to the strong biennial oscillation in ring widths seen in Phyllocladus in both countries, as noted by Ogden (this volume). A positive response to June through August temperature (austral

Table 5.3: Tasmanian temperature records incorporated in 15-station grid. Not all records are continuous from the date given.

SEQUENCE NUMBER	REFERENCE NUMBER	STATION NAME	LATITUDE S	LONGITUDE E	EARLIEST DATE
1	91022	Cressy Research	41°43′	147°05′	1940
2	91057	Low Head Lighthouse	41°03′	146°47′	1895
3	91094	Stanley Post Office	40°46′	145°18′	1892
4	91104	Launceston Airport	41°33′	147°13′	1939
5	92033	St. Helens Post Office	41°20′	148°14′	1910
6	92038	Swansea (Maria St.)	42°07′	148°04′	1896
7	92045	Eddystone Point Lighthouse	41°00′	148°21′	1908
8	93014	Oatlands Post Office	42°18′	147°22′	1884
9	94010	Cape Bruny Lighthouse	43°30′	147°09′	1924
10	94029	Hobart Regional Office	42°53′	147°20′	1841
11	94041	Maatsuyker Is.	43°40′	146°16′	1936
12	94056	Risdon	42°50′	147°19′	1923
13	97000	Cape Sorell Lighthouse	42°12′	145°10′	1900
14	97014	Waratah Post Office	41°26′	145°31′	1892
15	98001	Currie Post Office	39°56′	143°52′	1913

winter) is also seen in most Tasmanian _Phyllocladus_ response functions and is associated with the apparent negative effect of winter precipitation. It is possible that both temperature and precipitation variables are acting as proxies for available light, since it is known from our observations that _Phyllocladus_ responds strongly and positively to higher light levels. Thus, more light may be available for photosynthesis during clearer weather in the warmer and drier periods in winter. A similar explanation is offered by Brubaker (1980) for results obtained in mesic coniferous forest sites in the Pacific Northwest.

Although the direction of the temperature response is opposite in the two genera in most months, each function shows considerable seasonal coherence, with two or more successive months frequently showing significant departures in the same direction. Furthermore, these "natural seasons" correspond roughly in both genera. Therefore, we identified the eight-month warm season, October through May, as the period within which anomalous temperatures are most likely to produce annual rings that are wider or narrower than average. Although warm-season

temperature is not the only variable influencing ring width, the strength and seasonal coherence of the tree-growth response led us to expect that this parameter ought to contribute a dominant component of variance to the tree-growth records. Therefore, we selected the mean monthly average daily maximum temperature of the warm season as the single climate variable most likely to be successfully reconstructed using the available tree-ring data base in Tasmania.

Principal component analysis. Principal component (or eigenvector) analyses of the climate and dendrochronological data sets were carried out following procedures outlined by Sellers (1968) and LaMarche & Fritts (1971b). Our main objectives were to reduce the number of variables from the large number present in each of the original data sets, and to ensure that the new, reduced sets consisted of mutually uncorrelated variables.

The temperature data set consists of time series of mean daily maximum warm-season temperatures for the interval October through May for the period 1941 (October 1941-May 1942) to 1969 (October 1969-May 1970) at each of

Figure 5.14: Temperature portions of representative climate response functions for two coniferous genera.

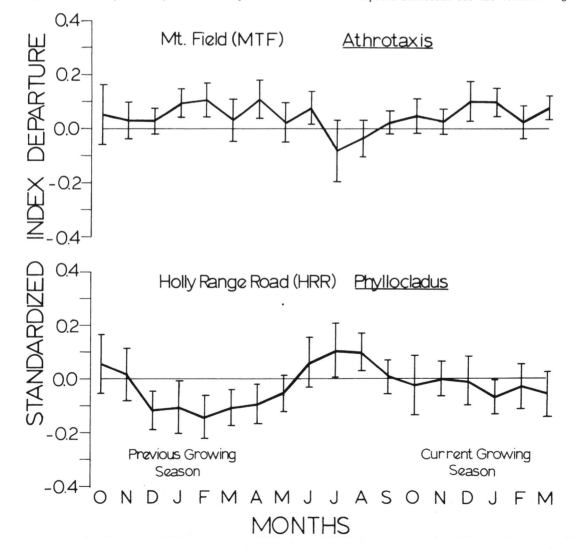

the 15 stations. Principal component analysis transformed these data to a set of 15 components, each with its associated 29-year amplitude series. Since the original variables are highly intercorrelated, the first few components account for most of the variance in the original data field. Here, components 1 through 4 explain 74%, 7%, 6%, and 4% of the variance, respectively, for a total of 91%. Most of the remaining components probably represent small-scale variability or errors in the original data. Maps of the component elements (not shown here) show that the first and most important component is associated with broad-scale temperature anomalies influencing the whole island in the same way. The second represents a pattern in which the eastern part of Tasmania shows an anomaly opposite in sign from that in the west. The third contrasts central Tasmania with the east and north. The fourth component emphasises only anomalies affecting the extreme northeast; it seemed relatively unimportant and was not included in further analyses.

The tree-ring data set includes the 11 site and composite chronologies previously described. The annual values are the mean ring-width indices (LaMarche et al., 1979d), and the period of analysis is 1776 to 1974. The analysis and the resulting reconstructions could have been extended to an earlier date if the Weindorfer Forest (WDF) Nothofagus series had been excluded, but we were interested in evaluating the performance of this species in comparison with the conifers. The first four (of 11) tree-ring components together explain 66% of the total variance in tree growth, accounting for 35%, 14%, 9%, and 8% of the variance, respectively. In contrast to the case of the temperature components the two most important tree-ring components do not represent broad scale spatial anomaly patterns. Instead, the first two are clearly

associated with differences between the chronologies of the two coniferous genera, seemingly independent of geographic location. As shown in Table 5.4, the elements of the first component are moderately to highly negative for Phyllocladus, and somewhat negative for Athrotaxis. Component 2 has near-zero to moderately positive elements for Phyllocladus and highly negative elements for Athrotaxis. Nothofagus has intermediate values. The elements of components 3 and 4 have larger ranges of values for all species. Their map patterns (not shown here) suggest that they do reflect spatial anomaly patterns, quite possibly linked to temperature anomalies, but a more extensive tree-ring data base may be required for further analysis of this possibility.

One of the main reasons for the sharp discrimination between genera shown in the results of the principal component analysis (especially components 1 and 2) may lie in the different time-series characteristics of the Athrotaxis and Phyllocladus chronologies. Figure 5.15 shows recent portions of representative tree-ring records for Athrotaxis selaginoides (CMT), Athrotaxis cupressoides (PNL and MTF), and Phyllocladus aspleniifolius (CLH, HRR). Visibly, there is much less high-frequency variation in both species of Athrotaxis than in Phyllocladus. This observation is borne out in the chronology statistics presented by Ogden (this volume), which show that three Tasmanian Athrotaxis cupressoides records and a single Athrotaxis selaginoides record have an average mean sensitivity (a measure of year-to-year variability) of only 0.13 and 0.14 respectively, whereas 10 Phyllocladus

Table 5.4: Tree growth anomaly patterns in Tasmania, 1776-1974. The sign and magnitude of the principal component elements indicates the importance of the associated anomaly pattern in explaining ring-width variations at each site.

TREE-RING SITE	MAP DESIGNATION	PRINCIPAL COMPONENT			
		1	2	3	4
Phyllocladus					
Dove River Road	DDR	-0.18	-0.04	-0.34	0.85
Pieman River	PIE	-0.37	-0.03	0.30	0.07
Lyell Highway	LYL	-0.40	0.08	0.05	-0.11
Holly Range Road	HRR	-0.40	0.00	0.09	-0.14
Clear Hill	CLH	-0.37	0.28	-0.08	0.09
Bruny Island	BIT	-0.36	0.15	-0.31	-0.31
St. Helens	SHT	-0.37	0.11	-0.13	-0.08
Athrotaxis					
Cradle Mountain	CMT	-0.12	-0.47	0.36	0.25
Pine Lake	PNL	-0.12	-0.61	-0.13	-0.22
Mount Field	MTF	-0.12	-0.51	-0.53	-0.02
Nothofagus					
Weindorfer Forest	WDF	-0.25	-0.16	0.48	0.12

Figure 5.15: Recent portions of tree-ring chronologies for three coniferous species showing contrast in high-frequency variability between Phyllocladus aspleniifolius (CLH, HRR), Athrotaxis cupressoides (PNL, MTF), and Athrotaxis selaginoides (CMT).

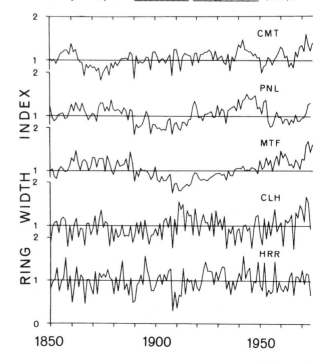

asplenifolius chronologies yield a mean value of 0.29. Corresponding values for the first order autocorrelation coefficient (a measure of persistence) are 0.64 and 0.26, respectively. To separate time/frequency-domain patterns of variability in data sets of this kind, pre-treatment with digital filters may be useful in future analyses incorporating a larger data base.

Canonical regression analysis. A stepwise canonical regression procedure similar to that described by Fritts et al. (1979) was used to derive a transfer function (Lofgren & Hunt, this volume) to be used in estimating past temperatures from tree-ring data. The predictand data set consists of the amplitudes of the three most important principal components of warm season temperatures in Tasmania, previously derived from analysis of the 15-station grid, which together account for 87% of the variance in the original data set. The predictor variables are the amplitudes of the four most important components (66% of the total variance) of tree growth from analysis of 11 chronologies, including current and two lagged values, for a total of 12 predictor variables. Tree-ring variables representing lagged relationships between climate and the growth were incorporated in the derivation of the calibration equation because the response functions had shown that the temperature of the current growing season (year t) has a marked effect on the width of the ring formed in the following year (t+1). Inclusion of an additional year's lag (t+2) was suggested by the strong biennial oscillation in Phyllocladus ring-width series, which we interpreted as an indication that lags of greater than one year may be important in climate-growth relationships in this genus. The importance of lagged relationships was also borne out by the structures of a series of preliminary multiple linear regression models that we had developed to test the predictability of an individual station record (Cape Sorell) from the tree-ring principal component data set, using different calibration periods and different combinations of predictor variables.

The canonical regression analysis yielded three canonical variates, corresponding to the number of predictand variables. These have associated canonical correlation coefficients of 0.87, 0.85, and 0.34, respectively. Only the first two canonical correlations were considered important enough to warrant incorporation in further analyses. Examination of the coefficients shows that the first canonical variate strongly emphasises the relationship between temperature and the second, third, and fourth components of current and lagged tree growth. The second canonical variate emphasises the importance of the first tree-growth component in the year concurrent with (t) and following (t+1) the temperature departure. The structure of the canonical model is similar to that of the best MLR model derived for Cape Sorell, evaluated on the basis of verification using independent data.

Finally, the canonical equations were solved to yield regression coefficients expressing the relationship between the 12 predictor variables and the three predictand variables. That is, the final result of this phase of the analysis is a set of three equations, one for each of the three temperature components, each of which expresses the estimated value of the amplitude of that component as a function of the value of the amplitudes of the four tree-ring components in the current and two following years.

Reconstruction and results. Estimates of the yearly values of the mean maximum October-May temperature were made for each of the 15 stations incorporated in the original temperature data grid (Table 5.3). These were obtained by first estimating the values of the amplitudes of the three most important temperature principal components using the canonical regression equations previously discussed, in which the tree-ring component amplitudes are the predictor variables. The resulting estimated yearly temperature amplitudes were then each multiplied by the elements of the corresponding temperature component for each station, and the respective products summed. Finally, because the estimates were in standard normal form, the mean and standard deviation of the observed seasonal temperature record for the calibration period were used to transform these normalised estimates to units of degrees Celsius. The estimated values for each station encompass the period 1776 to 1972, because this is the period for which the tree-ring principal component analysis was done, less the two years (1973 and 1974) lost because the t+1 and t+2 tree-ring values were included in the regression.

An important test of any scheme for estimating past values of a climate parameter from proxy data sources such as tree rings is its ability to approximate the observed values of the parameter in a data set independent of that used in developing the transfer function or calibration relationship (Gordon, this volume). We sought independent verification of some of our initial results by comparing the tree-ring based temperature estimates with observational data for periods preceding the beginning of the calibration period. We selected three stations for these tests on the basis of their location near the northern (Waratah), western (Cape Sorell), and southern (Cape Bruny) limits of the tree-ring data network, and the existence, for each station, of a reasonable length of record prior to 1941. The observed and reconstructed temperature series are plotted in Figure 5.16.

A variety of parametric and non-parametric tests were applied to compare quantitatively the reconstructed and observed temperature series. These are discussed in detail by Fritts et al. (1979), and are intended to test the reliability of a transfer function model in providing estimates of a climate parameter. Our results are summarised in Table 5.5. The general agreement between

Figure 5.16: Observed and reconstructed warm season (October–May) temperatures at three Tasmanian stations: A, Waratah Post Office; B, Cape Sorell Lighthouse; C, Cape Bruny Lighthouse.

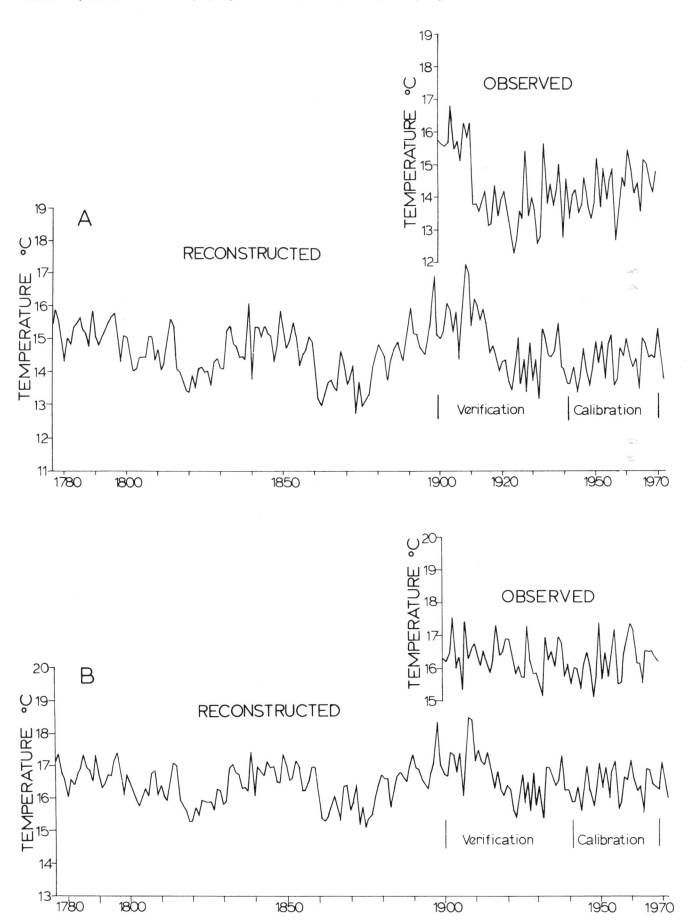

estimates and observations is good, as measured by the high simple linear correlation coefficients (r) for all stations in both the dependent and independent data periods, as well as by the significant results of all the product mean tests, and by the generally positive values of the more rigorous reduction of error (RE) statistic. Tests on the first differences of the series gave mixed results, indicating differences in the direction of year-to-year changes of estimated and observed values, although these differences may be small. It should be emphasised that the simple sign test and RE are both sensitive to departures of the mean of the estimated values from the mean of the observed values in the verification period (Fritts et al., 1979). Nonetheless, the overall results compare well with those from other tree-ring climate models, such as those used to estimate winter temperature and precipitation over North America (Fritts et al., 1979; their Table 3), and annual precipitation in California (Fritts & Gordon, this volume).

DISCUSSION

The results of this first attempt to reconstruct an important climate parameter such as surface temperature, together with other recent findings (Holmes et al., this volume; Campbell, this volume) show that dendroclimatic studies have great potential for future applications in Australia and in other areas of the Southern Hemisphere. Although we believe that the reconstructions presented here may be among the best available indicators of the recent climatic history of Tasmania prior to the start of

continuous instrumental records, evaluation of their potential and limitations must be made in light of some important qualifications regarding the data base and the methodology.

Examination of the estimates of warm-season temperatures in Tasmania since 1776 (Figure 5.16) reveal some important general points. Slow variations of the order of 1 to 2°C are apparent over time periods of up to several decades, and the means of different decades appear to have differed significantly. The warmest periods seem to have been the early 1830s through the late 1850s, and the late 1890s to mid-1910s. The reconstructions show cooler periods during the mid-1810s to late 1830s, and the late 1850s to late 1870s. The period since the late 1910s also appears somewhat below the long-term mean, except for the late 1930s. Although we have reconstructed temperatures cooler than recent normals for periods of up to a few decades, there is no evidence that temperatures in Tasmania were persistently below recent values over intervals of a century or more during the later part of Europe's Little Ice Age. This may mean that this part of the Southern Hemisphere did not follow the same trends as the North Atlantic sector, or could reflect the influence of the ocean surrounding Tasmania in moderating temperature extremes. Extreme temperatures, as are observed on the Australian mainland, are rare in Tasmania. The occasional continental influence is seen only when strong flow of heated air from the interior mainland in the north penetrates as far south as Tasmania, or when cold outbursts, greatly modified by the ocean, reach the island from the south.

Figure 5.16: Continued.

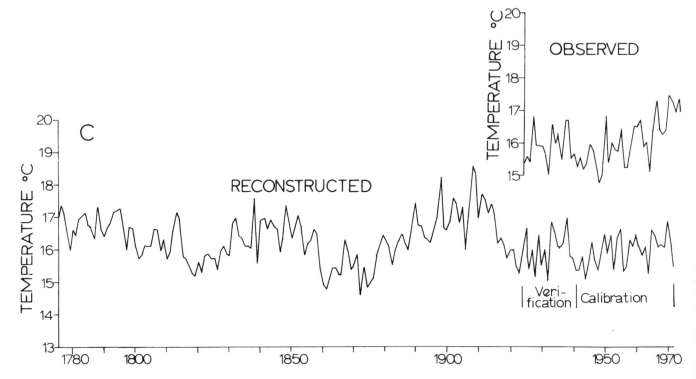

The apparent absence of very large, very low-frequency temperature variations during the past 200 years could also be in part an artificial result, caused by removal of certain long-term trends and fluctuations in tree growth as a result of the ways in which the raw tree-ring data are processed, but we think this less likely. We are also uncertain about the performance of the principal components and canonical regression procedures in the low-frequency domain, particularly where the period of calibration is short, and where the tree-ring and climate data series may not exhibit trends that are more characteristic of the longer reconstruction period. However, some degree of confidence in the fidelity of the lower-frequency components of our reconstructions is warranted by the good general agreement with the longer observational records presented here, as well as the smoothed record of February temperatures for six Tasmanian stations shown by Pearman et al. (1976) which illustrates the same general downward trend from the 1890s to about 1950 that we observe in our reconstructions.

The methodology employed has some important consequences for the characteristics of the reconstructed data and their relationships to the observational records. It is clear from Figure 5.16 that the three estimated series for Waratah, Cape Sorell, and Cape Bruny bear greater similarity to each other than do the respective observational records. This similarity results in part from the fact that the same set of tree-ring principal components were used for calibration purposes. More important, however, is the use of a limited number of components of temperature and the use of only two of the three canonical variates. Although the most important components preserve information on broad-scale anomaly patterns and island-wide trends, local temperature variability and possible errors or inhomogeneities in individual station records are largely eliminated. Better verification results would probably have been obtained if we had used spatially smoothed observational data for this purpose. Station-by-station reconstruction and verification might be improved if more principal

components of temperature were incorporated in the canonical analysis. For individual record reconstruction, other methods, such as multiple linear regression, might also be appropriately used. Possible problems with instrumental records are illustrated by Waratah (Figure 5.16a), where an abrupt decrease in temperature is apparent between 1910 and 1911, superimposed on a more general downward trend. We have endeavoured to discover the possible reasons for this anomalous discontinuity by research into station records, but can find only reference to a change in instrument location on completion of a new Post Office in 1913, which does not appear to account for the anomaly; a Stevenson Screen apparently was in use well before 1910. Unfortunately, site records here, as elsewhere, are far from complete.

Although artificial predictive skill does not appear to be a major problem in our analysis, the substantially poorer verification statistics for the independent data period in comparison with the calibration period pose some problems. Conservation of degrees of freedom through the inclusion of fewer tree-ring variables and fewer lags, as well as lengthening the calibration period might provide a better balance between calibration and verification statistics, and more reliable estimates.

RECONSTRUCTED ANNUAL PRECIPITATION FOR CALIFORNIA

H.C.Fritts & G.A.Gordon

INTRODUCTION

We have calibrated accurately dated tree-ring chronologies surrounding California, U.S.A., with instrumental precipitation data for the state, and the resulting transfer function was used to reconstruct past variations in precipitation from past variations in tree growth. The reconstructions are described for the years 1600 AD through 1961 AD and compared to 20th century experience.

PRECIPITATION DATA

When transfer function techniques are used to reconstruct variations in climate, a long instrumental

Table 5.5: Verification results for warm-season temperature estimates for Tasmania.

STATION	NO. YEARS	r	FIRST DIFF.	SIGN TEST#	FIRST DIFF.	PRODUCT MEANS TEST#	REDUCTION OF ERROR
CALIBRATION PERIOD (DEPENDENT DATA)							
Waratah	29	0.90*	0.90*	Yes	Yes	Yes	0.79
Cape Sorell	29	0.88*	0.87*	Yes	Yes	Yes	0.76
Cape Bruny	29	0.74*	0.64*	Yes	No	Yes	0.55
VERIFICATION PERIOD (INDEPENDENT DATA)							
Waratah	42	0.63*	0.60*	No	Yes	Yes	0.12
Cape Sorell	41	0.43*	0.48*	No	No	Yes	-0.58
Cape Bruny	17	0.56*	0.61*	No	No	Yes	0.04

* Significant at the 95% confidence level   # Yes (no) indicates result of test significant (not significant) at the 95% confidence level

record is required to allow both calibration and verification (Lofgren & Hunt, this volume; Gordon, this volume). Therefore, the most important criterion for selecting the instrumental data was the length of record. A second criterion was the representativeness of the data for the entire state.

Meteorological records, particularly earlier records, are subject to both systematic and random errors. A survey of 19th century records (Kennedy & Gordon, 1980) reveals several possible types of non-climatic influence on the data. Some of these influences can be detected as inhomogeneities in individual station records by comparing the data with those from surrounding stations. The data were first examined visually for unusual values. An automated computer routine was then used to check for years out of sequence, missing data, and unusually large (or small) monthly totals in normally dry (or wet) months. Problems were noted, and whenever possible, rectified by reference to the U.S. Weather Bureau's Climatic Summaries or by comparing the data with those from nearby stations.

Estimates for occasional missing values were obtained by a multiple regression which used data at surrounding stations as estimators. The number of missing monthly observations amounted to less than 0.5% of the total observations, and missing data estimation is expected to have little if any effect on the overall precipitation estimates. After missing observations were estimated, the stations were screened for spatial homogeneity. This was accomplished using double-mass plots and the Mann-Kendall test (Kohler, 1949; Bradley, 1976). Screening reduced the number of acceptable stations to 51. The 18 stations which had continuous records beginning in 1872 or earlier were then selected for the final data set (Table 5.6 and Figure 5.17). The 18 precipitation records were averaged together to form a sequence of statewide monthly average precipitation totals to be used in the calibration and verification.

SELECTION OF DENDROCLIMATIC DATA

The large-scale climatic patterns that affect the precipitation of California (Pyke, 1972) can be reflected in spatial patterns of tree-ring widths throughout the western United States (Fritts et al., 1979). The work of Fritts & Shatz (1975) indicated that an adequate number of western tree-ring chronologies extended back to 1600 AD. Therefore, we surveyed the available chronology data and selected those sites located within 970km of the California-Nevada border at Lake Tahoe that extended back to at least 1599 AD. A total of 63 chronologies were found which conformed to those initial criteria. Thirty-one of these chronologies which were from the network selected by Fritts & Shatz (1975) were combined with 21 additional chronologies, many of which had become available since 1975. When several highly correlated chronologies were available from a particular area, only those with the best overall statistics and incorporating the greatest number of cores (or samples) were selected. A total of 52 chronologies from 10 different coniferous species were used in the following analysis (Table 5.7, Figure 5.18).

The dominant spatial modes of variation in the dendroclimatic predictor data can be extracted by means of principal component analysis (Lofgren & Hunt, this volume). The principal components can then be used as predictors in place of the individual chronologies in the transfer function equations. We used only the first 15 principal components which accounted for 77% of the variance in the original tree-ring data set. This helps to eliminate small-scale variance that is most likely to be "noise" from the empirical climate-growth relationship and emphasises the large-scale dominant modes of ring-width variation in the chronology network (Fritts, 1976).

Table 5.6: The selected California precipitation stations used for calibration and verification.

NUMBER	ABBREVIATION	STATION NAME	FIRST YEAR	FIRST MONTH	US WEATHER BUREAU I.D. NUMBER
1	YREK	Yreka, Calif.	1872	10/1871	F2986600H
2	EUR	Eureka, Calif.	1872	10/1871	00412910H
3	RDG2	Redding, Calif.	1872	10/1871	004B7296H
4	REDB	Red Bluff, Calif.	1872	10/1871	004B7290H
5	CHCO	Chico, Calif.	1872	10/1871	00421715H
6	NEVA	Nevada City, Calif.	1872	10/1871	004B6136H
7	MARY	Marysville, Calif.	1872	10/1871	004B5385H
8	AUBR	Auburn, Calif.	1872	10/1871	00420383
9	RKLN	Rocklin, Calif.	1872	10/1871	004B7516H
10	DVS2	Davis Exp. Farm, Calif.	1872	10/1871	00422294H
11	SRNB	San Rafael, Calif.	1872*	10/1871	004C788BH
12	STK4	Stockton, Calif.	1872#	10/1871	00458560
13	LIV2	Livermore, Calif.	1872	10/1871	004C4997H
14	MER2	Merced, Calif.	1872	1/1872	004D5532G
15	SLOP	San Luis Obispo, Calif.	1872	10/1871	00447851H
16	SBAR	Santa Barbara, Calif.	1872	10/1872	004E7902H
17	LA	Los Angeles, Calif.	1872	10/1871	00465115H
18	SND	San Diego, Calif.	1872	10/1871	00467740H

* Three months of data estimated for 10-12/1963   # Ten months of data estimated for 11/1925; 9-12/1948; 1-5/1949

MODEL STRUCTURE, CALIBRATION, AND VERIFICATION

Model structure. The occurrence of precipitation in
most of California is highly seasonal, with a wet winter
and early spring and a dry summer. This seasonal pattern
is common to all the low-lying stations, but the spatial
pattern is of a much greater complexity (Granger, 1977).
It is generally believed that anomalous precipitation in
California is related to the anomalies in the general
circulation of the atmosphere (Granger, 1977). The
dominant factor controlling precipitation in California is
the Pacific subtropical high pressure cell which
fluctuates on a very large spatial scale. The southward
migration of this high pressure region during winter
allows the tracks of precipitation-bearing Pacific storms
to reach California. The timing of the winter wet season
has been shown by Pyke (1972) to be a function of the
latitude of the high pressure cell with the location of
maximum precipitation moving southward at a rate of about
one degree latitude each 4.5 days (Williams, 1948).

The element of control over precipitation exhibited
by the large-scale fluctuations and the nature of the
variability in the ring-width response of the trees
(Fritts, 1974) helped to determine the approach used to
model the relationship between growth in western trees and
California precipitation. Trees from different sites and
geographic locations usually respond differently to
climate at different times of the year (Fritts, this
volume), and the response of the trees may lag behind the
influence of climate. In past experiments (Fritts et al.,
1979), the lagging relationship was introduced on the
predictor side of the regression equation; the seasonal

Figure 5.17: California precipitation stations.

climatic predictand variables were used to represent a
spatial field of a single climatic variable. The present
problem required only a single statewide precipitation
index. Therefore, the predictand variables could represent
lagging relationships over time rather than a spatial
field, and the variability of precipitation from one month
to the next could be calibrated directly with the various
growth responses in the predictor grid.

The monthly precipitation data were prepared for
calibration as follows. The state average monthly values
for the driest months of June, July, and August were
combined into a single value denoted as "Summer". This
combination resulted in each tree-growth year being
represented by 10 "monthly" precipitation values, September through May, and Summer. Common logarithms were then
taken of each monthly value. The logarithm transformation
helps to normalise the skewed distribution of the monthly
precipitation data and eliminates the possibility of the
regression equations giving negative estimates when the
actual value should have been zero. Where individual
months (values in inches) were recorded as zero, the value
was changed to 0.001in, giving a logarithm of −3.0. The
monthly logarithms were then normalised by subtracting the
mean and dividing by the standard deviation for 1901
through 1963.

A predictand matrix appearing in the regression
equation consisted of 63 rows (years) and 20 columns
(months). The columns represent the 10 months concurrent
with the growth "year" of the tree t, and the 10 months
preceding the growth year t−1 (Figure 5.19). The growth
year is defined as September through Summer and the
January month determines its calendar year i. Thus, each
row of the predictand matrix corresponds to growth in the
calendar year i and consists of normalised logarithms of
monthly precipitation for the months of September through
December of year i−2, January through December of year
i−1, and January through May, and Summer of year i. The
predictor matrix consisted of the first 15 principal
components of tree growth in year i only. The predictor
and predictand matrices were then entered into a stepwise
canonical regression scheme (Glahn, 1968; Fritts et al.,
1971; Fritts, 1976; Blasing, 1978). The particular scheme
used here is a modification using a stepwise procedure
described by Fritts et al. (1979) and Lofgren & Hunt (this
volume). In this application, the canonical procedure
relates spatial patterns of ring-width response to
time-varying patterns of monthly precipitation amounts.
The regression equation, calibrated in the stepwise
canonical manner, is now called a transfer function and
can be used to convert the component amplitudes of tree
growth into estimates of the logarithms of monthly
precipitation. This transformation makes up the
reconstruction. For each year that there are tree-growth
data, the reconstruction consists of estimates of the
normalised logarithm of monthly precipitation for the 10

months concurrent with and 10 months preceding growth.
Therefore, the reconstruction contains two estimates for
each month in each calendar year i. The monthly estimates
concurrent with growth were labelled "the t estimates",
and those estimates corresponding to months preceding the
growth year were labelled "t-1 estimates". Each label
denotes whether the relationship behind the estimate was a
non-lagged or lagged one: t and t-1, respectively.

It was required that the precipitation index derived
from this study correspond to the California water year
defined to be October through September (Roos, 1973) with
the calendar year the same as January's. It was necessary,
therefore, to convert the normalised monthly estimates
into total precipitation for each water year. This was
accomplished by first denormalising the monthly estimates
by multiplying by the standard deviation and adding the
mean and taking the antilog. This procedure resulted in
estimates of statewide monthly precipitation which were
then summed over the water year.

At this point a small correction to the estimates
was made necessary by the logarithm transform. It is a
property of the regression scheme that the mean value of
the monthly estimates will equal the mean of the
logarithms of the instrumental observations. However, this
property does not hold true when the antilogs of the
estimated logarithms are taken. The means of the antilogs
are smaller than the mean monthly precipitation and when
these are totalled over the water year, the mean of the
estimated annual totals are too low. For example, the mean
of the instrumental data for the water years 1901 to 1961
was 599mm while the means for the t and t-1 estimates were
495 and 487mm, respectively. These differences represented
biases of 10.4mm and 11.2mm in the t and t-1 estimates.
Therefore, each t and t-1 estimate from 1600 to 1961 was
adjusted upward by this difference so that the 1901 to
1961 mean estimates equalled the mean of the instrumental
data. In this way the downward bias of the estimated water
year totals was removed without altering the relative
variation in the regression estimates.

Calibration and verification. Even though most of the
tree-ring sites on which the predictors are based are

Table 5.7: The number of selected chronologies listed by
species.

SPECIES NAME	NUMBER CHRONOLOGIES IN 52-CHRONOLOGY GRID
Abies concolor	2
Libocedrus decurrens	1
Pinus edulis	5
Pinus flexilis	4
Pinus jeffreyi	3
Pinus longaeva	6
Pinus monophylla	1
Pinus ponderosa	13
Pseudotsuga macrocarpa	1
Pseudotsuga menziesii	16

outside the State of California, the regression model
accounted for 22.1% of the total variance in the logarithms of monthly total precipitation. This value is
deceptively low; the integrated growth of the trees as
represented by the annual rings cannot be expected to
resolve monthly precipitation variations over the
preceding two-year period. We calibrated against the
monthly values so that seasonal variations in the
tree-growth responses could be accommodated in the
calibration. The canonical variates selected the most
important modes of covariation between the seasonal
precipitation regime over a two-year period and the
spatial anomalies in tree-ring width.

The calibration period was selected to reserve 29
years, 1872 through 1900, as independent data with which
to check the reliability of the estimates. The
verification procedures used here are described in detail
by Fritts et al. (1979) and Gordon (this volume).
Table 5.8 summarises the verification results obtained
with the monthly estimates for both the t and t-1
relationships along with correlation results for the
calibration (dependent) period. All but eight of the
correlations are positive averaging 0.47 and 0.45 for the
t and t-1 relationships. However, for the independent
period only 12 of the correlations are positive and

Figure 5.18: The selected dendroclimatic grid of 52
chronologies.

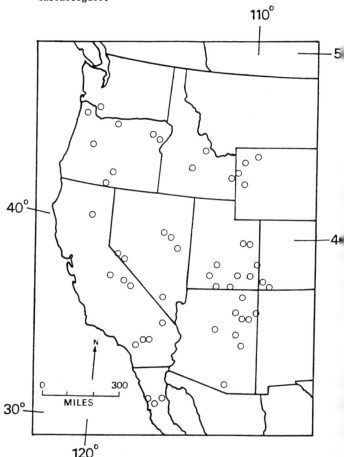

average −0.002 and 0.051 for the t and t-1 relationships, respectively. Few verification tests of the monthly estimates passed and only two of the reduction of error statistics are positive. Such a result is consistent with the results of Fritts (1976, pp.428-429) in that the response of widths to climate is too coarse to make monthly estimates of sufficient accuracy. A better test of the reconstruction is made by transforming the monthly estimates into annual totals and comparing the result with the actual annual totals (see also, Fritts, this volume).

When the estimates are combined over the entire water year (October through current September), which more or less corresponds to the seasons affecting each ring width, the verification statistics are substantially higher. The correlation coefficients for the independent data are closer to the values obtained for the dependent period. Two out of five and one out of five statistics pass for the t and t-1 relationships, respectively. The reduction of error statistics are positive, indicating useful information in the precipitation estimates. Results for the average of the t and t-1 estimates were also considered, but their statistics were no better than those for the t relationship.

RESULTS

Based largely on the verification results, the t reconstruction of the water year was chosen as the more reliable. The reconstruction covers the period 1600 to 1961 and was filtered with a low-pass digital filter (Fritts, 1976). This filter passes about 50% of the variance at a period of eight years. The filtered reconstruction focuses attention on the fluctuations at longer time scales.

The filtered and unfiltered reconstructions are plotted along with the instrumental data in Figure 5.20 to allow a visual comparison. The agreement is quite good even with more consistency than expected by chance for the independent data. It is apparent from the period prior to 1885, however, that the reconstruction is probably more reliable for deficit years than for surplus years, and that the filtered estimates derived from tree growth are below those for the instrumental data. This is consistent with the law of limiting factors which states that a biological process, such as growth, cannot proceed faster than is allowed by the most limiting factor. Growth was limited in dry years by water stress; but in wet years moisture was adequate, and growth was limited by factors unrelated to precipitation.

The entire reconstruction is illustrated in Figure 5.21 and shows a dominance of low-frequency variations in the fluctuations in statewide precipitation. However, Thomas (1959) and Granger (1977) report that there is a lack of synchroneity in the timing and intensity of precipitation anomalies between the northern and southern stations. The emphasis on the low frequencies could be in part the result of combining northern and southern California stations into a single record distorting the temporal pattern, lengthening anomalous periods, and reducing the magnitude of the anomalies. If such distortion occurs, individual precipitation anomalies noted at any one station could be obscured in the reconstruction. A second factor which might emphasise the low-frequency variations in the reconstruction is a possible lagging effect in the growth of the trees (Fritts, 1976), especially those from the high-altitude treeline (LaMarche, 1974a,b). The occurrence of a severe climatic event such as drought or low temperature can cause reduced growth for several growing seasons after the event. This would cause some carry-over greater than one year, which would exceed the capacity of the model used here. Some evidence of this effect can be seen in Figure 5.20 during the period 1935 to 1945. We did try lagging the predictors to accommodate first-order autoregression, but this did not improve the relationship.

The introduction of some highly autocorrelated high-altitude chronologies into the dendroclimatic grid and the lack of chronologies in central California might also contribute to a weakened and lagging response and to the low-frequency variations that were reconstructed. In addition the high-altitude chronologies respond primarily to low-frequency variations in temperature, but they do contain some high-frequency information on precipitation (LaMarche, 1974a,b; LaMarche & Stockton, 1974). It is assumed that the calibration separated the precipitation from the temperature information, but it may not have done the job adequately. Further experimentation with lagging models and with the inclusion or exclusion of these high-altitude chronologies in the dendroclimatic grid is

Figure 5.19: Schematic diagram illustrating the relationships between the calendar year, i, and the growth year, t. The diagram represents the 20 entries in a particular row of the predictand matrix.

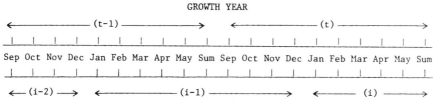

GROWTH YEAR

←———————— (t-1) ————————→   ←———————— (t) ————————→

Sep Oct Nov Dec Jan Feb Mar Apr May Sum Sep Oct Nov Dec Jan Feb Mar Apr May Sum

←— (i-2) —→   ←——————— (i-1) ———————→   ←——— (i) ———→

CALENDAR YEAR

needed before it can be resolved whether the dominance of low-frequency variations is real or an artefact of this particular calibration and dendroclimatic grid.

Because we combined the northern and southern climate data for the state, the reconstruction was examined and compared with records based on precipitation fluctuations on a large spatial scale rather than at individual station locations. Two of the more familiar sources of information are used here: a study of lake levels in the western United States (Harding, 1965), and a precipitation index for southern California (Lynch, 1931). Because the primary interest in a long-term record of precipitation fluctuations arises from the need to programme future water resource development, the focus of this discussion will be on deficiencies of precipitation rather than upon surpluses.

DISCUSSION

The most noteworthy anomaly in the reconstruction is the extensive period of deficient rainfall extending from about 1760 through 1820. Lynch's indices, based entirely on mission crop records during this period, agree except that the dry period according to Lynch commenced later, about 1780. A dry period of this length is unparalleled anywhere in our reconstruction. Not only is it of great length, but also the magnitude of the precipitation deficit is large. During the period 1761 to 1820, 87% of the reconstructed annual totals were below 508mm (20

inches) while a comparable period in the 20th century, 1901 to 1960, shows only 12% of the years to be reconstructed at below 508mm. The 20th century instrumental data show 28% of the years between 1901 and 1960 to have less then 508mm of precipitation. While the trees may have overemphasised the length of the period of precipitation deficit, it is likely that the severity of the 1760 to 1820 drought was even greater than indicated by individual yearly estimates of the reconstructions. Lynch's results were based on mission crop records in the Los Angeles area, and the interpretation of these records has recently come under some criticism because of the close association of variations in some crop data with anthropogenic parameters as well as with precipitation fluctuations (Rowntree & Raburn, 1980). It remains to be seen, however, if any revision of Lynch's indices would confirm our indication of an earlier commencement of this dry period.

Dry periods near the intensity of the 1760 to 1820 period are reconstructed at other times as well. These are 1600 to 1625, 1665 to 1670, 1720 to 1730, and 1865 to 1885. In all of these intervals of precipitation deficit, the annual rainfall fell below the reconstructed level of the 20th century dry period of 1924 to 1938. This 20th century drought was reconstructed with an average water year total of 531mm. The instrumental data gave an average of 551mm. However, the latter data show that rainfall had returned to more normal levels by 1935, three years earlier than did the more persistent reconstructed values.

Table 5.8: Verification of t and t-1 estimates with dependent data correlations.

| | INDEPENDENT PERIOD | | | DEPENDENT PERIOD |
	CORRELATION	TESTS PASSED*	RE	CORRELATION
		t estimates		
Sept.	0.202	0	-0.092	0.515
Oct.	0.059	0	-0.293	0.470
Nov.	0.065	0	-0.702	0.401
Dec.	0.199	0	-0.311	0.568
Jan.	-0.435	0	-1.389	0.643
Feb.	0.104	0	-0.434	0.450
Mar.	0.097	0	-0.754	0.560
Apr.	-0.120	0	-0.103	0.372
May	-0.063	0	-1.305	0.474
Sum.	-0.125	0	-0.405	0.274
Average	-0.002	0	-0.579	0.473
		t-1 estimates		
Sept.	0.415	3	0.165	0.425
Oct.	-0.451	0	-1.392	0.546
Nov.	0.342	2	-0.050	0.517
Dec.	-0.126	0	-0.200	0.401
Jan.	-0.048	0	-0.214	0.453
Feb.	-0.308	0	-0.269	0.487
Mar.	0.083	0	-0.079	0.321
Apr.	0.282	1	0.058	0.418
May	0.174	0	-1.082	0.575
Sum.	0.146	0	-0.324	0.392
Average	0.051	0.6	-0.339	0.454
Annual t #	0.402	2	0.065	0.543
Annual t-1	0.270	1	0.025	0.529

* Number out of 5 possible   # Annual water year totals, October to September

The deficit period reconstructed from 1865 to 1885 is covered by the records of both Harding and Lynch. Lynch indicates a dry period beginning around 1840 and lasting until about 1880. The reconstruction shows a dry sequence in the mid-1840s, but a recovery to normal levels during the 1850s and then a general deficit until the mid-1880s. Harding reports that the deficit in the 1840s was more severe in the southern Great Basin. Lynch's index applies exclusively to southern California, and this could help explain the lag in recovery of the statewide totals reconstructed in the 1850s which is not evident in Lynch's indices. Harding does not extend the dry period in the Great Basin beyond 1860. In fact, he considers 1860 to 1885 to be a period of runoff surplus. The actual data available from 1873 to 1880 show a near normal amount of rainfall averaging 574mm, while the reconstruction averaged 462mm. It appears possible that the more

persistent reconstruction overemphasised the duration of the late 19th century dry period, a result that could be caused by a delay in the trees' recovery from a prolonged period of stress. The agreement with Lynch's data late in the century does suggest, however, that California did experience some precipitation deficit beyond 1860.

CONCLUSIONS

It appears from this reconstruction that significant precipitation deficits have occurred in California. The mean reconstructed precipitation for 1600 to 1900 is 470mm (18.5 inches), which is 78% of the precipitation measured for the 1901-60 period. Fluctuations of sufficient intensity and duration to have economic and social impact have probably occurred six times since 1600. In the perspective of a 360-year reconstruction of precipitation, the period since 1890 has been one of precipitation surplus.

The effort reported upon here should not be considered to be the final result for a dendroclimatic reconstruction of California precipitation. Experiment-ation with and additions of more and different tree-ring sites, especially more located within the California borders, are essential. Subdivision of the climate data grid to reflect the variation between northern and southern stations is also desirable. Some variations in the model structure and different seasonal definitions should be studied. This study has indicated the value of calibrating and extending the reconstruction beyond the area covered by the trees, especially in regions where present day climate information may be inadequate.

AN OVERVIEW OF DENDROCLIMATIC TECHNIQUES, PROCEDURES, AND PROSPECTS

H.C.Fritts

INTRODUCTION

At the Second International Workshop on Global Dendroclimatology, we used a relatively new term, the signal-to-noise ratio (Graybill, this volume), in discussing tree-ring chronologies. The numerator of this ratio can be defined as the mean variance in common among

Figure 5.20: Unfiltered and filtered reconstructions (line) and instrumental data (dots), 1873-1961.

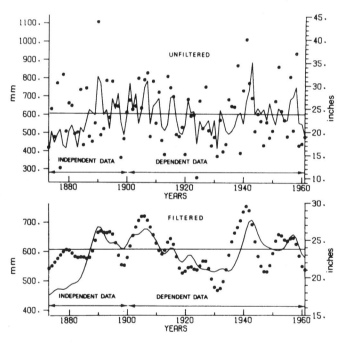

Figure 5.21: Reconstructed statewide precipitation index in inches for California. Mean line drawn for 1901 to 1963 water year total, 23.82in (605mm).

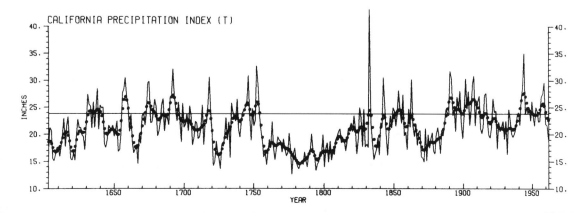

trees. The denominator is the mean information not in common, sometimes called the statistical error. Cropper (this volume) discusses the interpretation of this parameter.

While the term may be new to dendroclimatology, the general concept is not, for many of the dendroclimatic procedures and techniques were developed over the years to maximise dendroclimatic information content (Fritts, 1976). This is accomplished by either maximising the size and kinds of information in the climate signal, minimising the size of the noise, or simplifying and appropriately structuring the signal to facilitate its use. In this brief overview, I focus on this basic concept, which should receive more attention in future work, as well as illustrate how many dendroclimatic operations we have discussed in this volume help to maximise the climate information contained in tree-ring data. Detailed recommendations can be found in Hughes et al. (1980).

IMPROVING THE CLIMATE SIGNAL

The procedures for selecting suitable sites for dendroclimatology, not necessarily the traditional arid sites, ensure collection of the most responsive and statistically sensitive material with the largest climate signal (LaMarche, this volume). The more the growth processes are limited by climatic conditions, the more the variation in common in the chronologies, the higher the signal-to-noise ratios, and the greater the climate information contained in the chronology. The restriction of sampling only trees of a single species growing on similar habitats for a specific chronology keeps the climate signal clear and simple so it can be readily studied and analysed by objective statistical techniques. Tree-ring cross-dating prevents confounding of the climate information owing to improper identification of the growth year. In addition, cross-dating is the matching of the high-frequency climate signal among trees. It has been used from the beginning of the science and its effect is selection of tree-ring materials with similar information content. The fact that trees cross-date indicates that they do contain climate information.

Standardisation removes the effects of increasing tree age on ring width, the effects of varying site productivity among different trees, as well as many changes in the tree's environment which are unrelated to climate (Graybill, this volume). It helps to eliminate non-climatic variations and enhances the signal-to-noise ratio and relative climate information content of the chronology. Sample replication by collecting large numbers of trees and cores, measuring along two or more radii, and averaging the standardised results helps to minimise the between-tree and within-tree variations representing the noise component (Cropper, this volume). Increasing the sample size in this manner reduces the error (or noise) and increases the amount of common variation in the averaged chronology, up to a certain point.

The assembling of climatically sensitive chronologies from different species, exposures, altitudes, and habitats into a dendroclimatic grid (Fritts & Shatz, 1975) to be used for large-scale calibration and reconstruction is a procedure which maximises the total climate information content in a data set, although many differences will exist in the signals that the incorporated chronologies represent (Fritts, this volume). Multivariate calibration uses these differences, as well as the similarities, in the climate signals of the chronologies and extracts that information which is pertinent to the climate data set used for calibration (Lofgren & Hunt, this volume). Different climate data can be used for calibration and the results applied to different problems concerning variations in climate.

Measuring, processing, and calibrating a variety of tree-ring parameters, including earlywood and latewood widths, wood density (Schweingruber, this volume; Conkey, this volume), wood anatomy (Eckstein & Frisse, this volume), and isotopic ratios (Long, this volume), as well as ring width, add to a particular sample's information content. The different measurements can be used together to further maximise the climate signal in a total data set (Schweingruber et al., 1978a,b).

Grid size and density of the sites are important considerations in maximising the signal in a climate reconstruction (Kutzbach & Guetter, 1980, this volume). Generally, an increase in grid density enhances the climate signal, yet there is obviously a practical upper limit. It seems best from their study to use a greater spatial coverage of predictors than predictands where possible. It should be noted, however, that, in some cases, reconstruction of the climate at substantial distances beyond the area of such a tree-ring grid is possible (Fritts et al., 1979). Moreover, a regular network is not necessarily the best; it may be judicious to concentrate sampling effort, and grid density, in areas where the trees are climatically sensitive and climatic change is strong (Pittock, this volume; Kelly this volume).

Lagging and autocorrelated relationships in tree-ring chronologies can be dealt with to fully utilise the information in the chronology. This is a difficult statistical problem and has been discussed at length in Chapter 2 of this volume. Although recommended by some workers, it is not necessary to discard potentially useful information indiscriminately by removing (or correcting) all autocorrelated and moving-average variance before calibration. One can treat this variance in different ways, use these sets along with the untreated data, and solve for the differences in the predictors of climate variance. The calibration process (Lofgren & Hunt, this volume) then assists in selecting and properly weighting the autocorrelated and corrected chronology combinations needed for each reconstruction.

Calibration with data at different frequencies has been discussed (Guiot et al., this volume). This potentially powerful method could represent a major advance in the statistical treatment of dendroclimatological data and warrants careful evaluation and comparison with existing techniques.

The climate signal will be no stronger than the accuracy and representativeness of the data used to calibrate. The instrumental data, as well as tree-ring measurements, must be examined carefully and the apparent inhomogeneities, errors, and inappropriate values eliminated (Pittock, this volume; Kelly, this volume; Fritts & Gordon, this volume). Instrumental measurements of temperature, precipitation, pressure and humidity are perhaps poor estimates of the environmental factors directly affecting tree growth. More meaningful combinations of these measurements can be used such as calculated evapotranspiration, drought indices (Stockton & Meko, 1975; Guiot et al., this volume; Cook, this volume), and degree days (Jacoby, this volume). Such combinations may be of particular value in studies of crop production and energy demand.

The principal components can be calculated from each data set to reorder the information from large-scale, spatially coherent to small-scale and uncorrelated relationships. Apart from being convenient statistically, the low order principal components of the tree-ring data, which are likely to contain the largest proportion of the climate signal, may be selected for study and the smaller ones discarded to eliminate the small-scale, more or less random variance. Stepwise multiple regression or stepwise canonical analysis (Lofgren & Hunt, this volume) can be used to identify the most promising calibrated relationships. These procedures place statistical controls on the variance entered into regression so that large errors are not likely to dilute the signal and unduly reduce the climate information in a reconstruction.

Calibration models may involve different variables, different numbers of principal components, and different lags in the predictors. The reconstructions from different models can be averaged to enhance the climate signal and reduce the more or less random noise components. However, indiscriminate averaging and the use of too many variables are likely to lead to reduced information content. Care must be taken in assessing statistical significance as averaging can lead to a reduction in the number of degrees of freedom. We have found that when the number of climate principal components used in calibration is varied, one can obtain different results. In addition, the order of the principal component does not necessarily reflect its usefulness in reconstruction work. It can be argued that a reconstruction would be useful even if only a very small number of the available principal components of climate are calibrated, and not necessarily the lowest order ones

(Kelly, this volume). While this is true, I recommend that one should try to calibrate more than the minimum number of principal components up to a reasonable point if the available degrees of freedom are adequate.

A climate variable must often be calibrated one season at a time to accommodate the seasonal variability in the climate-growth response (Fritts, this volume). Sometimes monthly climatic values can be calibrated (Jacoby, this volume; Fritts & Gordon, this volume), but many trees respond to conditions integrated over more than one month (Fritts, 1976). The monthly and seasonal reconstructions are likely to be perturbed by climatic conditions in other months and seasons which have limited the growth process. These perturbations may be resolved in part by averaging the reconstructions of a climate variable for different months or seasons throughout the year to obtain annual estimates of that variable. The independent monthly or seasonal instrumental data can be averaged over the year and used to measure and verify that there is improvement. However, the reduced number of degrees of freedom in the averages must be carefully considered when calculating the statistics of the averaged series.

Climate data can be averaged over a region of homogeneous climate before reconstruction is attempted, or reconstructions can be averaged for several climate stations in such a region to eliminate of the local variations. In the latter case, these results can be compared with the regionally averaged climate data to measure improvements in the large-scale relationship (Conkey, this volume). Ring width contains a climate signal integrated over several years. Reconstructions may, therefore, be improved further by filtering or averaging annual estimates over longer time periods such as pentads or decades after calibrating on a seasonal basis. Alternatively, the data could be filtered before calibration, but such an approach is unproven.

A variety of climate variables can be calibrated and reconstructed. In this way, the maximum amount of information in a particular tree-ring data set can be extracted to provide the most complete picture of past climatic conditions. The soundness and the degree of consistency in the different reconstructions also provide a type of verification. Long, well-dated proxy climate and documentary records or phenological information can be used for calibration and verification. Long records from China, for example, could be calibrated along with the tree-ring records from North America to reconstruct climatic conditions over the North Pacific Ocean and East Asia. Verification with independent data is mandatory (Gordon, this volume). It is the statistical proof that reconstructions do in fact contain the desired climate information and are not the result of chance. Verification statistics can be used along with the calibration statistics to choose the calibration model structure that

CLIMATE RECONSTRUCTIONS   194

provides the optimum dendroclimatic reconstruction.

AN EXAMPLE

To illustrate the use of these techniques designed to maximise the climate information contained in, and extractable from, tree-ring data, I shall draw on my colleagues' and my own research. Anomalies in 65 well-replicated North American tree-ring chronologies spanning the interval 1602 to 1963 (Figure 5.22) were used by my group at the Laboratory of Tree-Ring Research, to reconstruct spatial anomalies in temperature, precipitation, and pressure for areas over North America and the North Pacific (Fritts et al., 1979, 1980). Many of the techniques discussed in the previous section helped to maximise the information in the tree-ring and climatic data sets. The multivariate calibration and verification techniques that were used have been described by Lofgren & Hunt (this volume) and Gordon (this volume). Different

Figure 5.22: The 65 replicated ring-width chronologies used to reconstruct past climate. From Fritts & Shatz (1975).

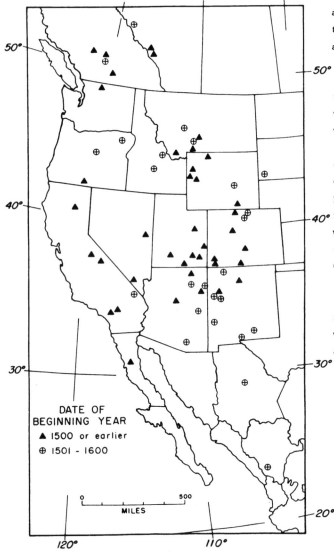

DATE OF
BEGINNING YEAR
▲ 1500 or earlier
⊕ 1501 - 1600

calibrations were run using different principal components and different lags were considered. Those reconstructions with the best calibration and verification statistics for each variable and season were selected. The reconstructions from two to three of the best calibrations for each season and variable were averaged, and the calibration and verification statistics were recalculated for the averaged reconstruction to measure any improvements. In all cases, at least one combination was better than the best single calibration. The combination with the best overall statistics was selected and the combinations from all four seasons were then averaged to obtain the annual estimate. Substantial improvements in the mean annual reconstruction over the seasonal reconstructions were evident (Table 5.9).

Figure 5.23 shows the spatial patterns of calibration percentages for dependent data and the reduction of error on independent data for the annual temperature and precipitation reconstructions. Note that the results are better for large areas downwind (to the east) of the tree-ring chronology grid over western North America than it is for many areas within the chronology grid. The annual reconstructions for individual stations within 11 different regions were also averaged, correlated with the averaged climate data for each region, and the square of these results compared to the square of the correlations averaged for the individual stations to document the improvement (Figure 5.24). All regions show some improvement. In addition, the regional climate reconstructions and the instrumental data were treated with a low-pass filter, and the square of the correlation was calculated to assess additional improvements (Figure 5.24). Each step of averaging resulted in some enhancement of the climate information in all regions except regions 4 and 11, where the filtered data for precipitation exhibited less agreement. The increase in the proportion of variance appears to be real, for, when we corrected the square of the correlation for loss of degrees of freedom using the equations of Kutzbach & Guetter (1980, this volume), the improvement, though sometimes reduced, continued to be present.

Figure 5.25 shows the regionally averaged and filtered reconstructions of temperature for the six western regions along with the instrumental data, plotted as departures from the 1901 to 1970 mean values. This was

Table 5.9: Calibrated variance and verification tests for the four seasonal reconstructions of temperature as compared to those for the annual averages.

	SEASONAL RECONSTRUCTIONS	ANNUAL AVERAGES
% Calibrated Variance	36	48
Verification Tests		
% Tests Significant	28	30
% Reduction of Error > 0	59	75

the period mean used to determine the principal components of the climate data. It is evident from all four figures that the tree ring-climate calibration and reconstruction in areas near and downwind from the trees were best. The calibration results for temperature were better over a wider area than precipitation, even though the sampled trees were from semi-arid sites, where drought was thought to be most limiting. This supports the proposition that a variety of climate variables can be recorded in the tree-ring data and different variables can be used for calibration, and hence reconstructed, if warranted by the response function analysis.

The averaged reconstructions for each decade are given in Figures 5.25 and 5.26 to portray our best reconstructions and to bring out the decade-long variance in the spatial anomalies that is evident. This takes advantage of the greater reliability of the filtered data (Figure 5.24). For many decades in the period 1602 to 1900, the climate was reconstructed to have been warmer and drier in the southwestern United States for many decades than it was in the 20th century, but the central and eastern areas of the nation were reconstructed to have been cooler and wetter (Figure 5.25). This was most obvious in the 1800s, 1820s, 1850s, and 1890s (Figures

5.26 and 5.27). The 1830s was unusually cool and moist especially in the central parts of the country. This was followed by general warming and drying especially in the west. There was severe drought in the Plains in the 1860s which persisted in the southwest through to the end of the century.

CONCLUSION

This volume illustrates the potential for continent- and world-wide dendroclimatic analysis, given suitable grids of tree-ring chronologies and appropriate methods for calibration. Although it has taken two decades to collect the tree-ring data and to learn how to obtain the reconstructions shown in this overview, the first large-scale reconstructions that have been attempted, it need not take that long before the first hemisphere-wide climatic variations are reconstructed. Viable methods are well researched, and the possibility for climate reconstruction has been demonstrated in this volume for several continents, even for areas at considerable distance from the tree grids.

New chronologies are being collected in eastern North America and these will allow improvement of our reconstructions in that area and extension of coverage

Figure 5.23: The percentage of variance calibrated (the square of the correlation) for the dependent data at each station in the grid, and the reduction of error for stations with seven or more years of independent data.

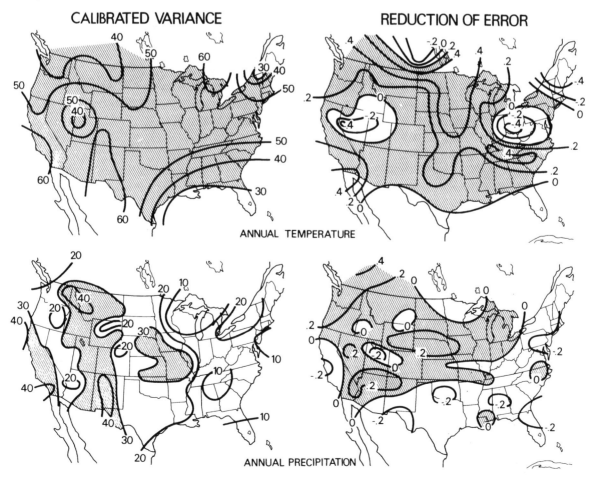

CALIBRATED VARIANCE   REDUCTION OF ERROR

ANNUAL TEMPERATURE

ANNUAL PRECIPITATION

within a few years to beyond the eastern tree-ring data network well into the North Atlantic. Appropriate European chronologies could be available in sufficient numbers by that time to attempt extending the coverage to western Europe. With tree-ring chronologies from eastern Asia, and possibly other proxy climate data and documentary information from China, the area of reliable North American climate reconstructions may also be extended westward as we have done for surface pressure over the North Pacific Ocean to Asia. When a sufficient number of useful chronologies become available for central Asia, eastern Europe, the Mediterranean, and the Arctic, we can attempt the first hemispheric-wide reconstructions of large-scale

climatic variations. With the chronology development and other research in the Southern Hemisphere, it may be that certain more or less world-wide changes in climate will be reconstructed and verified before 1990, given adequate funding and dedicated scientists.

Figure 5.24: Comparisons of the correlation coefficients squared x 100 for: A, the individual station reconstructions in each region compared to the station instrumental data for 1901-1961; B, the averaged reconstruction for all stations in the region compared to the averaged instrumental data; and C, the averaged reconstructions for all stations filtered compared to the averaged instrumental data treated with the same filter.

CORRELATION SQUARED FOR:

A. Individual stations,
B. Regional averages unfiltered,
C. Regional averages filtered

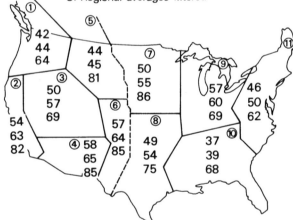

ANNUAL TEMPERATURE

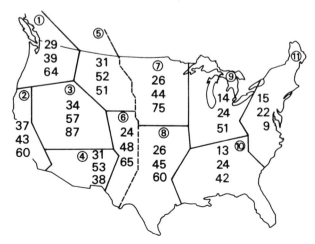

ANNUAL PRECIPITATION

Figure 5.25: The averaged temperature reconstructions for stations in each of the western regions (Figure 5.24) filtered and plotted as departures from the 1901-1970 averages. Note that this mean includes nine years not included in the calibration. Dots on the right are the filtered values of the corresponding instrumented record.

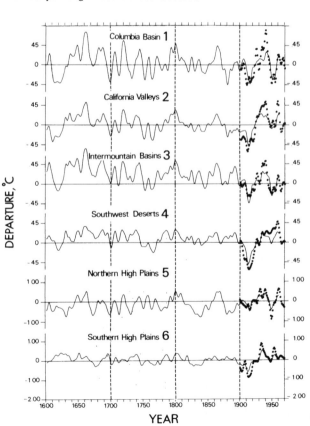

Figure 5.26: The spatial anomalies in reconstructed temperature averaged for each decade and plotted as departures (°C) from the 1901-1970 mean values.

Figure 5.27: The spatial anomalies in reconstructed precipitation averaged for each decade and plotted as percentages of the 1901-1970 mean values. Dashed lines indicate averages where the independent verification was poor (Figure 5.23), and the reconstructions are likely to be invalid.

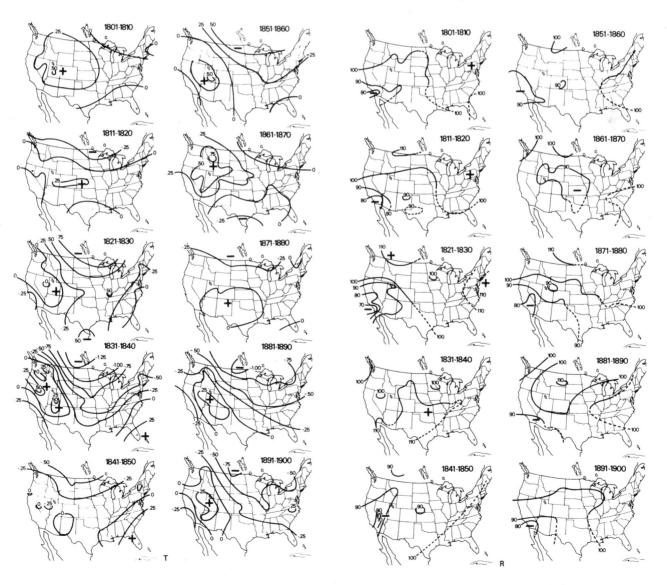

BIBLIOGRAPHY

Aanstadt, S. (1960). Daterte årringer i furu fra Solør. *Blyttia, 18,* 49-67.

Alestalo, J. (1971). Dendrochronological interpretation of geomorphic processes. *Fennia, 105,* 1-140.

Allan, H.H.B. (1961). *Flora of New Zealand, Vol.1.* Government Printer, New Zealand.

Aloui, A. (1978). *Quelques aspects de la dendroclimatologie en Khroumiries (Tunisie).* D.E.A. d'Ecologie méditerranéeanne, Université Aix-Marseille III.

American National Standards Institute (1966). American national standard programming language FORTRAN. *Bulletin X3.9-1966.* ANSI: New York.

American National Standards Institute (1978). American national standard programming language FORTRAN. *Bulletin X3.9-1978.* ANSI: New York.

Amos, G.L. & Dadswell, H.E. (1950). Wood structure in relation to growth of *Beilschmeida bancroftii.* Progress Report No. 1, Project W.S. 15/3. C.S.I.R.O., Division of Forest Products.

Anderson, T.W. (1958). *An Introduction to Multivariate Statistical Analysis.* Wiley: New York.

Anderson, R.L., Allen, D.M. & Cady, F.B. (1972). Selection of predictor variables in linear multiple regression. In: *Statistical Papers in Honor of George W. Snedecor,* ed., T.A. Bancroft. Iowa State University Press.

Andrews J.F. (1965). The weather and circulation of July 1965. *Monthly Weather Review, 93*(10), 647-654.

Andrews, T.J., Lorimer, G.H. & Tolbert, N.E. (1971). Incorporation of molecular oxygen into glycine and serine during photorespiration in spinach leaves. *Biochemistry, 10,* 4777-4782.

Arakawa, H. (1960). Japanese tree-ring analyses. *Arch. Met. Geophy. Biokl. (B), 10*(2), 210-212.

Arakawa, H. (1969). *Climates of Northern and Eastern Asia. World Survey of Climatology, Vol.8,* Elsevier: Amsterdam.

Arnold, L.D. (1979). *The climatic response in the partitioning of the stable isotopes of carbon in Juniper trees from Arizona.* Ph.D. Dissertation, University of Arizona.

Atwood, W.W. (1940). *Physiographic Provinces of North America.* Ginn: Boston.

Australian Academy of Science (1976). Report of a Committee on Climatic Change. *Report No.21,* Australian Academy of Science, Canberra.

Baillie, M.G.L. (1977a). The Belfast oak chronology to AD 1001. *Tree-Ring Bulletin, 37,* 1-12.

Baillie, M.G.L. (1977b). An oak chronology for South Central Scotland. *Tree-Ring Bulletin, 37,* 33-44.

Baillie, M.G.L. (1977c). Dublin medieval dendrochronology. *Tree-Ring Bulletin, 37,* 13-20.

Baillie, M.G.L. (1980a). Dendrochronology - the Irish view. *Current Archaeology, 7,* 61-63.

Baillie, M.G.L. (1980b). Some observations of gaps in tree-ring chronologies. In: *Symposium on Archaeological Sciences,* ed. H. Aspinall & S.E. Warren, 4-7 January 1978. University of Bradford Press.

Bär, O. (1977). *Geographie Europas.* Lehrmittelverlag Kantons Zurich, 315 S.

Barefoot, A.C. (1975). A Winchester dendrochronology for 1635-1972 AD. Its validity and possible extension. *Journal of the Institute of Wood Science, 7,* 25-32.

Barnett, T.P. (1978). Estimating variability of surface air temperatures in the Northern Hemisphere. *Monthly Weather Review, 106*(9), 1353-1367.

Barnett, T.P. & Hasselman, K. (1979). Techniques of linear prediction, with application to oceanic and atmospheric fields in the tropical Pacific. *Reviews of Geophysics and Space Physics, 17,* 949-968.

Barry, R.G. (1978). Climatic fluctuations during periods of historical and instrumental record. In: *Climatic Change and Variability. A Southern Perspective,* ed. A.B. Pittock, L.A. Frakes, D. Jenssen, J.A. Peterson & J.W. Zillman, pp.150-166. Cambridge University Press.

Barry, R.G. & Perry, A.H. (1973). *Synoptic Climatology. Methods and Applications.* Methuen: London.

Barry, R.G. & Ives, J.D. (1974). Introduction. In: *Arctic and Alpine Environments,* ed. J.D. Ives & R.G. Barry, pp.1-13. Methuen: London.

Barry, R.G. & Chorley, R.J. (1976). *Atmosphere, Weather and Climate, 3rd Edition.* Methuen: London.

Barry, R.G., Fritts, H.C., Imbrie, J., Kutzbach, J., Mitchell, J.M. & Savin, S.M. (1979). Paleoclimatic research: status and opportunities, ed. A.D. Hecht. *Quaternary Research, 12*(1), 6-17.

Bartholin, T.S. (1973). Undersøgelse af muligheden for dendrokronologisk datering af egetrae i Danmark, specielt Sønderjylland. *Det forstlige Forsøgsvaesen i Danmark, 33,* 217-241.

Bartholin, T.S. (1975). Dendrochronology of oak in Southern Sweden. *Tree-Ring Bulletin, 35,* 25-29.

Batley, R.A.L. (1956). Some practical aspects of dendrochronology in New Zealand. *Journal of the Polynesian Society, 65,* 232-244.

Bauch, J. (1978). Tree-ring chronologies for the Netherlands. In: *Dendrochronology in Europe,* ed. J. Fletcher, British Archaeological Research Series, No.51, pp.133-137.

Bauch, J. & Eckstein, D. (1978). Holzdatierungen im Kloster Wienhausen (Celle). Ein Beitrag zu seiner Baugeschichte. *Niedersächsische Denmalpflege, 9,* 79-92.

Bauch, J., Liese, W. & Eckstein, D. (1967). Ueber die Altersbestimmung von Eichenholz in Norddeutschland mit Hilfe der Dendrochronologie. *Holz als Roh- und Werkstoff, 25,* 285-291.

Beaulieu de, J.L. (1977). *Contribution pollenanalytique à l'histoire tardi-glaciaire et holocène de la végétation des Alpes méridionnales françaises.* Thesis, Université Aix-Marseilles III.

Becker, B. (1978). Dendrochronological evidence for Holocene changes in the drainage system of southern central Europe. In: *Dendrochronology in Europe,* ed. J. Fletcher, British Archaeological Research Series, No.51, pp.289-290.

Becker, B. (1979). Die postglaziale Eichenjahrringchronologie Süddeutschlands und ihre Bedeutung für die Datierung neolithischer Chronologien aus der Schweiz. *Zeitschrift Schweiz. Archäologie und Kunstgeschichte, 36,* 91-92.

Becker, B. & Delorme, A. (1978). Oak chronologies for Central Europe and their extension from medieval to prehistoric times. In: *Dendrochronology in Europe,* ed. J. Fletcher, British Archaeological Research Series, no.51, pp.59-64.

Becker, B. & Frenzel, B. (1977). Palaeookologische Befunde zur Geschichte Postglazialer Flussauen im Sudlichen Mitteleuropa. In: *Dendrochronologie und Postglaziale Klimaschwankungen in Europa,* ed. B. Frenzel. Franz Steiner Verlag GMBH: Wiesbaden.

Becker, B. & Giertz-Siebenlist, V. (1970). Eine über 1100 jährige mitteleuropäische Tannenchronologie. *Flora, 159,* 310-346.

Bednarz, Z. (1975). Geographical range of similarities of annual growth curves of stone pine (*Pinus cembra* L.) in Europe. In: *Bioecological Fundamentals of Dendrochronology. Symp. Mat. XII Int. Bot. Congr., Leningrad,* pp.75-83.

Bednarz, Z. (1976). Wpływ klimatu na zmienność szerokości słojów rocznych limby (*Pinus cembra* L.) w Tatrach. *Acta Agraria et Silv., Ser. Silv., 16,* 17-34.

Bednarz, Z. (1978). Tree-ring research in Poland: a progress report. *International Tree-Ring Data Bank Newsletter, 3*(2), 1-2. Laboratory of Tree-Ring Research, University of Arizona, Tucson.

Bell, R.E. (1958). Dendrochronology. *New Zealand Science Review, 16,* 13-17.

Bell, V. & Bell, R.E. (1958). Dendrochronological studies in New Zealand. *Tree-Ring Bulletin, 22*(1), 7-11.

Bell, W.T. & Ogilvie, A.E.J. (1978). Weather compilations as a source of data for the reconstruction of European climate during the medieval period. *Climatic Change, 1,* 331-348.

Berger, A. (1979). Spectrum of climatic variations and

their causal mechanisms. *Geophysical Surveys, 3,* 351-402.

Berger, A., editor. (1981). *Climatic Variations and Variability: Facts and Theories.* D. Reidel: Dordrecht, Boston, London.

Berger, A., Guiot, J., Mathieu, L. & Munaut, A. (1979). Cedar tree-rings and climate in Morocco. *Tree-Ring Bulletin, 39,* 61-75.

Berry, J.A., Osmond, C.B. & Lorimer, G.H. (1978). Fixation of $^{18}O_2$ during photorespiration. *Plant Physiology, 62,* 954-967.

Berthet, P., Feytmans, E., Stevens, D. & Genette, A. (1976). A new divisive method of classification illustrated by its application to ecological problems. In: *Proceedings of the 9th International Biometric Conference,* Boston, August 22-27, 1976. Biometric Society: Boston.

Bertrand, Fr. (1979). Reconstruction mathématique du climat au Maroc. *Mémoire Lic. Sc. Math.,* Institut d'Astronomie et de Géophysique, Université de Louvain-la-Neuve.

Beschel, R.E. & Webb, D. (1963). Growth ring studies on arctic willows. *Axel Heiberg Research Reports:* Preliminary Report 1961-1962, pp.189-198.

Bigeleisen, J. (1949). The relative reaction velocities of isotopic molecules. *Journal of Chemical Physics, 17,* 675-678.

Bitvinskas, T.T. (1961). Regularities of the stands growth. *Musugirius, No.9.* In Lithuanian.

Bitvinskas, T.T. (1964). Dynamics of the pine stands increment and the means of prognosis (in Lithuanian SSR). *Reports of TSChA,* 1964, 497-503. In Russian.

Bitvinskas, T.T. (1968). Object and problems of the Dendroclimatochronological Laboratory, Institute of Botany of the Academy of Sciences of the Lithuanian SSR. *Reports of the all-Union Scientific Conference on the Problem of Dendrochronology and Dendroclimatology, Vilnius,* 1968, 144-147. In Russian.

Bitvinskas, T.T. (1974). *Dendroclimatological Research.* Gidroteometizdat: Leningrad. In Russian.

Bitvinskas, T.T., Dergachev, V.A., Kairaitis, J.J. & Zakarka, R.A. (1972). About the possibility of constructing long-term dendroscales in the south of the Baltic area. In: *Dendroclimatochronology and Radiocarbon, Kaunas,* 1972, 69-75. In Russian.

Bitvinskas, T.T. & Kairaitis, J.J. (1975). The radial growth variation of oak stands and its relationship with the environmental conditions, climate and solar activity. In: *Bioecological Basis of Dendrochronology, Vilnius,* 1975, 69-74. In Russian.

Bitvinskas, T.T., Savukynene, N. & Grigelyte, M. (1976). Application of complex research methods in the study of environmental conditions. In: *Indication of Natural Processes and the Environment, Vilnius,* 1976, 31-33. In Russian.

Bitvinskas, T.T. & Kairaitis, J.J. (1978). Dendrochronological scales of the profile Murmansk - the Carpathians. In: *Dendroclimatological Scales of the Soviet Union, Kaunus,* 1978, 52-78. In Russian.

Bjerknes, J. (1969). Atmospheric teleconnections from the equatorial Pacific. *Tellus, 18,* 820-829.

Blasing, T.J. (1975). *Methods for analysing climatic variations in the North Pacific Sector and Western North America for the last few centuries.* Ph.D. Thesis, University of Wisconsin, Madison.

Blasing, T.J. (1978). Time series and multivariate analysis in paleoclimatology. In: *Time Series and Ecological Processes,* ed. H.H. Shuggart, Jr., SIAM-SIMS Conference Series, No.5, pp.211-226. Society for Industrial and Applied Mathematics: Philadelphia.

Blasing, T.J. (1981). Characteristic anomaly patterns of summer sea-level pressure for the Northern Hemisphere. *Tellus, 33,* 428-437.

Blasing, T.J. & Fritts, H.C. (1975). Past climate of Alaska from tree rings. In: *Climate of the Arctic,* ed. G. Weller & S. Bowling, 24th Alaska Science Conference, 1973, pp.48-58. Geophysical Institute, University of Alaska: Fairbanks, Alaska.

Blasing, T.J. & Fritts, H.C. (1976). Reconstruction of past climatic anomalies in the North Pacific and

Western North America from tree-ring data. *Quaternary Research, 6,* 563-579.

Blasing, T.J., Duvick, D.N. & West, D.C. (1981). Dendroclimatic calibration and verification using regionally-averaged and single-station precipitation data. *Tree-Ring Bulletin,* in press.

Boninsegna, J.A. & Homes, R.L. (1977). Estudios dendrocronologicos en los Andes Centrales y Pategonico-Fueguinos. *Memoria Anual 1976 (III),* 75-91. Instituto de Nivologia y Glaciologia: Mendoza, Argentina.

Boninsegnas, J.A. & Holmes, R.L. (1978a). Distribucion geografica de *Araucaria araucana* (Moll.) C. Koch en relacion a sus caracteristics dendrocronologicas. *Annales 1977, 4,* 107-113. Instituto Argentino de Nivologia y Glaciologia: Mendoza, Argentina.

Boninsegna, J.A. & Holmes, R.L. (1978b). Breve description de un relicto de *Austrocedrus chilensis* (D.Don) Endl. en Huiganco (Pcia. del Neuquen). *Annales 1977, 4,* 115-123. Instituto Argentino de Nivologia y Glaciologia: Mendoza, Argentina.

Bonneman, A. & Röhrig, E. (1971). *Waldbau auf ökolgischer Grundlage.* Vol.1. P. Parey Verl: Hamburg & Berlin.

Borisov, A.A. (1965). *Climates of the U.S.S.R.* Aldine: Chicago.

Borel, L. & Serre, F. (1967). Phytosociologie et analyse des cernes ligneux: l'example de trois fôrets du Huat Var (France). *Oecologia Plantarum, 4,* 1955-1976.

Bowler, J., editor (1981). *Proceedings of the CLIMANZ Conference,* 8-13 February 1981, Howmans Gap, Australia. Australian Academy of Science: Canberra.

Bowler, J.M., Hope, G.S., Jennings, J.N., Singh, G. & Walker, D. (1976). Late Quaternary climates of Australia and New Guinea. *Quaternary Research, 6,* 359-394.

Box, G.E.P. & Jenkins, G.H. (1976). *Time Series Analysis, Forecasting and Control.* Holden-Day: San Francisco.

Bradley, R.S. (1976). *Precipitation History of the Rocky Mountain States.* Westview Press: Boulder.

Brathen, A. (1978). A tree-ring chronology for oak from the Gotha river area, Western Sweden. In: *Dendrochronology in Europe,* ed. J. Fletcher, British Archaeological Research Series, no.51, p.131.

Braun, E.L. (1950). *Deciduous Forests of Eastern North America.* Hafner Press: New York.

Brebme, K. (1951). Jahrringchronologische und -klimatologische Untersuchungenan Hochebirgslarchen des Berchtesgadner Landes. *Zeitschrift für Weltforstwirtsch, 14(2/3),* 65-80.

Brett, D.W. (1978). Dendroclimatology of elm in London. *Tree-Ring Bulletin, 38,* 35-44.

Brinkmann, W.A.R. (1976). Surface temperature trend for the northern hemisphere - updated. *Quaternary Research, 6(3),* 355-358.

Brubaker, L.B. (1980). Spatial patterns of tree-growth anomalies in the Pacific Northwest. *Ecology, 61,* 798-807.

Bryson, R.A. (1966). Air masses, streamlines, and the boreal forest. *Geographical Bulletin, 8(3),* 228-269.

Bryson, R.A. & Hare, F.K. (1974). *Climates of North America.* World Series of Climatology, Vol. 11. Elsevier: Amsterdam.

Budyko, M.J. (1971). *Climate and Life.* Gidroteometizdat: Moscow. In Russian. (Published in English, 1974, Academic: London.)

Burk, R.L. (1979). *Factors affecting $^{18}O/^{16}O$ ratios in cellulose.* Dissertation, University of Washington, Seattle.

Burk, R.L. & Stuiver, M. (1981). Oxygen isotope ratios in trees reflect mean annual temperature and humidity. *Science, 211,* 1417-1419.

Burrows, C.J. & Greenland, D.E. (1979). An analysis of the evidence for climatic change in New Zealand in the last thousand years: Evidence from diverse natural phenomena and from instrumental records. *Journal of the Royal Society of New Zealand, 9,* 321-373.

Bussell, W.T. (1968). The growth of some New Zealand trees. 1. Growth in natural conditions. *New Zealand Journal of Botany, 6,* 63-75.

Cain, W.F. & Suess, H.E. (1976). Carbon-14 in tree rings. *Journal of Geophysical Research, 81,* 3688-3694.

Cameron, R.J. (1957). Lake shore forest as an indicator of past rainfall. *New Zealand Journal of Forestry, 7,* 104.

Cameron, R.J. (1960). Dendrochronology in New Zealand. *Journal of the Polynesian Society, 69,* 37-38.

Cameron, J.F., Berry, P.F. & Phillips, E.W.J. (1959). The determination of wood density using beta-rays. *Holzforschung, 13,* 78-84.

Campbell, D.A. (1980). *The feasibility of using tree-ring chronologies to augment hydrologic records in Tasmania, Australia.* M.S. Thesis, University of Arizona, Tucson.

Carnahan, J.A. (1976). Natural vegetation: A commentary to supplement the map sheet "Natural Vegetation". In: *Atlas of Australian Resources (Second Series),* p.26. Division of National Mapping, Department of National Resources: Canberra.

Carter, C.M. (1971). *Studies in dendrochronology, North Island, New Zealand.* B.Sc. (Hons.) Thesis, Victoria University of Wellington.

Cerskiene, J.J. (1972). Correlation of the tree-ring width of fir tree and climatic factors in the west of Lithuania. In: *Dendrochronology and Radiocarbon, Kaunas,* 1972, 49-54. In Russian.

Chalabi, M.N. (1980). *Analyse phytosociologique, dendrométrique et dendroclimatologique des forêts de Quercus cerris subsp. pseudocerris et contribution à l'étude taxinomique du genre Quercus L. en Syrie.* Thesis, Université Aix-Marseille III.

Chang, J.-H. (1972). *Atmospheric Circulation Systems and Climates.* Oriental: Honolulu.

Cheng, T.C. (1935). On the relation between tree rings and rainfall in Peiping. *Fangzhi Yueken, 8*(6), 13-16. In Chinese.

Christeller, J.T., Lang, W.A. & Troughton, J.H. (1976). Isotope discrimination by ribulose 1, 5-diphosphate carboxylae. *Plant Physiology, 57,* 580-582.

Clark, D. (1975). Understanding canonical correlation analysis. *Concepts and Techniques in Modern Geography, No.3,* Geo Abstracts: Norwich.

Clark, N., Blasing, T.J. & Fritts, H.C. (1975). Influence of interannual climate fluctuations on biological systems. *Nature, London, 256,* 302-304.

Clayton-Greene, K.A. (1977). Structure and origin of *Librocedrus bidwillii* stands in the Waikato District, New Zealand. *New Zealand Journal of Botany, 15,* 19-28.

Cleaveland, M.K. (1975). *Dendroclimatic relationships of shortleaf pine (Pinus echinata Mill.) in the South Carolina Piedmont.* M.S. Thesis, Clemson University.

Cochrane, G.R. (1973). The general environment and biogeography. In: *The Natural History of New Zealand, An Ecological Survey,* ed. G.R. Williams, pp.1-27. A.H. & A.W. Reed: Wellington, New Zealand.

Cole, M.M. (1961). *South Africa.* Methuen: London.

Conkey, L.E. (1979a). *Dendroclimatology in the northeastern United States.* M.S. Thesis, University of Arizona.

Conkey, L.E. (1979b). Response of tree-ring density to climate in Maine, U.S.A. *Tree-Ring Bulletin, 39,* 29-38.

Cook, E.R. (1976). *A tree-ring analysis of four tree species growing in southeastern New York State.* M.S. Thesis, University of Arizona.

Cook, E.R. & Jacoby, G.C. (1977). Tree-ring - drought relationships in the Hudson Valley, New York. *Science, 198,* 399-401.

Cook, E.R. & Jacoby, G.C. (1979). Evidence for quasi-periodic July drought in the Hudson Valley, New York. *Nature, London, 282,* 390-392.

Cook, E.R. & Peters, K. (1979). The smoothing spline: A new approach to standardizing forest interior tree-ring width series for dendroclimatic studies. *Lamont-Doherty Geological Observatory, Contributions.* L-DGO: New York.

Corona, E. (1975). Indagine dendrocronologica in Val di Tovel. *Esperienze e Ricerche, nuova serie, 5,* 261-274.

Corona, E. (1979). Le segnature di Landshut nell'abete rosso cisalpino. *L'Italia Forestale e Montana, 34,* 37-42.

Coulter, J.D (1967). Mountain Climate. *Proceedings of the New Zealand Ecological Society, 14,* 40-57.

Coulter, J.D. (1973). Ecological aspects of the climate. In: *The Natural History of New Zealand,* ed. G.R. Williams, pp.28-60. A.H. & A.W. Reed: Wellington, New Zealand.

Craddock, J.M. & Flood, C.R. (1969). Eigenvectors for representing the 500 mb geopotential surface over the Northern Hemisphere. *Quarterly Journal of the Royal Meteorological Society, 95,* 576-593.

Craig, H. & Gordon, L.I. (1965). Deuterium and oxygen 18 variations in the ocean and marine atmosphere. In: *Stable Isotopes in Oceanographic Studies and Palaeotemperatures,* ed. E. Tongiorgi, pp.9-130. Laboratoria Geologia Nucleare: Pisa.

Crawford, J.P. (1975). The availability of climatological data. In: *Proceedings of a Symposium on Meteorology and Forestry,* October 1974, pp.5-10. New Zealand Meteorological Service: Wellington, New Zealand.

Cropper, J.P. (1981). *Reconstruction of North Pacific surface pressure anomaly types from Alaskan and Western Canadian tree-ring data.* M.S. Thesis, University of Arizona, Tucson.

Cropper, J.P. & Fritts, H.C. (1981). Tree-ring width chronologies from the North American Arctic. *Arctic and Alpine Research, 13*(3), 245-260.

Cunningham, A. (1964). Notes on carbonised wood and leaf fragments occurring in Taupo pumice in the vicinity of the Kaweka Range. *New Zealand Journal of Botany, 2,* 107-119.

Dallimore, W. & Jackson, A.B. (1954). *A Handbook of Coniferae and Ginkgoaceae, (4th Edition, revised by S.G. Harrison, 1966).* Edward Arnold: London.

Damon, P.E. (1968). Radiocarbon and climate. *Meteorological Monographs, 8,* 151-154.

Dansgaard, W., Johnsen, S.J., Clausen, H.B. & Langway, C.C., Jr. (1971). Climatic record revealed by the Camp Century ice core. In: *The Late Cenozoic Glacial Ages,* ed. K.K. Turekian, pp.37-56. Yale University Press.

Dansgaard, W., Johnsen, S.J., Reeh, N., Gundestrup, N., Clausen, H.B. & Hammer, C.U. (1975). Climatic changes, Norsemen and modern man. *Nature, London, 255,* 24-28.

Davis, R.E. (1976). Predictability of sea surface temperature and sea level pressure anomalies over the North Pacific Ocean. *Journal of Physical Oceanography, 6*(3), 249-266.

Dean, J.S. & Robinson, W.J. (1978). Expanded tree-ring chronologies for the southwestern United States. *Chronology Series III,* Laboratory of Tree-Ring Research, University of Arizona, Tucson.

De Boer, H.J. (1951). Tree-ring measurements and weather fluctuations in Java from AD 1514. *Nederland Akad. van Wetensch Proc. Ser.B, 54,* 194-209.

De Boer, H.J. (1952). Tree-ring measurements in Java and the sunspot cycles from AD 1514. *Nederland. Akad. van Wetensch Proc. Ser. B: Phys. Sci., 55,* 386-394.

De Corte, C. (1979). Première reconstitution du climat au Maroc à partir de la dendrochronologie. *Mémoire Lic. Sc. Geogr.,* Institut d'Astronomie et de Géophysique, Université de Louvain-La-Neuve.

Delorme, A. (1973). Aufbau einer Eichenjahrringchronologie fur das südliche Weser- und Leinebergland. *Forstarchiv, 44,* 205-209.

DeNiro, M.J. & Epstein, S. (1979). Relationship between the oxygen isotope ratios of terrestrial plant cellulose. *Science, 205,* 51-53.

Dergachev, V.A. & Sanadze, A.A. (1974). Concentration of ^{14}C in dendrochronologically dated specimen. In: *Reports of the 5th all-Union Conference on the problem "Astrophysical Phenomena and Radiocarbon", Tbilisi,* 1974, 63-71. In Russian.

de Swardt, S.J. & Burger, O.E. (1941). Natural provision against drought. *Forestry in S.A., 16,* 193-196.

de Winter, B. & Vahrmeijer, J. (1972). *The National List of Trees.* van Schaik: Pretoria, South Africa.

Devaux, J.P., Le Bourhis, M. & Moutte, P. (1975).
   Structure et croissances comparées de quelques
   peuplements de pins d'Alep dans l'île de Port-Cros
   (Parc National). *Biologie et Ecologie Méditerrané-
   enne*, *2*, 15-31.
de Vries, Hl. (1958). Variation in concentration of rad-
   iocarbon with time and location on the Earth.
   *Koninkl. Nederl. Akademie Van Wetenschappen,
   Amsterdam, Prc. Series B61*, 94-102.
DeWitt, E. & Ames, M. (1978). Tree-ring chronologies of
   eastern North America. *Chronology Series IV, Vol.
   1*, Laboratory of Tree-Ring Research, University of
   Arizona, Tucson.
Dietz, R. & Holden, J.C. (1970). The breakup of Pangaea.
   Scientific American, 223, 30-41.
Diller, O.D. (1935). The relation of temperature and pre-
   cipitation to the growth of beech in northern Indi-
   ana. *Ecology, 16*, 72-81.
Douglass, A.E. (1914). A method of estimating rainfall by
   the growth of trees. In: *The Climatic Factor*, ed.
   E. Huntington, Carnegie Institute of Washington
   Publications, 192, pp.101-122.
Douglass, A.E. (1919). *Climatic Cycles and Tree Growth,
   Vol.I.* Carnegie Institute of Washington Publications,
   289.
Douglass, A.E. (1921). Dating our prehistoric ruins. *Nat-
   ural History, 21*(1), 27-30.
Douglass, A.E. (1936). *Climatic cycles and tree growth.
   Vol.III.* Carnegie Institute of Washington Public-
   ations, 289.
Douglass, A.E. (1937). Tree-rings and chronology. *Univ-
   ersity of Arizona Bulletin, Phys. Sci. Ser. 1, 8*(4).
Douglass, A.V. (1976). Past air-sea interactions over the
   eastern North Pacific as revealed by tree-ring data.
   Ph.D. Dissertation, University of Arizona, Tucson.
Draper, N.R. & Smith, H. (1966). *Applied Regression Anal-
   ysis.* Wiley: New York.
Drew, L.G., editor (1974). Tree-ring chronologies of
   Western America, Vol.IV: Colorado, Utah, Nebraska
   and South Dakota. *Chronology Series I*, Laboratory
   of Tree-Ring Research, University of Arizona, Tucson.
Drew, L.G., editor (1975a,b). Tree-ring chronologies of
   Western America, a. Vol.V: Washington, Oregon, Idaho,
   Montana and Wyoming. b. Vol.VI: Western Canada and
   Mexico. *Chronology Series I*, Laboratory of Tree-Ring
   Research, University of Arizona, Tucson.
Drew, L.G., editor (1976). Tree-ring chronologies for
   dendroclimatic analysis: an expanded Western North
   American grid. *Chronology Series II*, Laboratory of
   Tree-Ring Research, University of Arizona, Tucson.
Druce, A.P. (1966). Tree-ring dating of recent volcanic
   ash and lapilli, Mt. Egmont. *New Zealand Journal of
   Botany, 4*, 3-41.
Duke, N.C., Birch, W.R. & Williams, W.T. (1981). Growth
   rings and rainfall correlations in a Mangrove tree
   of the genus *Diosyros* (Ebebaceae). *Australian Jour-
   nal of Botany*, in press.
Dunwiddie, P.D. (1978). Recent dendrochronological samp-
   ling in New Zealand. A preliminary note. *New Zealand
   Journal of Botany, 16*, 409-410.
Dunwiddie, P.D. (1979). Dendrochronological studies of indigenous New Zealand trees. *New Zealand Journal of
   Botany, 17*, 251-266.
Dunwiddie, P.D. & LaMarche, V.C., Jr. (1980a). A climatically responsive tree-ring record for *Widdringtonia
   cedarbergensis*, Cape Province, South Africa. *Nature,
   London, 286*, 796-797.
Dunwiddie, P.D. & LaMarche, V.C., Jr. (1980b). Dendrochronological characteristics of some native Australian
   trees. *Australian Forestry, 43*, 124-135.

Eckstein, D. (1969). *Entwicklung und Anwendung der Dendro-
   chronologie zur Altersbestimmung der Siedlung Haitabu.*
   Dissertation, University of Hamburg.
Eckstein, D. (1972). Tree-ring research in Europe. *Tree-
   Ring Bulletin, 32*, 1-18.
Eckstein, D. (1976). Die absolute Datierung der wikinger-
   zeitlichen Siedlung Haithabu. *Naturwissenschaft
   Rundschau, 29*, 81-84.
Eckstein, D. (1978). Regional tree-ring chronologies along

parts of the North Sea coast. In: *Dendrochronology
   in Europe*, ed. J. Fletcher, British Archaeological
   Research Series, no.51, pp.117-124.
Eckstein, D., Mathieu, K. & Bauch, J. (1972). Jahrringan-
   alyse und Baugeschichtsforschung - Aufbau einer
   Jahrring chronologie für die Vier- und Marscheland
   bei Hamburg. *Abhandlungen und Verhandlungen des
   Naturwissenschaftlichen Vereins in Hamburg, 16*, 73-
   100.
Eckstein, D., Brongers, J.A. & Bauch, J. (1975). Tree-ring
   research in the Netherlands. *Tree-Ring Bulletin, 35*,
   1-13.
Eckstein, D., Frisse, E. & Quiehl, F. (1977). Holzanatom-
   ische Untersuchungen zum Nachweis anthropegener Ein-
   flüsse auf die Umweltbedingungen einer Rotbuche.
   Angewandte Botanik, 51, 47-56.
Eckstein D. & Schmidt, B. (1974). Dendroklimatologische
   Untersuchungen an Stieleichen aus dem maritem Klima-
   gebiet Schleswig-Holsteins. *Angewandte Botanik, 48*,
   371-383.
Eckstein, D., Schwab, F. & Zimmerman, W.H. (1979). Aufbau
   und Anwendung einer Jahrringchronologie im nieder-
   sächsische Küstenraum. *Probleme der Küstenforschung
   im südlichwen Nordseegebiet, 13*, 99-121.
Eddy, J.A. (1977). Climate and the changing sun. *Climatic
   Change, 1*, 173-190.
Eidem, P. (1953). Om svingninger i tykkelsestilveksten hos
   gran (*Picea abies*) og furu (*Pinus silvestris*) i
   Trøndelag. *Medd. Norske Skogforsøksvesen, 12*, 1-155.
Eidem, P. (1959). En grunnskala til tidfesting av trevirke
   fra Flesburg i Numedal. *Blyttia, 17*, 69-84.
Elliot, D. (1979). The current regenerative capacity of
   the northern Canadian trees, Keewatin, N.W.T.,
   Canada: Some preliminary observations. *Arctic and
   Alpine Research, 11*(2), 243-251.
Ellis, J.C. (1971). The use of X-rays in measuring ring
   widths from increment borings. *New Zealand Journal
   of Forestry Science, 1*(2), 223-230.
Emberger, L. (1939). Aperçu général sur la végétation du
   Maroc. Commentaires de la carte phytogéographique
   du Maroc 1:1,500,000. *Veröffentlichungen des Geobot-
   anischen Instituts, Eidgenössische Technische Hoch-
   schule Rübel in Zürich, 14*, 40-147.
Enmoto, K. (1937). The tree rings of cryptomeria at Yak-
   ushima. *Tenki to Kiko, 4*(5), 211-215. In Japanese.
Enright, N.J. (1978). *The comparative ecology and population dynamics of Araucaria species in Papua New
   Guinea.* Ph.D. Thesis, Australian National University,
   Canberra.
Epstein, S., Thompson, P. & Yapp, C.J. (1977). Oxygen and
   hydrogen isotopic ratios in plant cellulose. *Science,
   198*, 1209-1215.
Epstein, S. & Yapp, C.J. (1976). Climatic implications of
   the D/H ratio of hydrogen in C-H groups in tree
   cellulose. *Earth & Planetary Science Letters, 30*,
   252-261.
Epstein, S., Yapp, C.J. & Hall, J.H. (1976). The determination of the D/H of non-exchangeable hydrogen in
   cellulose extracted from aquatic and land plants.
   Earth & Planetary Science Letters, 30, 241-251.
Estes, E.T. (1969). *The dendrochronology of three tree
   species in the central Mississippi Valley.* Ph.D.
   Dissertation, Southern Illinois University.

Fabijanowski, J. (1962). Lasy tatrzańskie. In: *Tatrzański
   Park Narodowy*, ed. W. Szafer, pp.240-304. Kraków.
Farmer, J.G. (1979). Problems in interpreting tree-ring
   records. *Nature, London, 279*, 229-231.
Feliksik, E. (1972). Studia dendroklimatologiczne nad
   świerkiem (*Picea excelsa* L.). *Acta Agraria et Sil-
   vestra, 12*, 41-70.
Feliksik, E. (1975). Present state of dendrochronological
   investigation in Poland. In: *Bioecological Fundamentals of Dendrochronology. Symp. Mat. XII Int. Bot.
   Cong.*, Leningrad, pp.21-25.
Ferguson, C.W. (1968). Bristlecone pine: science and esth-
   etics. *Science, 159*(3817), 839-846.
Ferguson, C.W. (1969). A 7104-year annual tree-ring chronology for bristlecone pine, *Pinus aristata*, from the
   White Mountains, California. *Tree-Ring Bulletin, 29*
   (3-4), 1-29.

Ferguson, C.W. (1970). Dendrochronology of bristlecone pine, *Pinus aristata*: establishment of a 7484-year chronology in the White Mountains of Eastern-Central California, U.S.A. In: *Radiocarbon Variations and Absolute Chronology*, ed. I.U. Olsson, pp.237-259. Wiley: New York.

Ferhi, A., Letolle, R. & Lerman, J.C. (1975). Oxygen isotope ratios of organic matter: Analyses of natural composition. In: *Proceedings of the 2nd International Conference on Stable Isotopes*, U.S. ERDA Conference 741027, Oak Brook, Illinois, pp.716-724.

Ferhi, A. & Letolle, R. (1977). Transpiration and evaporation as the principal factors in oxygen isotope variations of organic matter in land plants. *Physiol. Veg.*, *15*, 107-117.

Ferhi, A., Long, A. & Lerman, J.C. (1977). Stable isotopes of oxygen in plants: a possible paleohygrometer. *Hydrology and Water Resources, Arizona and Southwest*, *7*, 191-198.

Ferhi, A. & Letolle, R. (1979). Relation entre le milieu climatique et les teneurs en oxygen-18 de la cellulose des plantes terrestres. *Physiol. Veg.*, *17*, 107-117.

Fleming, C.A. (1962). New Zealand biogeography: A paleontologist's approach. *Tuatara*, *10*, 53-108.

Fleming, C.A. (1979). *The Geological History of New Zealand and its Life*. Auckland University Press: Auckland.

Fletcher, J. (1977). Tree-ring chronologies for the 6th to 16th centuries for oaks of southern and eastern England. *Journal of Archaeological Science*, *4*, 335-352.

Flohn, H., editor (1969). *General Climatology, 2. World Survey of Climatology, Vol.2.* Elsevier: Amsterdam.

Flohn, H. (1980). Possible climatic consequences of a man-made global warming. *Research Report RR-80-30*, International Institute for Applied Systems Analysis, Laxenburg, Austria.

Förstel, H. (1978). The enrichment of ^{18}O in leaf water under natural conditions. *Radiocarbon and Environmental Biophysics*, *15*, 323-344.

Förstel, H. (1979). The enrichment of $H_2^{18}O$ under field and under laboratory conditions. In: *Proceedings of the 3rd International Conference on Stable Isotopes*, ed. E.R. Klein & P.D. Klein, pp.191-203. Academic: New York.

Forsythe, G.E. (1957). Generation and use of orthogonal polynomials for fitting data with a digital computer. *Journal of the Society of Industrial and Applied Mathematics*, *5*(2), 74-88.

Fowells, H.A. (1965). *Silvics of Forest Trees of the United States*. Agricultural Handbook No.271, U.S. Department of Agriculture, Washington DC.

Francey, R.J. (1981). Tasmanian tree rings belie suggested anthropogenic $^{14}C/^{13}C$ trends. *Nature, London, 290*, 232-235.

Franklin, D.A. (1968). Biological flora of New Zealand. 3. *Dacrydium cupressinum* Lamb. (Podocarpaceae) Rimu. *New Zealand Journal of Botany*, *6*, 493-513.

Franklin, D.A. (1969). Growth rings in rimu from South Westland terrace forest. *New Zealand Journal of Botany*, *7*, 177-188.

Franklin, J.F. & Dyerness, C.T. (1973). *Natural Vegetation of Oregon and Washington*. U.S.D.A. Forest Service General Technical Report PNW-8. U.S. Department of Agriculture: Washington DC.

Freyer, H.D. (1979a). On the ^{13}C record in tree rings. I. ^{13}C variations in northern hemisphere trees during the last 150 years. *Tellus*, *31*, 124-138.

Freyer, H.D. (1979b). On the ^{13}C record in tree rings. II. Registration of microenvironmental CO_2 and anomalous pollution effect. *Tellus*, *31*, 308-312.

Freyer, H.D. (1980). Record of environmental variables by ^{13}C measurements in tree rings. In: *Proceedings of the International Meeting on Stable Isotopes in Tree-Ring Research*, New Paltz, New York, ed. G.C. Jacoby, U.S. Department of Energy Publ. No.12, CONF-7905180, UC-11, pp.13-21.

Freyer, H.D. (1981). ^{13}C record in tree rings during the past half millenium - climatic fluctuations and anthropogenic impact. Paper presented at *IAMAP Third Scientific Assembly*, Hamburg FRG.

Frezet, D., Tessier, L., Guiot, J., Pons, A. & Serre, F. (1979). Dendroclimatologie de *Quercus pubescens* de la montagne St-Maurice et del la fôret de Bordeaux-Bois-de-Vache (Drôme). *Rapport de l'exercise 1978-79 Action concerte D.G.R.S.T. "Structure dynamique et utilisation des formations à Chene pubescent"*.

Frisse, E. (1977). *Xylometriche und dendroklimatologische Untersuchungen uber den Einfluss von Temperatur und Niederschlag auf Eichen und Buchen*. Dissertation, University of Hamburg.

Fritts, H.C. (1958). An analysis of radial growth of beech in a central Ohio forest during 1954-1955. *Ecology*, *39*, 705-720.

Fritts, H.C. (1959). The relation of radial growth to maximum and minimum temperatures in three tree species. *Ecology*, *40*, 261-265.

Fritts, H.C. (1960). Multiple regression analysis of radial growth in individual trees. *Forest Science*, *6*, 334-349.

Fritts, H.C. (1962). The relations of growth ring variations in American beech and white oak to variations in climate. *Tree-Ring Bulletin*, *25*, 2-10.

Fritts, H.C. (1963). Computer programs for tree-ring research. *Tree-Ring Bulletin*, *25*(3-4), 2-7.

Fritts, H.C. (1965). Tree-ring evidence for climatic changes in Western North America. *Monthly Weather Review*, *93*, 421-443.

Fritts, H.C. (1966). Growth-rings of trees: their correlation with climate. *Science*, *154*(3752), 973-979.

Fritts, H.C. (1967). Growth rings of trees: A physiological basis for their correlation with climate. In: *Ground Level Climatology*, ed. R.A. Shaw, A.A.A.S. Publ. No.86, pp.45-65.

Fritts, H.C. (1969). Bristlecone pine in the White Mountains of California; growth and ring-width characteristics. *Papers of the Laboratory of Tree-Ring Research*, *4*, University of Arizona, Tucson, Arizona.

Fritts, H.C. (1974). Relationships of ring widths in arid-site conifers to variations in monthly temperature and precipitation. *Ecological Monographs*, *44*(4), 411-440.

Fritts, H.C. (1976). *Tree Rings and Climate*. Academic: London.

Fritts, H.C., Smith, D.G., Cardis, J.W. & Budelsky, C.A. (1965a). Tree-ring characteristics along a vegetation gradient in Northern Arizona. *Ecology*, *46*, 393-401.

Fritts, H.C., Smith, D.G. & Stokes, M.A. (1965b). The biological model for paleoclimatic interpretation of Mesa Verde tree-ring series. *American Antiquities*, *31*(2), Part 2, 101-121.

Fritts, H.C., Mosimann, J.E. & Botterff, C.P. (1969). A revised computer program for standardizing tree-ring series. *Tree-Ring Bulletin*, *29*, 15-20.

Fritts, H.C., Blasing, T.J., Hayden, B.P. & Kutzbach, J.E. (1971). Multivariate techniques for specifying tree-growth and climate relationships and for reconstructing anomalies in paleoclimate. *Journal of Applied Meteorology*, *10*(5), 845-864.

Fritts, H.C. & Shatz, D.J. (1975). Selecting and characterizing tree-ring chronologies for dendroclimatic analysis. *Tree-Ring Bulletin*, *35*, 31-40.

Fritts, H.C. & Lofgren, G.R. (1978). *Patterns of climatic change revealed through dendroclimatology*. U.S. Army Coastal Engineering Research Center Requisition Purchase Request IWR-B-78-119. Laboratory of Tree-Ring Research, University of Arizona, Tucson.

Fritts, H.C., Lofgren, G.F. & Gordon, G.A. (1979). Variations in climate since 1602 as reconstructed from tree rings. *Quaternary Research*, *12*(1), 18-46.

Fritts, H.C., Lofgren, G.R. & Gordon, G.A. (1980). Past climate reconstructed from tree rings. *Journal of Interdisciplinary History*, *10*(4), 773-793.

Fritts, P. & Fontes, J., editors. (1980). *Handbook of Environmental Isotope Geochemistry Vol.1, The Terrestrial Environment*, A. Elsevier: Amsterdam.

Galazii, G.I. (1972). The annual increment of trees in relation to changes in climate, water level and relief on the NW shore of Lake Baikal. *Tr. Limnol. In - ta Sib. Otd. AN SSR*, *13*(4), 71-214. In Russian.

Garfinkel, H.L. (1979). *Climate-growth relationships in*

white spruce (Picea glauca) (Moench) Voss) in the south central Brooks range of Alaska. Ph.D. Thesis, University of Washington, Seattle.

Garfinkel, H.L. & Brubaker, L.B. (1980). Modern climate-tree-growth relationships and climatic reconstruction in sub-Arctic Alaska. *Nature, London, 286,* 872-874.

Garner, B.J. (1958). *The Climate of New Zealand.* Edward Arnold: London.

Garnier, M. (1974). Longues séries de measures de précipitations en France, Zone 4 (Méditérranéene). *Mémorial de la Meteorologie Nationale, Fasc.4,* No.53.

Gassner, A. & Christiansen, W.Fr. (1943). Dendroklimatologische Untersuchungen über die Jahrringentwicklung der Kiefer in Anatolien. *Nova Acta Leopold, 12,* 80-137.

Gentilli, J., editor. (1971). *Climate of Australia and New Zealand. World Survey of Climatology, Vol.13.* Elsevier: Amsterdam.

Gentilli, J. (1972). *Australian Climate Patterns.* Thomas Nelson (Australia): Adelaide.

Giddings, J.L., Jr. (1938). Buried wood from Fairbanks, Alaska. *Tree-Ring Bulletin, 4*(4), 3-5.

Giddings, J.L., Jr. (1941). Dendrochronology in northern Alaska. *University of Alaska Publication, Vol.IV; Laboratory of Tree-Ring Research Bulletin No.1,* University of Arizona, Tucson.

Giddings, J.L., Jr. (1943). Some climatic aspects of tree growth in Alaska. *Tree-Ring Bulletin, 4,* 26-32.

Giddings, J.L., Jr. (1947). Mackenzie River Delta chronology. *Tree-Ring Bulletin, 13,* 26-29.

Giddings, J.L., Jr. (1948). Chronology of the Kobuk-Kotzebue series. *Tree-Ring Bulletin, 14*(4), 26-32.

Giddings, J.L., Jr. (1951). The forest edge at Norton Bay, Alaska. *Tree-Ring Bulletin, 18*(1), 2-8.

Giddings, J.L., Jr. (1953). Yukon River spruce growth. *Tree-Ring Bulletin, 20*(11), 2-5.

Glahn, H.R. (1968). Canonical correlation and its relationship to discriminant analysis and multiple regression. *Journal of the Atmospheric Sciences, 25*(1), 23-31.

Glock, W.S. & Agerter, S.R. (1962). Rainfall and tree growth. In: *Tree Growth,* ed. T.T. Kozlowski. Ronald Press: New York.

Godley, E.J. (1976). Flora. In: *New Zealand Atlas,* ed. I. Wards, pp.108-113. Government Printer: Wellington.

Gong Gaofa, Chen Enhiu & Wen Huanran (1979). The climatic fluctuations in Heilongjiang province, China. *Acta Geographica Sinica, 34*(2), 129-138. In Chinese.

Gordon, G.A. (1980). Verification tests for dendroclimatological reconstructions. *Technical Note No.19,* Northern Hemisphere Climate Reconstruction Group, Laboratory of Tree-Ring Research, University of Arizona, Tucson.

Graf-Henning, H.-J. (1960). *Statistische Methoden bei textilen Untersuchungen.* Springer: Berlin, Gottingen, Heidelberg.

Granger, O.E. (1977). Secular fluctuations of seasonal precipitation in lowland California. *Monthly Weather Review, 105*(4), 386-397.

Gray, J. (1981). The use of stable-isotope data in climate reconstruction. In: *Climate and History: Studies in Past Climate and their Impact on Man,* ed. T.M.L. Wigley, M.J. Ingrams & G. Farmer. Cambridge University Press.

Gray, J. & Thompson, P. (1976). Climatic information from $^{18}O/^{16}O$ ratios of cellulose in tree rings. *Nature, London, 262,* 481-482.

Gray, J. & Thompson, P. (1977). Climatic information from $^{18}O/^{16}O$ analysis of cellulose, lignin, and whole wood from tree-rings. *Nature, London, 270,* 708-709.

Graybill, D.A. (1979a). *Program operating manual for RWLIST, INDEX and SUMAC.* Laboratory of Tree-Ring Research, University of Arizona, Tucson.

Graybill, D.A. (1979b). Revised computer programs for tree-ring research. *Tree-Ring Bulletin, 39,* 77-82.

Green, J.W. (1963). Wood cellulose. In: *Methods in Carbohydrate chemistry, Vol.3,* ed. R.L. Whistler, pp.9-21. Academic: New York.

Green, H.V. (1965). The study of wood characteristics by means of a photometric technique. In: *Proceedings; Meeting of the Working Groups on Wood Quality, Sawing*

and Machining, Wood and Tree Chemistry, ed. Section 41, IUFRO, Melbourne, 4-15 October 1965, Vol.2. C.S.I.R.O., Division of Forest Products: Melbourne.

Gribbin, J. editor. (1978). *Climatic Change.* Cambridge University Press.

Griffiths, J.F., editor. (1971). *Climates of Africa. World Survey of Climatology, Vol.10.* Elsevier: Amsterdam.

Grinsted, M.J. (1977). *A study of the relationships between climate and stable isotope ratios in tree rings.* Ph.D. Thesis, University of Waikato, Hamilton, New Zealand.

Guiot, J. (1980a). Spectral multivariate regression in dendroclimatology. *Institute d'Astronomie et de Géophysique, Université Catholique de Louvain-la-Neuve, Contribution No. 21.*

Guiot, J. (1980b). Response function after extracting principal components and persistence in dendroclimatology. *Institut d'Astronomie et de Géophysique, Université Catholique de Louvain-La-Neuve, Contribution No. 22.*

Guiot, J., Berger, A.L. & Munaut, A. (1978). La fonction réponse en dendroclimatologie. *Institut d'Astronomie et de Géophysique, Université Catholique de Louvain-la-Neuve, Scientific Report 1978/4.*

Guiot, J., Berger, A.L. & Munaut, A. (1981). Some new mathematical procedures in dendroclimatology with examples in Switzerland and Morocco. *Tree-Ring Bulletin,* submitted.

Hadley, E.B. (1969). Physiological ecology of *Pinus ponderosa* in southwestern North Dakota. *The American Midland Naturalist, 81*(2), 289-314.

Hardcastle, K.G. & Friedman, I. (1974). A method for oxygen isotope analysis of organic material. *Geophysical Research Letters, 1,* 165-167.

Harding, S.T. (1965). *Recent Variations in the Water Supply of the Western Great Basin.* Archive Series Report No.16, Water Resources Center Archives, University of California.

Harlow, W.M. & Harrar, E.S. (1969). *Textbook of Dendrology (5th edition).* McGraw-Hill: New York.

Harlow, W., Harrar, E.S. & White, F.M. (1979). *Textbook of Dendrology (6th edition).* McGraw-Hill: New York.

Harshberger, J.W. (1970). *The Vegetation of the New Jersey Pine-Barrens.* Dover: New York.

Harvey, L.D.D. (1980). Solar variability as a contributory factor to Holocene climatic change. *Progress in Physical Geography, 4,* 487-530.

Haugen, R.K. (1967). Tree ring indices: a circumpolar comparison. *Science, 158,* 773-775.

Hawley, F. (1941). *Tree-Ring Analysis and Dating in the Mississippi Drainage.* Chicago University Press: Chicago.

Helleputte, Cl. (1976). Contribution à l'étude dendrochronologique du cèdre dans le Rif et le Moyen Atlas (Maroc). *Memoire Lic. Sc. Géogr.,* Institut d'Astronomie et de Géophysique, Université Catholique de Louvain-la-Neuve.

Helley, E.J. & LaMarche, V.C., Jr. (1968). A 400-year flood in northern California. *United States Geological Survey Professional Paper 600-D,* U.S. Geological Survey, Washington D.C.

Herbert, J. (1972). Growth of silver beech in northern Fiordland. *New Zealand Journal of Forestry Science, 3,* 131-171.

Herbert, J. (1977). Growth pattern of planted tanehaka *(Phyllacladus trichomanides D.Don). New Zealand Journal of Botany, 15,* 513-515.

Hessel, J.W.D. (1980). Apparent trend of mean temperatures in New Zealand since 1930. *New Zealand Journal of Science, 23,* 1-9.

Heuberger, H. (1974). Alpine Quaternary glaciation. In: *Arctic and Alpine Environments,* ed. J.D. Ives & R.G. Barry. Methuen: London.

Heusser, C.J. (1964). Historical variations of Lemon Creek Glacier, Alaska and their relationships to the climatic record. *Journal of Glaciology, 5,* 77-86.

Heusser, C.J. (1966). Late-Pleistocene pollen diagrams from the Province of Llanquihue, southern Chile. *Proceedings of the American Philosophical Society, 110*(4), 269-304.

Hirano, R. (1920). The tree rings of cryptomeria and the

Bruckner cycle. *Journal of the Meteorological Society of Japan, 39*(10), 276-281. In Japanese.

Hoeg, O.A. (1956). Growth-ring research in Norway. *Tree-Ring Bulletin, 21*(1-4), 2-15.

Hoinkes, H.C. (1968). Glacier variation and the weather. *Journal of Glaciology, 7*(49), 1-13.

Hollstein, E. (1963). Jahrringchronologie mosel- und saarländischer Rotbuchen. *Mitteilungen der Deutschen dendrologischen Gesellschaft, No.66,* 165-172.

Hollstein, E. (1980). *Mitteleuropäische Eichenchronologie.* Zabern-Verlag: Mainz.

Holloway, J.T. (1954). Forests and climates in the South Island of New Zealand. *Transactions of the Royal Society of New Zealand, 82,* 329-410.

Holmes, R.L. (1978). Informe preliminar sobre el trabajo en terreno en Argentina y Chile 1975-1978. *Annales 1977, 4,* 187-197. Instituto Argentino de Nivologia y Glaciologia, Mendoza, Argentina.

Holmes, R.L., Stockton, C.W. & LaMarche, V.C. Jr. (1978). Extensión de registros de caudales a partire de cronologias de anillos de arboles en Argentina. *Annales 1977, 4,* 97-105. Instituto Argentino de Nivologia Y Glaciologia, Mendoza, Argentina.

Holmes, R.L., Stockton, C.W. & LaMarche, V.C., Jr. (1979). Extension of river flow records in Argentina from long tree-ring chronologies. *Water Resources Bulletin, 15*(4), 1081-1085.

Hotelling, H. (1936). Relations between two sets of variates. *Biometrika, 28,* 139-142.

Hough, A.F. & Forbes, R.D. (1943). The ecology and silvics of forests in the high plateaux of Pennsylvania. *Ecological Monographs, 13,* 301-320.

Hoyanagi, M. (1940). Tree-ring analysis and climatic change. *Geographical Review of Japan, 16*(12), 801-817.

Howe, S. & Webb, III, T. (1977). Testing the assumptions of paleoclimatic calibration functions. In: *Preprint Volume, Fifth Conference on Probability and Statistics,* 15-18 November 1977. American Meteorological Society: Boston.

Huber, F. (1976). Problèmes d'interdatation chez le pin sylvestre et influence du climat sur la structure de ses accroissements annuels. *Annales des Sciences Forestières, 33*(2), 61-86.

Huber, B. & Giertz-Siebenlist, V. (1969). Unsere tausendjährige Eichenchronologie durchschnittlich 57 (10-150) fach belegt. *Sitzungsberichte der Österreichischen Akademie der Wissenschaften, 178,* 37-42.

Hueck, K. (1951). Eine biologische method zum messen der erodieren Tatigkeit des Windes und des Wassers. *Deutsche Botannische Gesellschaft Berlin, 64,* 53-56.

Hughes, M.K., Leggett, P., Milsom, S.J. & Hibbert, F.A. (1978a). Dendrochronology of oak in North Wales. *Tree-Ring Bulletin, 38,* 15-23.

Hughes, M.K., Gray, B., Pilcher, J., Baillie, M.G.L. & Leggett, P. (1978b). Climatic signals in British Isles tree-ring chronologies. *Nature, London, 272,* 605-606.

Hughes, M.K., Kelly, P.M., Pilcher, J.R. & LaMarche, V.C., Jr., (1980). *Report and Recommendations of the Second International Workshop on Global Dendroclimatology.* The Organising Committee of the Second International Workshop on Global Dendroclimatology: Belfast.

Hustich, I. (1956). Notes on the growth of Scotch Pine in Utsjokl in northernmost Finland. *Acta Botanica Fennica, 56,* 3-13.

Hutson, W.H. (1977). Transfer functions under no-analog conditions: experiments with Indian Ocean planktonic foraminifera. *Quaternary Research, 8*(3), 355-367.

I.A.E.A. (1981). Statistical treatment of environmental isotope data in precipitation. *Technical Report Series, No.206,* I.A.E.A., Vienna.

Imbrie, J. & Kipp, N.G. (1971). A new micropaleontological method for quantitative palaeoclimatology: an application to a late Pleistocene Carribean core. In: *The Late Cenozoic Glacial Ages,* ed. K. Turekian, pp.71-182. Yale University Press: New Haven, Connecticut.

Ingram, M.J., Underhill, D.J. & Farmer, G. (1981). The use of documentary sources for the study of past climates. In: *Climate and History,* ed. T.M.L. Wigley, M.J.

Ingram & G. Farmer. Cambridge University Press.

Isachenko, T.E. (1955). *Vegetation Map of U.S.S.R., 1: 4,000,000.* G.U.G.K., Scientific and Editorial Section: Moscow.

Jackson, W.D. (1965). Vegetation. In: *Atlas of Tasmania,* ed. J.L. Davies, pp.30-35. Lands & Survey Department: Hobart, Tasmania.

Jackson, W.D. (1968). Fire, air, water and earth. An elementary ecology of Tasmania. *Proceedings of the Ecological Society of Australia, 3,* 9-16.

Jacoby, G.C. (1979). Tree-ring analysis to date terrace events at Icy Bay, Alaska. In: *Crustal Deformation Measurement Near Yakatuga, Gulf of Alaska,* Summary of Semi-annual report for USGS 14-08-0001-18272, U.S. Geological Survey, pp.22-27.

Jacoby, G.C. (1980a). X-ray densitometry in tree-ring studies. In: *Proceedings of Dendrology of the Eastern Deciduous Forest Biome.* Virginia Polytechnic Institue: Blacksburg, Virginia.

Jacoby, G.C. editor. (1980b). *Proceedings of the International Meeting on Stable Isotopes in Tree-Ring Research,* New Paltz, New York, Carbon Dioxide Effects Research and Assessment Program, Publication No.12, U.S. Dept. of Energy, CONF-790518, UC-11, Department of Energy: Washington D.C.

Jacoby, G.C. & Cook, E.R. (1981). High-latitude temperature tends as inferred from a 400-year tree-ring chronology from Yukon Territories, Canada. *Arctic and Alpine Research, 13,* 409-418.

Jährig, M. (1976). Zum Stand der dendrochronologischen Arbeiten. *Ausgrabungen und Funde, 21,* 191-194.

Jansen, H.S. (1962). Comparison between ring-dates and ^{14}C dates in a New Zealand kauri tree. *New Zealand Journal of Science, 5,* 74-84.

Jansen, H.S. & Wardle, P. (1971). Comparisons between ring age and ^{14}C age in rimu trees from Westland and Auckland. *New Zealand Journal of Botany, 9,* 215-216.

Jenne, R. (1975). Data sets for meteorological research. NCAR-TN/IA-111, National Center for Atmospheric Research, Boulder, Colorado.

Jennings, J.N. (1972). Some attributes of Torres Strait. In: *Bridge and Barrier: The Natural and Cultural History of Torres Strait,* ed. D. Walker, pp.29-38. Australian National University Publication BG/3, Canberra.

Jensen, Ad.S. (1939). Concerning a change of climate during recent decades in the arctic and subarctic regions, from Greenland in the west to Eurasia in the east, and contemporary biological and geophysical changes. *Det Kongelige Danske Videnskabernes Selskabs Biologiske Meddelelser, 14* (8), 1-75.

Johnson, S.J. (1977). Stable isotope profiles compared with temperature profiles in firn with historical temperature records. In: *Isotopes and Impurities in Snow and Ice,* Proceedings of the Grenoble Symposium, August/September 1975. *IAHS Publication No.118.*

Johnston, J. (1972). *Econometric Methods (2nd edition).* McGraw-Hill: New York.

Johnston, T.N. (1975). Thinning studies in Cypress Pine in Queensland. *Research Paper No.7,* pp.1-87, Dept. of Forestry, Queensland.

Jones, P.D. Wigley, T.M.L. & Kelly, P.M. (1982). Variations in surface air temperatures: Pt.1, the Northern Hemisphere, 1881-1980. *Monthly Weather Review,* in press.

Jonsson, B. (1969). Studier över den av väderleken orsakade variationen i årsringbredderna hos tall och gran i Sverige. *Royal College of Forestry, Stockholm, Research Notes, 16.*

Julian, P.R. (1970). An application of rank-order statistics to the joint spatial and temporal variations of meteorological elements. *Monthly Weather Review, 98* (2), 142-153.

Kaiser, F. (1979). *Ein spateiszeitlicher Wald im Dattau bei Winterthur/Schwiez.* Dissertation, Universität Zurich, Zeigler Winterthur.

Kärenlampi, L. (1972). On the relationships of the Scots pine annual ring width and some climatic variables at the Kevo Subarctic Station. *Reports of the Kevo Subarctic Station, 9,* 78-81.

Karpavicius, J.A. (1976). Sensitivity coefficient as an indicator of tree response to environmental conditions.
In: *Indication of the Natural Processes and Environment, Vilnius*, 1976, 45-47. In Russian.

Kay, P.A. (1978). Dendroecology in Canada's forest-tundra
transition zone. *Arctic and Alpine Research, 10*(1),
133-138.

Keeling, C.D., Mook, W.G. & Tans, P.P. (1979). Recent trend
in the $^{13}C/^{12}C$ ratio of atmospheric carbon dioxide.
Nature, London, 277, 121-123.

Keeling, C.D., Bacastow, R.B. & Tans, P.P. (1980). Predicted
shift in the $^{13}C/^{12}C$ ratio of atmospheric carbon dioxide. *Geophysical Research Letters, 7*, 505-508.

Kellogg, W.W. & Schware, R. (1981). *Climatic Change and Society. Consequences of Increasing Atmospheric Carbon
Dioxide*. Westview Press: Boulder.

Kelly, P.M. (1979). Towards the prediction of climate.
Endeavour, New Series, 3(4), 176-182.

Kelly, P.M., Jones, P.D., Sear, C.B., Cherry, B.S.G. &
Tavakol, R.K. (1982). Variations in surface air temp-
erature: Pt.2, Arctic regions, 1881-1980. *Monthly
Weather Review*, in press.

Kendall, M.G. & Stuart, A. (1979). *The Advanced Theory of
Statistics, Vol.2 (4th edition)*. Griffin: London.

Kennedy, E.A. & Gordon, G.A. (1980). Characteristics of
nineteenth century climatic data used in verification
of dendroclimatic reconstructions. *Technical Note 16*,
Laboratory of Tree-Ring Research, University of Arizona, Tucson.

Kidson, J.W. (1975a). Eigenvector analysis of monthly mean
surface data. *Monthly Weather Review, 103*(3), 177-186.

Kidson, J.W. (1975b). Tropical eigenvector analysis and the
southern Oscillation. *Monthly Weather Review, 103*(3),
187-196.

Klein, P. (1979). Alte Gemälde auf Holz gemalt. Jahhringanal-
ytische Untersuchungen an Gemaldetafeln. *Holz-Zentral-
blatt, 105*, 2287-2288.

Klemmer, I. (1969). *Die Periodik des Radialzuwashses in einen
Fichtenwald und deren meteorologische Steuerung*. Dissertation, Lud. -Max. -Universität München.

Knigge, W. & Schultz, H. (1961). Einfluss der Jahreswitterung
1959 auf Zellartverteilung, Fäserlange und Gefässweite
verschiedener Holzarten. *Holz als Roh- und Werkstoff,
22*, 293-303.

Koch, L. (1945). The East Greenland ice. *Meddelelser om Grønland, 130*(3).

Kohler, M.A. (1949). On the use of double-mass analysis for
testing the consistency of meteorological records.
Bulletin of American Meteorological Society, 30, 188-
189.

Kuniholm, P.I. & Striker, C.L. (1977). The tie-beam system
in the Nave Arcade of St. Eirene: structure and denro-
chronology. *Istanbuler Mitteilungen, Beiheit, 18*, 229-
240.

Kutzbach, J.E. (1970). Large-scale features of monthly mean
Northern Hemisphere anomaly maps of sea-level pressure.
Monthly Weather Review, 98(9), 708-716.

Kutzbach, J.E. & Guetter, P.J. (1980). On the design of pal-
eoenvironmental data networks for estimating large-
scale patterns of climate. *Quaternary Research, 14*(2),
169-187.

Kyncl, J. (1976). Prispevek k metodice dendrochronologie.
Archeologia historica, 2, 317-310.

LaMarche, V.C., Jr. (1961). Rate of slope erosion in the
White Mountains, California. *Geological Society of
America Bulletin, 72*, 1579.

LaMarche, V.C., Jr. (1968). Rates of slope degradation as
determined from botanical evidence, White Mountains,
California. *U.S. Geological Survey Professional Paper,
352-1*, Washington D.C.

LaMarche, V.C., Jr. (1969). Environment in relation to age
of bristlecone pines. *Ecology, 50*, 53-59.

LaMarche, V.C., Jr. (1973). Holocene climatic variations inferred from treeline fluctuations in the White Mountains, California. *Quaternary Research, 3*, 632-660.

LaMarche, V.C., Jr. (1974a). Paleoclimatic inferences from
long tree-ring records. *Science, 183*, 1043-1048.

LaMarche, V.C., Jr. (1974b). Frequency-dependent relationships between tree-ring series along an ecological
gradient and some dendroclimatic implications. *Tree-

Ring Bulletin, 34*, 1-20.

LaMarche, V.C., Jr. (1975). Potential of tree-rings for re-
construction of past climatic variations in the
Southern Hemisphere. *Proceedings of the WMO/IAMAP
Symposium on Long-Term Climatic Fluctuations*, Norwich,
England, W.M.O. No.421, pp.21-30. World Meteorologic-
al Organisation: Geneva.

LaMarche, V.C., Jr. (1978). Tree-ring evidence of past
climatic variability. *Nature, London, 276*, 334-338.

LaMarche, V.C., Jr. & Mooney, H.A. (1967). Altithermal
timberline advance in Western United States. *Nature,
London, 218*, 980-982.

LaMarche, V.C., Jr. & Fritts, H.C. (1971a). Tree rings,
glacial advance and climate in the Alps. *Zeitschrift
fur Gletscherkunde und Glaziologie, VII*, 125-131.

LaMarche, V.C., Jr. & Fritts, H.C. (1971b). Anomaly
patterns of climate over the Western United States,
1700-1930, derived from principal component analysis
of tree-ring data. *Monthly Weather Review, 99*, 138-
142.

LaMarche, V.C., Jr. & Mooney, H.A. (1972). Recent climatic
change and development of the bristlecone pine (*P.
Longaeva* Bailey), Krumholz zone, Mt. Washington,
Nevada. *Arctic and Alpine Research, 4*(1), 61-72.

LaMarche, V.C., Jr. & Harlan, T.P. (1973). Accuracy of
tree-ring dating of bristlecone pine for calibration
of the radiocarbon time scale. *Journal of Geophysical
Research, 78*(36), 8849-8858.

LaMarche, V.C., Jr. & Stockton, C.W. (1974). Chronologies
from the temperature-sensitive bristlecone pines at
upper treeline in western United States. *Tree-Ring
Bulletin, 34*, 21-45.

LaMarche, V.C., Jr., Holmes, R.L., Dunwiddie, P.W. & Drew,
L.G. (1979a,b,c,d,e). Tree-ring chronologies of the
Southern Hemisphere, a. Vol.1: Argentina, b. Vol.2:
Chile, c. Vol.3: New Zealand, d. Vol.4: Australia,
e. Vol.5: South Africa. *Chronology Series V*, Labor-
atory of Tree-Ring Research, University of Arizona,
Tucson.

Lamb, H.H. (1966). *The Changing Climate*. Methuen: London.

Lamb, H.H. (1970). Volcanic dust in the atmosphere with
chronology and assessment of its meteorological significance. *Philosophical Transactions of the Royal
Society, A, 226*, 425-533.

Lamb, H.H. (1972). *Climate: Present, Past and Future, Vol.
1: Fundamentals and Climate Now*. Methuen: London.

Lamb, H.H. (1977). *Climate: Present, Past and Future, Vol.
2: Climate History and the Future*. Methuen: London.

Lamb, H.H. & Johnson, A.I. (1959). Climatic variation and
observed changes in the general circulation. *Geografiska Annaler, 41*, 94-134.

Lamprecht, A. (1978). *Die Beziehungen zwischen Holzdichte-
werten von Fichten aus subalpinen Lagen des Tirols
und Witterungsdaten aus Chroniken im Zeitraum 1370-
1800 AD*. Dissertation, Universität Zürich.

Lange, R.T. (1965). Growth characteristics of an arid-zone
conifer. *Transactions of the Royal Society of South
Australia, 89*, 133-137.

Langford, J. (1965). Weather and climate. In: *Atlas of
Tasmania*, ed. J.L. Davies, pp.21-11. Lands & Survey
Department: Hobart.

Larsen, J.A. (1974). Ecology of the northern continental
forest border. In: *Arctic and Alpine Environments*,
ed. J.D. Ives & R.G. Barry, pp.341-369. Methuen:
London.

Larson, S.C. (1931). The shrinkage of the coefficient of
multiple correlation. *Journal of Educational Psychology, 22*, 45-55.

Lawrence, J. (1980). Stable isotopes in precipitation. In:
*Proceedings of the International Meeting on Stable
Isotopes in Tree-Ring Research*, ed. G.C. Jacoby,
Carbon Dioxide Effects Research and Assessment Pro-
gram, Publication No.12, U.S. Department of Energy:
Washington D.C.

Lawson, M.P. (1974). The climate of the great American
desert, reconstruction of the climate of the western interior United States, 1800-1850. *University
of Nebraska Studies: New Series No.46*, University
of Nebraska, Lincoln, Nebraska.

Lawson, C.L. & Hanson, R.J. (1974). *Solving Least Squares
Problems*. Prentice-Hall: New Jersey.

Leeper, W.G. (1970). Climates. In: *The Australian Environment (4th edition)*, ed. W.G. Leeper, pp.12-20. C.S.I.R.O. and Melbourne University Press: Melbourne.

Leggett, P., Hughes, M. & Hibbert, F.A. (1978). A modern oak chronology from North Wales and its interpretation. In: *Dendrochronology in Europe*, ed. J. Fletcher, British Archaeological Research series, no.51, pp. 187-194.

Lefébure, L. (1980). Modèles de la tendance en dendrochronologie. *Mémoire de Lic. Sc. Math.*, Institut d'Astronomie et de Géophysique, Université Catholique de Louvain-la-Neuve.

Lentner, M. (1975). *Introduction to Applied Statistics*. Prindle, Weber & Schmidt: Boston.

Lenz, O., Schär, E. & Schweingruber, F.H. (1976). Methodische Probleme bei der radiographisch-densitometrischen Bestimmung der Dichte und der Jahrringbreiten von Holz. *Holzforschung*, *30*, 114-123.

Leopold, L.B., Wolman, M.G. & Miller, J.P. (1964). *Fluvial Processes in Geomorphology*. W.H. Freeman: San Francisco.

Lerman, J.C. (1974). Isotopic "paleothermometers" on continental matter: assessment. In: *Les Methods Quantitative d'Etude des Variations du Climate au Cours du Pleistocene*, International C.N.R.S., no.219, pp.163-191, C.N.R.S.: Paris.

Libby, L.M. (1972). Multiple thermometry in paleoclimate and historic climate. *Journal of Geophysical Research*, *77*, 4310-4317.

Libby, L.M. & Pandolfi, L.J. (1974). Temperature dependence of isotopic ratios in tree rings. *Proceedings of the National Academy of Science*, *71*, 2482-2486.

Libby, L.M., Pandolfi, L.J., Payton, P.H., Marshall, J., Becker, B. & Giertz-Siebenlist, W. (1976). Isotopic tree thermometers. *Nature, London*, *261*, 284-288.

Liese, W. & Meyer-Uhlenried, K.H. (1957). Zur quantitativen Bestimmung der verschiedenen Zellarten im Holz. *Zeitschrift für Wissenschaftliche Mikroskopie und Mikroskopische Technik*, *63*, 269-275.

Liese, W. (1978). Bruno Huber - the pioneer of European dendrochronology. In: *Dendrochronology in Europe*, ed. J. Fletcher, British Archaeological Research Series, No.51, pp.1-10.

Lilly, M.A. (1977). An assessment of the dendroclimatological potential of indigenous tree species in South Africa. *Occasional Paper No.18*, Department of Geography and Environmental Studies, University of Witwatersrand, Johannesburg.

Linacre, E. & Hobbs, J. (1977). *The Australian Climatic Environment*. Wiley: Brisbane.

Little, E.L. (1971). *Atlas of the United States Trees. Vol.1, Conifers and Important Hardwoods*. U.S. Dept. of Agriculture Miscellaneous Publications No.1146, Washington D.C.

Lloyd, R.C. (1963). Indigenous tree rings. *New Zealand Journal of Forestry*, *8*, 824-825.

Lockerbie, L. (1950). Dating the moa hunter. *Journal of the Polynesian Society*, *59*, 78-82.

Lockwood, J.G. (1974). *World Climatology - An Environmental Approach*. Edward Arnold: London.

Lockwood, J.G. (1979). *Causes of Climate*. Edward Arnold: London.

Long, A., Arnold, L.D., Damon, P.E., Ferguson, C.W., Lerman, J.C. & Wilson, A.T. (1976). Radial translocation of carbon in bristlecone pine. *Proceedings of the 9th International ^{14}C Conference*, pp.532-537.

Long, A., Ferhi, A., Letolle, R. & Lerman, J.C. (1980). Potential use of δ^{18}O and δD together in plant cellulose as a paleoenvironment indicator. In: *Proceedings of the International Meeting on Stable Isotopes in Tree-Rings*, ed. G.C. Jacoby, New Paltz, New York, Carbon Dioxide Effects Research and Assessment Program Publication No.12. U.S. Department of Energy, Washington D.C.

Lorenz, E.N. (1956). Empirical orthogonal functions and statistical weather prediction. *M.I.T. Statistical Forecasting Project, Scientific Report No.1*, Contract No. AF 19 (604)-1566, Massachussets Institute of Technology.

Lorenz, E.N. (1977). An experiment in nonlinear statistical weather forecasting. *Monthly Weather Review*, *105*(5),

590-602.

Lutz, H.J. (1944). Swamp-grown eastern white pine and hemlock in Connecticut as dendrochronological material. *Tree-Ring Bulletin*, *10*, 26-28.

Lydolph, P.E. (1977). *Climates of the Soviet Union. World Survey of Climatology, Vol.7*. Elsevier: Amsterdam.

Lynch, H.B. (1931). *Rainfall and stream run-off in Southern California since 1769*. Metropolitan Water District of Southern California: Los Angeles.

Lyon, C.J. (1935). Rainfall and hemlock growth in New Hampshire. *Journal of Forestry*, *33*, 162-168.

Lyon, C.J. (1936). Tree-ring widths as an index of physiological dryness in New England. *Ecology*, *17*, 457-478.

Lyon, C.J. (1939). Objectives and methods in New England tree-ring studies. *Tree-Ring Bulletin*, *5*, 27-30.

Lyon, C.J. (1949). Secondary growth of white pine in bog and upland. *Ecology*, *30*, 549-552.

Lyon, C.J. (1953). Vertical uniformity in three New England conifers. *Tree-Ring Bulletin*, *21*, 10-16.

Mabbutt, J.A. (1970). Landforms. In: *The Australian Environment (4th edition)*, ed. G.W. Leeper, pp.1-11. C.S.I.R.O. and Melbourne University Press.

Macphail, M.K. (1979). Vegetation and climates in Southern Tasmania since the last glaciation. *Quaternary Research*, *11*, 306-341.

Maleckas, E.P. (1972). Algorithms for the statistical processing of information in dendroclimatochronological research. In: *Dendroclimatochronology and Radiocarbon, Kaunas*, 1972, 159-164. In Russian.

Maleckas, E.P., Kotcharov, G.J. & Bitvinskas, T.T. (1975). Automised system of dendroclimatological research. *Reports of the 6th all-Union Conference on the problem "Astrophysical Phenomena and Radiocarbon", Tbilisi*, 1975. In Russian.

Mark, A.F., Scott, G.A.M., Sanderson, F.R. & James, P.W. (1964). Forest succession on landslides above Lake Thompson, Fiordland. *New Zealand Journal of Botany*, *2*, 60-89.

Mark, A.F. & Adams, N.M. (1973). *New Zealand Alpine Plants*. A.H. & A.W. Reed: Wellington, New Zealand.

Marr, J.W. (1948). Ecology of the forest-tundra ecotone on the east coast of Hudson Bay. *Ecological Monographs*, *18*, 118-144.

Mason, H.L. & Langenheim, J.H. (1957). Language analysis and the concept Environment. *Ecology*, *38*(2), 325-340.

Matagne, Fr. (1978). Analyse statistique en dendrochronologie. *Mémoire Lic. Sc. Math.*, Institut d'Astronomie et de Geophysique, Universite Catholique de Louvain-la-Nueve.

Matthews, J.A. (1976). "Little Ice Age" paleotemperatures from high altitude tree growth in S. Norway. *Nature, London*, *264*, 243-245.

Mattox, W.G. (1973). Fishing in west Greenland 1910-1966: the development of a new native industry. *Meddelelser om Grønland*, *197*(1).

Mayer-Wegelin, H. (1950). Der Härtetaster. Ein neues Gerät zur Untersuchung von Jahrringaufbau und Holzegeflüge. *Allgemeine Forst- und Jagdzeitung*, *122*, 12-23.

Maunder, W. (1970). Chapters 8, 9 and 10. In: *Climate of Australia and New Zealand. World Survey of Climatology, Vol.1*, ed. J. Gentilli, pp.213-268. Elsevier: Amsterdam.

Mazany, T., Lerman, J.C. & Long, A. (1978). Climatic sensitivity in ^{13}C/^{12}C values in tree rings from Chaco Canyon. *American Association for the Advancement of Science poster session*, 12-17 February 1978, Washington D.C.

McCarthy, P.J. (1976). The use of balance half-sample replication in cross-validation studies. *Journal of the American Statistical Association*, *71*(355), 596-604.

McGlone, M.S. & Topping, W.W. (1977). Auruian (post-glacial) pollen diagrams from the Tongariro region, North Island, New Zealand. *New Zealand Journal of Botany*, *11*, 283-290.

Meko, D.M., Stockton, C.W. & Boggess, W.R. (1980). A tree-ring reconstruction of drought in southern California. *Water Resources Bulletin*, *16*(4), 594-600.

Metro, A. & Destremau, D.X. (1968/69). Essai en dendrochronologie. *Rapport Station de Recherche Forestière de Rabat, 11,* 1-20.

Meyer, H.A. (1941). Growth fluctuations of virgin hemlock from northern Pennsylvania. *Tree-Ring Bulletin, 7,* 20-23.

Meyer, B.S., Anderson, D.B., Bohning, R.H. & Fratianne, D.G. (1973). *Introduction to Plant Physiology (2nd edition).* Van Nostrand: Princeton.

Meylan, B.A. & Butterfield, B.G. (1978). The structure of New Zealand Woods. *DSIR Bulletin 222,* Wellington, New Zealand.

Miertsching, J.A. (1854). Travel Narrative of a North Pole Expedition in search of Sir John Franklin and for the Discovery of the Northwest Passage in the Years 1850-1854. *(Frozen Ships:* a translation by L.H. Neatby, 1967). Macmillan of Canada: Toronto.

Miessen, H. (1975). Contribution à la méthodologie dendrochronologie appliquée au pin sylvestre en Belgique. *Mémoire Lic. Sc. Geogr.,* Institut d'Astronomie et de Géophysique, Université Catholique de Louvain-la-Neuve.

Mikola, P. (1962). Temperature and tree growth near the northern timber line. In: *Tree Growth,* ed. T.T. Kowlowski, pp.265-274. Ronald Press: New York.

Mikola, P. (1971). Reflexion of climatic fluctuation in the forestry practices in Northern Finland. *Reports of the Kero Subarctic Research Station, 8,* 116-121.

Milsom, S.J. (1979). *Within- and between-tree variation in certain properties of annual rings of sessile oak, Quercus petraea (Mattuschka) Liebl., as a source of dendrochronological information.* Ph.D. Thesis, Council for National Academic Awards (Liverpool Polytechnic).

Mitchell, J.M., Jr. (1953). On the causes of instrumentally observed secular temperature trends. *Journal of Meteorology, 10(4),* 244-261.

Mitchell, J.M., Jr. (1961). Recent secular changes in global temperature. *Annals of the New York Academy of Sciences, 95(1),* 235-250.

Mitchell, J.M., Jr., Dzerdzeevskii, B., Flohn, H., Hofmeyr, W.L., Lamb, H.H., Rao, K.N. & Wallén, C.C. (1966). Climatic Change. *World Meteorological Organization Technical Note 79,* World Meteorological Organization, Geneva.

Moar, N.T. (1966). Post-glacial vegetation and climate in New Zealand. In: *World Climate from 8000 to 0 B.C.,* ed. J.S. Sawyer, pp. 155-156. Royal Meteorological Society: London.

Moar, N.T. (1971). Contributions to the Quaternary history of the New Zealand flora. 6. Aranuian pollen diagrams from Canterbury, Nelson and North Westland, South Island. *New Zealand Journal of Botany, 9,* 80-145.

Molchanov, A.A. (1976). *Dendrochronological Principles of Long Range Forecasting.* Gidroteometizdat: Moscow. In Russian.

Molley, B.P.J. (1969). Evidence for post-glacial climatic changes in New Zealand. *Journal of Hydrology, 8,* 56-67.

Moodie, D.W. & Catchpole, A.J.W. (1975). Environmental data from historical documents by content analysis: freeze-up and break-up of estuaries on Hudson Bay, 1714-1871. *Manitoba Geographical Studies, No. 5,* University of Manitoba, Winnipeg.

Moore, D.P. & Mathews, W.H. (1978). The Rubble Creek landslide, southwestern British Columbia. *Canadian Journal of Earth Sciences, 15,* 1039-1052.

Morrison, D.F. (1976). *Multivariate Statistical Methods (2nd edition).* McGraw-Hill: New York.

Morrow, P.A. & LaMarche, V.C., Jr. (1978). Tree-ring evidence for chronic insect suppression of productivity in subalpine *Eucalyptus. Science, 201,* 1244-1245.

Mosteller, F. & Tukey, J.W. (1968). Data analysis, including statistics. In: *Handbook of Social Psychology, Vol. 2.* ed. G. Lindzey & E. Aronson, Addison-Wesley: Reading, Massachusetts.

Mosteller, F. & Tukey, J.W. (1977). *Data Analysis and Regression.* Addison-Wesley: Reading, Massachusetts.

Mucha, S.B. (1979). Estimation of tree ages from growth rings of eucalypts in northern Australia. *Australian Forestry, 42,* 13-16.

Mukhamedshin, K.D. (1974). Variation in the increment of Juniper in the high mountains of the Tien-Shan over the last Millenium of the Holocene. In: *Tr. 5-go vses. Sovesheh po Probl. Astrofiz Yavlemya i radiouglerod 1973, Tbilisi ISSR,* 1974, 149-161. In Russian.

Mukhamedshin, K.D. & Sartbaev, S.K. (1972). The increment cycle of Juniper in the high mountain conditions of the Tien-Shan. *Isvestia Akademia Nauka Kirgiz SSR, 2,* 55-60. In Russian.

Müller-Stoll, W.R. (1947). Photometrische Holzstrukturuntersuchungen. 1. Mitteilung: Ueber die Ermittlung von Jahrringaufbau und Spätholzanteil auf photometrischem Wege. *Planta, 35,* 397-426.

Müller-Stoll, W.R. (1949). Photometrische Holzstrukturuntersuchungen. 2. Mitteilung: Ueber die Beziehungen der Lichtdurchlässigkeit von Holzschnitten zu Rohwichte und Wichtekontrast. *Forstwirtschaftliches Centralblatt, 68,* 21-63.

Munaut, A.V. (1966). Récherches dendrochronologiques sur *Pinus silvestris.* I. Etude de 45 pins sylvestres récents originaires de Belgique. *Agricultura, 2nd Series, 14(2),* 193-232.

Munaut, A.V., Berger, A.L., Guiot, J. & Mathieu, L. (1978). Dendroclimatological studies on cedars in Morocco. In: *Evolution des Atmospheres Planétaires et Climatologie de la Terre.* C.N.E.S: Nice.

National Academy of Sciences. (1975). *Understanding Climatic Change.* NAS: Washington D.C.

Neter, J. & Wasserman, W. (1974). *Applied Linear Statistical Models.* Irwin: Homewood, Illinois.

Newell, R.E., Kidson, J.W., Vincent, P.G. & Boer, G.J. (1972). *The General Circulation of the Tropical Atmosphere, Vol. 1.* MIT: Cambridge, Massachusetts.

Newell, R.E., Kidson, J.W., Vincent, P.G. & Boer, G.J. (1974). *The General Circulation of the Tropical Atmosphere, Vol. 2.* MIT: Cambridge, Massachusetts.

Nichols, G.E. (1913). The vegetation of Connecticut. II. Virgin forests. *Torreya, 13,* 199-215.

Nichols, H. (1975). Palynological and paleoclimatic study of the Late Quaternary displacements of the boreal forest-tundra ecotone in Keewatin and Machenzie, N.W.T., Canada. *Occasional Paper No. 18,* Institute of Arctic and Alpine Research, Boulder, Colorado.

Nichols, H., Kelly, P.M. & Andrews, J.T. (1978). Holocene palaeo-wind evidence from palynology in Baffin Island. *Nature, 273,* 140-142. London.

Nie, N.N., Hull, C.H., Jenkins, J.G., Steinbrenner, K. & Bent, D.H. (1975). *Statistical Package for the Social Sciences (2nd Edition).* McGraw-Hill: New York.

Norton, D.A. (1979). *Dendrochronological studies with Nothofagus solandri (Fagaceae) and Libricedrus bidwillii (Cupressaceae).* B.Sc. Hons. Thesis, University of Canterbury, Christchurch, New Zealand.

Ogden, J. (1978a). On the dendrochronological potential of Australian trees. *Australian Journal of Ecology, 3,* 339-356.

Ogden, J. (1978b). Investigations of the dendrochronology of the genus *Athrotaxis* D. Don (Taxodiaceae) in Tasmania. *Tree-Ring Bulletin, 38,* 1-13.

Ogden, J. (1978c). On the diameter growth rates of red beech *(Nothofagus fusca)* in different parts of New Zealand. *New Zealand Journal of Ecology, 1,* 16-18.

Ogden, J. (1981). Dendrochronological studies and the determination of tree age in the Australian tropics. *Journal of Biogeography,* in press.

Oliver, J. (1974). Tropical cyclones in N.E. Australia. *Climatic Research Unit Monthly Bulletin, 3,* 8-9. Climatic Research Unit, Norwich, England.

Oliver, W.R.B. (1931). An ancient maori oven on Mt. Egmont. *Journal of the Polynesian Society, 40(2),* 70-73.

Ording, A. (1941). Årringanalyser på gran og furn. *Meddedelser Norske Skogforsøksvesen, 7,* 105-354.

Orlicz, M. (1962). Klima Tatr. In: *Tatrzanski Parl Narodowy,* ed. W. Szafer. Kraków, Poland.

Orvig, S., editor (1970). *Climate of the Polar Regions. World Survey of Climatology. Vol. 14.* Elsevier: Amsterdam.

Oswalt, W.H. (1950). Spruce borings from the lower Yukon River, Alaska. *Tree-Ring Bulletin, 16*(4), 26-30.

Oswalt, W.H. (1952). Spruce samples from the Copper River drainage, Alaska. *Tree-Ring Bulletin, 19*(1), 5-10.

Oswalt, W.H. (1958). Tree-ring chronologies in south central Alaska. *Tree-Ring Bulletin, 22*(1-4), 16-22.

Paijmans, K. (1975). Explanatory note to the vegetation map of Papua New Guinea. *Land Research Series No. 35*, C.S.I.R.O., Melbourne.

Paijmans, K., editor. (1976). *New Guinea Vegetation*. C.S.I.R.O. and Australian National University Press: Canberra.

Palmén, E. & Newton, C.W. (1969). *Atmospheric Circulation Systems*. Academic: New York.

Palmer, E. & Pitman, N. (1972). *Trees of Southern Africa*. A.A. Balkema: Cape Town.

Palmer, W.C. (1965a). Meteorological drought. *U.S. Weather Bureau Research Paper 45*, U.S. Weather Bureau, Washington D.C.

Palmer, W.C. (1965b). Drought. *Weekly Weather and Crop Bulletin, 32*(30), 8.

Parker, M.L. (1971). Dendrochronological techniques used by the Geological Survey of Canada. In. *Paper 71-25*, *1-30*, Geological Survey of Canada, Edmonton, Canada.

Parker, M.L. & Meleskie, K.R. (1970). Preparation of X-ray negatives of tree-ring specimens for dendrochronological analysis. *Tree-Ring Bulletin, 30*, 11-22.

Parker, M.L. & Henoch, W.E.S. (1971). The use of Engelmann spruce latewood density for dendrochronological purposes. *Canadian Journal of Forest Research, 1*(2), 90-98.

Paterson, J.H. (1979). *North America; a Geography of Canada and the United States*. Oxford University Press.

Pearman, G.I. (1971). An exploratory investigation of the growth rings of *Callitris preissii* trees from Garden Island and Naval Base. *Western Australian Naturalist, 12*, 12-17.

Pearman, G.I., Francey, R.J. & Fraser, P.J.B. (1976). Climatic implications of stable carbon isotopes in tree rings. *Nature, 260*, 771-773. London.

Peattie, D.C. (1953). *A Natural History of Western Trees*. Houghton Miloflin: Boston.

Peters, K. & Cook, E.R. (1979). The cubic smoothing spline as a digital filter. *Lamont-Doherty Geological Observatory Contributions*, New York.

Pettersen, S. (1969). *An Introduction to Meteorology (3rd edition)*. McGraw-Hill: New York.

Peyre, C. (1979). *Recherches sur l'étagement de la végétation dans le Massif du Bou Iblane (Moyen Atlas oriental, Maroc)*. Thèse, Universite de Droit, D'Econ. et des Sc. d'Aix-Marseilles.

Phillips, E.W.J. (1965). Methods and equipment for determining the specific gravity of wood. In: *Proceedings; Meeting of working groups on wood quality, sawing and machining, wood and tree chemistry*, ed. Section 41, I.U.F.R.O., Melbourne, 4-15 October 1965, Vol. 2. C.S.I.R.O., Division of Forest Products: Melbourne.

Phipps, R.L. (1961). Analysis of five years dendrometer data obtained within three deciduous forest communities of Neotoma. *Special Report No. 3*, Ohio Agricultural Experimental Station, Wooster, Ohio.

Phipps, R.L. (1967). Annual growth of suppressed chestnut oak and red maple, a basis for hydrologic inference. *U.S. Geological Survey Professional Paper 485-C*, U.S. Geological Survey, Washington D.C.

Phipps, R.L. (1970). The potential use of tree rings in hydrologic investigations in eastern North America with some botanical considerations. *Water Resources Research, 6*(6), 1634-1640.

Phipps, R.L. (1974). The soil creep-curved tree fallacy. *United States Geological Survey Journal of Research, 3*, 371-377.

Phipps, R.L., Ierley, D.L. & Baker, C.P. (1979). Tree rings as indicators of hydrologic change in the Great Dismal Swamp, Virginia and North Carolina. *U.S. Geological Survey Water Resources Investigations 78-136*, U.S. Geological Survey, Washington D.C.

Pilcher, J.R. (1973). Tree-ring research in Ireland. *Tree-Ring Bulletin, 33*, 1-28.

Pilcher, J.R., Hillam, J., Baillie, M.G.L. & Pearson, G.W. (1977). A long sub-fossil oak tree-ring chronology from the north of Ireland. *New Phytologist, 79*, 713-729.

Pilcher, J. & Baillie, M.G.L. (1980a). Eight modern oak chronologies from England and Scotland. *Tree-Ring Bulletin, 40*, 45-58.

Pilcher, J. & Baillie, M.G.L. (1980b). Six modern oak chronologies from Ireland. *Tree-Ring Bulletin, 40*, 23-34.

Pittock, A.B. (1975). Climatic change and patterns of variation in Australian rainfall. *Search, 6*, 498-504.

Pittock, A.B. (1978). A critical look at long-term sun-weather relationships. *Reviews of Geophysics and Space Physics, 16*, 400-420.

Pittock, A.B., Frakes, L.A., Jensen, D.R., Peterson, J.A. & Zillman, J.W., editors. (1978). *Climatic Change and Variability. A Southern Perspective*. Cambridge University Press.

Pittock, A.B. (1980a). Patterns of climatic variation in Argentina and Chile, I, precipitation, 1931-60. *Monthly Weather Review, 108*(9), 1347-1361.

Pittock, A.B. (1980b). Patterns of climatic variation in Argentina and Chile, II, temperature, 1931-1960. *Monthly Weather Review, 108*(9), 1362-1369.

Polge, H. (1963). Une nouvelle méthode de détermination de la texture du bois: l'analyse densitométrique de clichés radiographiques. *Annales de l'école nat. eaux et forêts et de la station de recherche et exper., 20*, 531-581.

Polge, H. (1966). Etablissement des courbes de variation de la densité du bois par l'exploration densitométrique de radiographies d'échantillons prélevés à la tarière sur des arbres vivants. *Annales des sciences forestières, 23*, 1-206.

Portois, A.M. (1978). Influence climatique sur les cernes annuels de croissance des arbres. Une compariason entre le Maroc, la Belgique et la Suisse. *Mémoire Lic. Sc. Géogr.*, Institut d'Astronomie et de Géophysique, Université Catholique de Louvain-la-Neuve.

Potter, N. (1969). Tree-ring dating of snow avalanche tracks and the geomorphic activity of avalanches, northern Absaroka Mountains, Wyoming. In: *INQUA Volume, Geological Society of America Special Paper 123*, pp. 141-165.

Preisendorfer, R.W. & Barnett, T.P. (1977). Significance tests for empirical orthogonal functions. In: *Preprint Volume, Fifth Conference on Probability and Statistics*, 15-18 November 1977. American Meteorological Society: Boston.

Preston, R. (1966). *North American Trees*. M.I.T. Press: Cambridge, Massachussetts.

Pyke, C.B. (1972). Some meteorological aspects of the seasonal distribution of precipitation in the Western United States and Baja California. *Water Resources Center, Contribution 139*, University of California.

Raper, S.C.B. (1978). *Variations in the incidence of tropical cyclones and relationships with the global circulation*. Ph.D. Thesis, University of East Anglia.

Readshaw, J.L. & Mazanec, Z. (1969). Use of growth rings to determine past phasmatid defoliations of Alpine Ash forest. *Australian Forestry, 33*, 29-36.

Reille, M. (1975). *Contribution pollenanalytique à l'histoire tardiglaciaire et holocene de la végétation de la montagne corse*. Thèse, Université d'Aix-Marseilles III.

Reille, M. (1977). Contribution pollenanalytique à l'histoire holocène de la végétation des montagnes du Rif (Maroc septenrional). In: *Recherches françaises sur la Quaternaire*, INQUA, 1977. Supplement Bull. AFEQ, 1(50), 53-76.

Reinsch, C.H. (1967). Smoothing by spline functions. *Numerische Mathematik, 10*, 177-183.

Rex, D.F., editor. (1969). *Climate of the Free Atmosphere*.

World Survey of Climatology, *Vol.4.* Elsevier: Amsterdam.

Rittenbrug, D. & Ponticorvo, L. (1956). A method for determination of the ^{18}O concentration of the oxygen of organic compounds. *International Journal of Applied Radiation and Isotopes, 1,* 208-214.

Robinson, W.J. (1976). Tree-ring dating and archaeology in the American Southwest. *Tree-Ring Bulletin, 36,* 9-20.

Rodriguez-Iturbe, I. (1969). Estimation of statistical parameters for annual river flows. *Water Resources Research, 5,* 1418-1426.

Rome, C. (1980). Analyse cospectrale et applications en dendroclimatologie. *Mémoire Lic. Sc. Math.,* Institut d'Astronomie et de Géophysique, Université Catholique de Louvain-la-Neuve.

Romeder, C.R. (1973). *Méthodes et Programmes d'Analyse Discriminante.* Dunod: Paris.

Roos, M. (1973). *Drought Probability Study, Sacramento River Basin.* Department of Water Resources, State of California.

Röthlisberger, F. (1976). Gletscher und Klimaschwankungen im Raum Zermatt, Ferpècle und Arolla. *Die Alpen, 52,* 59-152.

Rowe, J.S (1972). Forest regions of Canada. *Canadian Forestry Service Publication No.1300,* Department of Fisheries and the Environment, Ottawa, Canada.

Rowntree, L. & Raburn, R. (1980). Rainfall variability and California mission agriculture: an analysis from harvest and tree-ring data. In: *Yearbook of the Association of Pacific Coast Geographer, Volume 42.*

Ryan, P., editor. (1972). *Encyclopaedia of Papua and New Guinea.* Melbourne University Press and the University of New Guinea.

Saito, H. (1950). An exploration about tree-rings in Hokkaido. *Journal of Meteorological Research, 2*(2), 42-50. In Japanese.

Salinger, M.J. (1976). New Zealand temperature since 1300 A.D. *Nature, London, 260,* 310-311.

Salinger, M.J. (1979). New Zealand climate: the temperature record, historical data and some agricultural implications. *Climatic Change, 2,* 109-126.

Salinger, M.J. (1980). New Zealand climate: 1. Precipitation patterns. *Monthly Weather Review, 108,* 1892-1904.

Schiegl, W.E. (1970). *Natural deuterium in biogenic materials - influence of environment and geophysical applications.* Ph.D. Dissertation, University of South Africa, Pretoria.

Schiegl, W.E. (1974). Climatic significance of deuterium abundance in growth rings of *Picea. Nature, London, 251,* 582-584.

Schmidt, B. (1977a). Der Aufbau von Jahrringchronologien im Holozän mit Eichen (*Quercus* sp.) aus dem Rhein-, Weser- und Werragebiet. *Erdwiss. Forschung, 13,* 91-98.

Schmidt, B. (1977b). *Dendroklimatologische Untersuchungen an Eichen nordwestdeutscher Standorte.* Dissertation, University of Hamburg.

Schmidt, B. & Aniol, R.W. (1981). Die Arbeitsweise der Dendrochronologie und ihre Verbessung durch Berucksichtigung von Weiserjahren. *Kölner Jahrbuch für Vor- und Fruhgeschichte, 16,* in press.

Schmidt, H.L., Winkler, F.J., Latzko, E. & Wirth, E. (1978). Carboxylation reactions and ^{13}C-Kinetic isotope effects in photosynthetic ^{13}C-values of plant material. *Israel Journal of Chemistry, 17,* 223-224.

Schneider, S.H. with Mesirow, L.E. (1976). *The Genesis Strategy.* Plenum: New York.

Schneider, W.J. & Conover, W.J. (1964). Tree growth proves nonsensitive indicator of precipitation in central New York. *U.S. Geological Survey Professional Paper 501-B,* B185-B187, U.S. Geological Survey, Washington D.C.

Schove, D.J. (1954). Summer temperatures and tree-rings in North Scandinavia A.D. 1461-1950. *Geografiska Annaler, 35,* 40-80.

Schüepp, M. (1961). Lufttemperaturen, 2 Teil. *Beiheft zu den Annalen der Schweizerischen MZA (Jahrgang 1960),* Zürich.

Schulman, E. (1942). Dendrochronology in the pines of Arkansas. *Ecology, 23,* 309-318.

Schulman, E. (1951). Tree-ring indices of rainfall, temperature, and river flow. In: *Compendium of Meteorology,* pp.1024-1029. American Meteorological Society: Boston.

Schulman, E. (1954). Longevity under adversity in conifers. *Science, 119,* 1396-1399.

Schulman, E. (1956). *Dendroclimatic Change in Semiarid America.* University of Arizona Press.

Schulman, E. (1958). Bristlecone pine, oldest known living thing. *National Geographic, 113*(3), 355-372.

Schuster, R.M. (1972). Continental movements, "Wallace's line" and Indomalayan-Australiasian dispersal of land plants: Some eclectic concepts. *The Botanical Review, 38,* 3-86.

Schweingruber, F.H. & Schär, E. (1976). Röntgenuntersuchungen an Jahrringen. *Neue Zürcher Zeitung, 180,* 33.

Schweingruber, F.J., Fritts, H.C., Bräker, O.U., Drew, L.G. & Schär, E. (1978a). The x-ray technique as applied to dendroclimatology. *Tree-Ring Bulletin, 38,* 61-91.

Schweingruber, F.H., Bräker, O.U. & Schär, E. (1978b). Dendroclimatic studies in Great Britain and in the Alps. In: *Evolution of Planetary Atmospheres and the Climatology of the Earth,* pp.369-372. C.E.R.N.: Paris.

Schweingruber, F.H. & Ruoff, U. (1979). Stand und Anwendung der Dendrochronologie in der Schweiz. *Archäologie und Kunstgeschichte, 36,* 69-90.

Schweingruber, F., Bräker, O.U. & Schär, E. (1979). Dendroclimatic studies on conifers from central Europe and Great Britain. *Boreas, 8,* 427-452.

Schwerdtfeger, W., editor. (1976). *Climates of Central and South Africa. World Survey of Climatology, Vol.12.* Elsevier: Amsterdam.

Scott, S.D. (1964). Notes on archaeological tree-ring dating in New Zealand. *New Archaeological Association Newsletter, 7*(1), 34-35.

Scott, S.D. (1972). Correlation between tree-ring width and climate in two areas in New Zealand. *Journal of the Royal Society of New Zealand, 2*(4), 545-560.

Seber, G.A.F. (1977). *Linear Regression Analysis.* Wiley: New York.

Sell, J. (1978). Quantitative automatische Bildanalyse in der Materialprüfung. Prinzip, Systemaufbau, Anwendung. *Material und Technik, 6,* 79-83.

Sellers, W.D. (1968). Climatology of monthly precipitation patterns in the western United States, 1931-1966. *Monthly Weather Review, 96*(9), 585-595.

Senter, F.H. (1938). Dendrochronology in two Mississippi drainage tree-ring areas. *Tree-Ring Bulletin, 5,* 3-6.

Serre, F. (1969). Variations de l'épaisseur des anneaux ligneux chez le Thuya de Barbarie (*Tetraclinis articulata*) et le climat en Tunisie. *Ann. Fac. Sc. Marseille, 42,* 193-204.

Serre, F. (1973). *Contribution à l'étude dendroclimatologique du pin d'Alep (Pinus halepensis Mill.).* Thèse Dr. des Sciences Nat., Marseille.

Serre, F. (1977). A factor analysis of correspondence applied to ring widths. *Tree-Ring Bulletin, 37,* 21-32.

Serre, F. (1978a). The dendroclimatological value of the European Larch (*Larix decidua* Mill.) in the French Maritime Alps. *Tree-Ring Bulletin, 38,* 25-34.

Serre, F. (1978b). Résultats dendroclimatiques pour les Alpes meridionales francaises. In: *Evolution des Atmosphères Planetaires et Climatologie de la Terre,* ed. C.N.E.S., pp.381-385. C.N.E.S.: Toulouse.

Serre, F., Lück, H.B. & Pons, A. (1966). Premières recherches sur les relations entre les variations des anneaux ligneux chez *Pinus halepensis* Mill. et les variations annuelles du climat. *Oecologia Plantarum, 1,* 117-135.

Sherwood, J.A., editor. (1978). Holdings of the International Tree-Ring Data Bank. *International Tree-Ring Data Bank Newsletter, 3,* 7-16, University of Arizona, Tucson.

Shida, J. (1935). Climatic change and historical vicissitude in the Far East. *Kagakuchishiki, 15*(1), 12-19. In Japanese.

Shroder, J.F., Jr. (1978). Dendrogeomorphological analysis of mass movement on Table Cliffs Plateau, Utah. *Quaternary Research, 9,* 168-185.

Shroder, J.F., Jr. (1980). Dendrogeomorphology: review and new techniques of tree-ring dating. *Progress in Physical Geography, 4*(2), 161-188.

Siegenthaler, U. & Oeschger, H. (1980). Correlation of ^{18}O in precipitation with temperature and altitude. *Nature, London, 285,* 314-335.

Simpson, C. (1966). *The Viking Cycle.* William Morrow: New York.

Singh, G., Kershaw, A.P. & Clark, R.A. (1980). Quaternary vegetation and fire history in Australia. In: *Fire and Australian Biota,* ed. A.M. Gill, R.A. Groves & I.R. Noble. Australian Academy of Science: Canberra.

Sirèn, G. (1961). Skogsgranstalleṣom indikator for klimafluktuationaerna i norra Fennoskandien under historisk tid. *Communicationes Instituti Forestalis Fenniae, Helsinki, 54(2).*

Smed, J. (1946-67). Monthly anomalies of the surface temperature in areas of the northern North Atlantic. *International Council for the Exploration of the Sea,* 1-22.

Smith, B.N., Herath, H.M.W. & Chase, J.B. (1973). Effect of growth temperature on carbon isotopic ratios in barley, pea and rape. *Plant and Cell Physiology, 14,* 177-182.

Smith, B.N., Oliver, J. & McMillan, C. (1976). Influence of carbon source, oxygen concentration, light intensity and temperature on ^{13}C/^{12}C ratios in plant tissue. *Botanical Gazette, 37,* 99-104.

Smith, D.C., Borns, H. & Baron, R. (1981). Climate stress and Maine agriculture, 1785-1885. In: *Climate and History,* ed. T.M.L. Wigley, M.J. Ingram & G. Farmer. Cambridge University Press.

Smith, D.G. (1974). Remarks on: The effects of ice jams on Alberta rivers (abstract). *Great Plains - Rocky Mountains Geographical Journal, 3,* 122.

Sneyers, R. (1979). Homogénéité et stabilité des éléments météorologique à Uccle (Belgique). *Il Nuovo Cimento, 20(1),* 101-113.

Sokal, R.R. & Rohlf, J. (1969). *Biometry: the Principles and Practice of Statistics in Biological Research.* W.H. Freeman: San Francisco.

Soons, J.M. (1979). Late Quaternary environments in the Central South Island of New Zealand. *New Zealand Geographer, 35,* 16-23.

Sparrow, C.J., Healy, T.R., editors (1968). *Meteorology and Climate of New Zealand: A Bibliography.* Oxford University Press.

Specht, R., Roe, E.M. & Broughton, V.H., editors. (1974). Conservation of major plant communities in Australia and Papua New Guinea. *Australian Journal of Botany, Supplementary Series, 7.*

Spiegel, M.R. (1961). *Schaum's Outline of Theory and Problems of Statistics.* Schaum: New York.

Stahle, D.W. (1978). *Tree-ring dating of selected Arkansas log buildings.* M.S. Thesis, University of Arkansas.

Stevens, D. (1975). A computer program for stimulating cambium activity and ring growth. *Tree-Ring Bulletin, 35,* 49-56.

Stockton, C.W. (1971). *The feasibility of augmenting hydrologic records using tree-ring data.* Ph.D. Thesis, University of Arizona, Tucson.

Stockton, C.W. (1975). Long-term streamflow records reconstructed from tree-rings. *Papers of the Laboratory of Tree-Ring Research, No.5,* University of Arizona, Tucson.

Stockton, C.W. (1976). Long-term streamflow reconstructions in the upper Colorado river basin using tree rings. In: *Colorado River Basin Modelling Studies,* ed. G.C. Clyde, D.H. Falkenborg & J.P. Riley, pp.401-441. Utah State University: Logan, Utah.

Stockton, C.W. & Fritts, H.C. (1971a). *An empirical reconstruction of water levels for Lake Athbasca (1810-1967) by analysis of tree rings.* Laboratory of Tree-Ring Research, University of Arizona, Tucson.

Stockton, C.W. & Fritts, H.C. (1971b). Conditional probability of occurrence for variations in climate based on width of annual tree rings of Arizona. *Tree-Ring Bulletin, 31,* 3-24.

Stockton, C.W. & Fritts, H.C. (1971c). Augmentation of hydrologic records using tree-ring data. In: *Hydrology and Water Resources in Arizona and the Southwest, Vol.1.* Arizona Academy of Sciences: Tucson, Arizona.

Stockton, C.W. & Fritts, H.C. (1973). Long-term reconstr-uction of water level changes of Lake Athbasca by analysis of tree rings. *Water Resources Bulletin, 9,* 1006-1027.

Stockton, C.W. & Meko, D.M. (1975). A long-term history of drought occurrence in Western United States as inferred from tree-rings. *Weatherwise, 28(6),* 244-249.

Stockton, C.W., Meko, D.M. & Mitchell, J.M., Jr. (1978). Tree-ring evidence of a 22-year rhythm of drought area in western United States and its relation to the Hale solar cycle. Paper presented to *Working Group VII under the U.S./U.S.S.R. Agreement of Protection of the Environment,* Crimean Astrophysical Observatory and Kislovosk Observatory, U.S.S.R., September 12-23.

Stockton, C.W. & Boggess, W.R. (1980). Augmentation of hydrologic records using tree rings. In: *Improved Hydrologic Forecasting,* pp.239-265. American Society of Civil Engineers.

Stokes, M.A. & Smiley, T.L. (1968). *An Introduction to Tree-Ring Dating.* University of Chicago Press.

Stokes, M.A. & Drew, L.G. & Stockton, C.W. editors. (1973). Tree-ring chronologies of Western America. I. Selected tree ring stations. *Chronology Series I,* Laboratory of Tree-Ring Research, University of Arizona, Tucson.

Stone, M. (1974). Cross-validatory choice and assessment of statistical predictions. *Journal of the Royal Statistical Society, Series B, 36(2),* 11-147.

Strahler, A.N. (1960). *Physical Geography (2nd edition),* Wiley: New York.

Streten, N.A. (1980). Some synoptic indices of the Southern Hemisphere mean sea level circulation in 1972-77. *Monthly Weather Review, 108,* 18-36.

Stuiver, M. (1978). Atmospheric carbon dioxide and carbon reservoir changes. *Science, 199,* 245-249.

Stuiver, M. (1980). Solar variability and climatic change during the current millenium. *Nature, London, 286,* 868-871.

Stuiver, M. & Quay, P.D. (1980). Changes in atmospheric carbon-14 attributed to a variable Sun. *Science, 207,* 11-19.

Sturman, A.P. (1979). Aspects of the synoptic climatology of the southern South American and Antarctic peninsula. *Weather, 34(6),* 210-223.

Suess, H.E. (1968). Climatic change, solar activity, and the cosmic-ray production rate of natural radiocarbon. *Meteorological Monographs, 8,* 146-150.

Taljaard, J.J. (1972a). Physical features of the Southern Hemisphere. In: *Meteorology of the Southern Hemisphere,* ed. C.W. Newton, *Meteorological Monographs, 35,* 1-8.

Taljaard, J.J. (1972b). Synoptic meteorology of the Southern Hemisphere. In: *Meteorology of the Southern Hemisphere,* ed. C.W. Newton, *Meteorological Monographs, 35,* 139-211.

Taljaard, J.J., Schmitt, W. & van Loon, H. (1969). Frontal analysis with application to the Southern Hemisphere. *Notos, 10,* 25-58, Weather Bureau, Pretoria.

Tans, P.P. (1978). *Carbon 13 and carbon 14 in trees and the atmospheric CO_2 increase.* Ph.D. Dissertation, Rijksuniversiteit te Groningen, Netherlands.

Tarankov, V.I. (1973). Introduction to dendroclimatology in the Far East. *Tr. Biol. pochv. in-ta Dal'nevost. navchtsentr AN SSSR, 12(115),* 7-23. In Russian.

Tatsuoka, M.M. (1971). *Multivariate Analysis.* Wiley: New York.

Ten Brink, N.W. & Wedick, A. (1974). Greenland ice sheet history since the last glaciation. *Quaternary Research, 4,* 429-440.

Teng, S.C. (1948). Tree ring and climate in Kansu. *Botanical Bulletin of the Academia Sineca, 2(3),* 211-214. In Chinese.

Tessier, L. (1978). *Contribution dendroclimatologique à l'étude du peuplement forestier des environs des Chalets de l'Orgère (Savoie).* Thèse, Université d'Aix Marseille III.

Thie, J. (1974). Distribution and thawing of permafrost in the southern part of the discontinuous permafrost zone in Manitoba. *Arctic, 27,* 189-200.

Thiercelin, F. (1970). Tardivité du débourrement et densité du bois dans une population adulte de *Picea abies.* *Ann. Sci. Forest, 3,* 243-254.

Thomas, H.E. (1959). Reservoirs to match our climatic fluctuations. *Bulletin of the American Meteorological Society, 40*(5), 240-249.

Thornthwaite, C.W. (1948). An approach towards a national classification of climate. *Geographical Review, 38*, 35-94.

Tickell, C. (1973). *Climatic Change and World Affairs*. Pergamon: Oxford.

Tomlinson, A.I. (1976). Climate. In: *New Zealand Atlas*, ed. I. Wards, pp.82-89. Government Printer: Wellington, New Zealand.

Thompson, P. & Gray, J. (1977). Determination of $^{18}O/^{16}O$ ratios in compounds containing C, H and O. *International Journal of Applied Radiation and Isotopes, 28*, 411-415.

Topping, W.W. (1971). Some aspects of Quaternary eruptives of the Tongagiro volcanic centre, New Zealand. Ph.D. Thesis, Victoria University of Wellington, Wellington, New Zealand.

Tortorelli, L.A. (1956). *Maderas y Bosques Argentinos*. Editorial Acme: Buenos Aires.

Tranquillini, Q. (1979). Physiological ecology of the Alpine timberline tree existence at high altitudes with special reference to the European Alps. *Ecological Studies, 31*.

Trenberth, K.E. (1976). Fluctuations and trends in indices of the Southern Hemisphere circulation. *Quarterly Journal of the Royal Meteorological Society, 108*, 18-36.

Trenberth, K.E. & Paolino, D.A., Jr. (1980). The Northern Hemisphere sea-level pressure data set: trends, errors and discontinuities. *Monthly Weather Review, 108*(7), 855-872.

Trendelenburg, R. & Mayer-Wegelin, H. (1955). *Das Holz als Rohstoff*. Carl Hanser: Munich.

Trewartha, G.T. (1961). *The Earth's Problem Climates (2nd edition, 1981)*. Methuen: London.

Triat-Laval, H. (1979) *Contribution pollenanalytique à l'histoire tardi et postglaciaire de la végétation de la bassevallée du Rhône*. Thèse, Université d'Aix Marseille III.

Troughton, J.H. & Card, K.A. (1975). Temperature effects on the carbon-isotope ratio of C_3 and C_4 and crassulacean-acid metabolism (CAM) in plants. *Planta, 123*, 185-190.

Troup, A.J. (1965). The Southern Oscillation. *Quarterly Journal of the Royal Meteorological Society, 91*, 490-506.

Tyson, P.D., Dyer, T.G.J. & Mametse, M.N. (1975). Secular changes in South African rainfall: 1910-1972. *Quarterly Journal of the Royal Meteorological Society, 101*, 104-110.

Ulam, S.M. (1966). On general formulations of simulation and model construction. In: *Prospects for Simulation and Simulations of Dynamic Systems*, ed. G. Shapiro & M. Rogers. Spartan Books: New York.

Urey, H.C. (1947). The thermodynamic properties of isotopic substances. *Journal of the Chemical Society of London, 1947*, 562-581.

Uttinger, H. (1965). Niederschlag, 1-3, Teil. *Beiheft zu den Annalen der Schweizerischen MZA (Jahrgang 1964)*, Zurich.

van Loon, H. (1972a). Temperature in the Southern Hemisphere. *Meteorological Monographs, 13*(35), 25-28.

van Loon, H. (1972b). Pressure in the Southern Hemisphere. *Meteorological Monographs, 13*(35), 87-100.

van Loon, H. & Madden, R. (1981). The Southern Oscillation: Part I. Global associations with pressure and temperature in the Northern Hemisphere. *Monthly Weather Review, 109*, 1150-1169.

van Wyk, J.H. (1972). *Trees of the Kruger National Park*. Purnell: Cape Town, South Africa.

Velimský, T. (1976). K dendrochronologickému výzkumu středovekých nálezů z Mostu. *Archaeologia historica, 2*, 299-306.

Vibe, C. (1967). Arctic animals in relation to climatic fluctuations. *Meddelelser om Grønland, 170*(5), 1-227.

Viereck, L.A. (1965). Relationship of white spruce to lenses of perennially frozen ground, Mount McKinley National Park, Alaska. *Arctic, 18*(4), 262-267.

Viereck, L.A. & Little, E.L. (1972). *Alaska Trees and Shrubs. Agriculture Handbook No.410*, Forest Service, U.S. Department of Agriculture, Washington D.C.

von Breitenbach, F. (1974). *Southern Cape Forests and Trees*. Government Printer: Pretoria.

von Jazewitsch, W. (1961). Zur klimatologischen Auswertung von Jahrringkurven. *Fortw. Obl., 80*(5/6), 175-190.

von der Kall, T. (1978). Der Jahrringaufbau der Kiefer (*Pinus silv.* L.) und seine jährliche Variationen. *Mitt. d. Forstl. Vers. -u. Bundesanstalt Baden-Württemberf, Freiburg i. Br., 89*.

Vowinkel, E. (1955). Southern Hemisphere weather map analysis: Five-year mean pressures; Part II. *Notos, 4*, 204-216, Weather Bureau, Pretoria.

Wahl, E.W. (1968). A comparison of the climate of the eastern United States during the 1830's with the current normals. *Monthly Weather Review, 96*, 73-82.

Walker, D. (1978). Quaternary climate os the Australian region. In: *Climatic Change and Variability. A Southern Perspective*, ed. A.B. Pittock, L.A. Frakes, J.A. Peterson & J.W. Zillman, pp.82-97. Cambridge University Press.

Walker, D. & Guppy, J.C. (1976). Generic plant assemblages in the highland forests of Papua New Guinea. *Australian Journal of Ecology, 1*, 203-212.

Walker, G.T. & Bliss, E.W. (1932). World Weather V. *Memoirs of the Royal Meteorological Society, 4*(36), 53-84.

Walker, G.T. & Bliss, E.W. (1937). World Weather VI. *Memoirs of the Royal Meteorological Society, 4*(39), 119-139.

Wallén, C.C. (1970). *Climates of Northern and Western Europe. World Survey of Climatology, Vol.5*. Elsevier: Amsterdam.

Wallén, C.C. (1977). *Climates of Central and Southern Europe. World Survey of Climatology, Vol.6*. Elsevier: Amsterdam.

Wallis, J.R. & O'Connell, P.E. (1973). Firm reservoir yield; How reliable are historic hydrological records ? *Hydrological Sciences Bulletin, 18*, 347.

Wallis, A.W. & Roberts, H.V. (1956). *Statistics: A New Approach*. The Free Press: Glencoe, Illinois.

Walsh, J.E. (1977). The incorporation of ice station data into a study of recent Arctic temperature fluctuations. *Monthly Weather Review, 105*, 1527-1535.

Walsh, J.E. (1978). Temporal and spatial scales of the Arctic circulation. *Monthly Weather Review, 106*, 1532-1544.

Wardle, P. (1963a). The regeneration gap of New Zealand gymnosperms. *New Journal of Botany, 3*, 301-315.

Wardle, P. (1963b). Vegetation studies on Secretary Island. Part 5. Population structure and growth of rimu (*Dacrydium cupressinum*). *New Zealand Journal of Botany, 1*, 208-214.

Wardle, P. (1963c). Growth habits of New Zealand subalpine shrubs and trees. *New Zealand Journal of Botany, 1*, 18-47.

Wardle, P. (1969). Biological flora of New Zealand. 4. *Phyllocladus alpinus* Hock F. (Podocarpaceae) Mountain Toatoa, Celery Pine. *New Zealand Journal of Botany, 7*, 76-95.

Wardle, P. (1970). The ecology of *Nothofagus solandri*. 4. Growth, and general discussion to parts 1 to 4. *New Zealand Journal of Botany, 8*, 609-646.

Wardle, P. (1973a). Native vegetation. In: *The Natural History of New Zealand. An Ecological Survey*, ed. G.R. Williams, pp.155-169. A.H. & A.W. Reed: Wellington, New Zealand.

Wardle, P. (1973b). Variations in the glaciers of Westland National Park and the Hooker Range, New Zealand. *New Zealand Journal of Botany, 11*, 349-388.

Wardle, P. (1978). Regeneration status of some New Zealand conifers with particular reference to *Libocedrus bidwillii* in Westland National Park. *New Zealand Journal of Botany, 16*, 471-477.

Watson, B. & Wylie, J. (1972). An investigation into the properties of long term rainfall records in Tasmania. *Report of the Hydro-Electric Commission*, Hobart, Tasmania.

Webb, L.J. (1978). A structural comparison of New Zealand and south-east Australian rain forests and their tropical affinities. *Australian Journal of Ecology, 3*, 7-21.

Webb, T, III, & Bryson, R.A. (1972). Late- and post-glacial climatic change in the northern midwest, USA: quantitative estimates derived from fossil pollen spectra by multivariate statistical analysis. *Quaternary Research, 2,* 70-111.

Webb, T., III, & Clarke, D.R. (1977). Calibrating micropaleontological data in climatic terms: a critical review. *Annals of the New York Academy of Sciences, 288,* 93-118.

Weidick, A. (1975). A review of Quaternary investigations in Greenland. *Institute of Polar Studies Report No.55,* Ohio State University.

Weissel, J.K., Hayes, D.E. & Herron, E.M. (1977). Plate tectonics synthesis. The displacements between Australia, New Zealand and Antarctica since the Late Cretaceous. *Marine Geology, 25,* 231-277.

Wellington, A.B., Polach, H.A. & Noble, I.R. (1979). Radiocarbon dating of lignotubers from male forms of *Eucalyptus. Search, 10,* 282-283.

Wells, J.A. (1972). Ecology of *Podocarpus hallii* in central Otago, New Zealand. *New Zealand Journal of Botany, 10,* 399-426.

Wendelken, W.J. (1976). Forests. In: *New Zealand Atlas,* ed. I. Wards, pp.98-107. Government Printer: Wellington, New Zealand.

Whelan, T., Sackett, W.M. & Benedict, C.R. (1973). Enzymatic fractionation of carbon isotopes by phosphoenolpyruvate carboxylase from C_4 plants. *Plant Physiology, 51,* 1051-1054.

Wherry, R.J. (1931). A new formula for predicting the shrinkage of the multiple correlation coefficient. *Annals of Mathematical Statistics, 2,* 440-457.

Wieser, E. & Becker, B. (1975). Die Entwicklung des spätmittelalterlichen Säulenbaues in Bad Windsheim und Uffenheim. *Jahrbuch Bayer. Denmalpflege, 29,* 35-78.

Wigley, T.M.L. (1981). Climate and paleoclimate: what can we learn about solar luminosity variations. *Solar Physics, 74,* 435-471.

Wigley, T.M.L., Gray, B.M. & Kelly, P.M. (1978). Climatic interpretation of $\delta^{18}O$ and δD in tree rings. *Nature, London, 271,* 92-93.

Wigley, T.M.L., Jones, P.D. & Kelly, P.M. (1980). Scenario for a warm, high-CO_2 world. *Nature, London, 283,* 17-21.

Wigley, T.M.L., Ingram, M.J. & Farmer, G., editors. (1981). *Climate and History.* Cambridge University Press.

Willett, H.C. (1950). Temperature trends of the past century. *Centenary Proceedings of the Royal Meteorological Society,* London, pp.195-206.

Williams, L.D., Wigley, T.M.L. & Kelly, P.M. (1981). Climatic trends at high northern latitudes during the last 4,000 years compared with ^{14}C fluctuations. In: *Sun and Climate.* C.N.E.S.: Toulouse.

Williams, P., Jr. (1948). The variation of the time of maximum precipitation along the west coast of North America. *Bulletin of the American Meteorological Society, 29*(4), 143-145.

Williams, R.W.M. & Chavasse, C.G.R. (1951). The silviculture of silver beech in Southland. *New Zealand Journal of Forestry, 2*(3), 219-235.

Wilson, A.T. & Grinsted, M.J. (1975). Palaeotemperatures from tree rings and the D/H ratio of cellulose as a biochemical thermometer. *Nature, London, 257,* 387-388.

Wilson, A.T. & Grinsted, M.J. (1976). The possibilities of deriving past climatic information from stable isotope studies on tree-rings. In: *Proceedings of the International Conference on Stable Isotopes,* August 1976, pp.1-12. Institute of Nuclear Sciences: Lower Hutt, New Zealand.

Wilson, A.T. & Grinsted, M.J. (1977a). The D/H ratio of cellulose as a biochemical thermometer: a comment on "Climatic implications of the D/H ratio of hydrogen in C-H groups in tree cellulose" by S. Epstein and C.J. Yapp. *Earth and Planetary Science Letters, 37* (1977), 246-248.

Wilson, A.T. & Grinsted, M.J. (1977b). $^{12}C/^{13}C$ in cellulose and lignin as palaeothermometers. *Nature, London, 265,* 133-135.

World Meteorological Organization. (1981). On the assessment of the role of CO_2 on climate variations and their impact. *Proceedings of a Joint WMO/ICSU/UNEP Meeting of Experts.* W.M.O. (World Climate Programme): Geneva.

Wu Xiangding & Lin Zhenyao (1978). A preliminary analysis of climatic variation during the last hundred years and its outlook on Tibetan Plateau. *Kexue Tongbao, 23*(12), 746-750. In Chinese.

Yamamoto, T. (1949). Tree-ring and climate. *Tenbun to Kisho, 15*(5), 12-17. In Japanese.

Yamazawa, K. (1929). About the tree growth since 1628 in region of Hida Mts. *Journal of the Meteorological Society of Japan, Series II, 7*(6), 186-190. In Japanese.

Yevjevich, V.M. (1967). An objective approach to definitions and investigations of continental hydrologic droughts. *Colorado State University Hydrologic Papers No.23.*

Zoltai, S.C. (1975). Tree ring record of soil movements on permafrost. *Arctic and Alpine Research, 7,* 331-340.

SPECIES INDEX

219

SUBJECT INDEX

Aerial photographs, 86
Africa, 65, 105, 151
Age effects, on response functions, 58
Age structure, 3
Age-trend, see growth-trend
Air pollution, 135, 136
Alaska, 21, 105, 108, 109, 111, 114
Algeria, 155
Alluvia, 11
Alpha-sources, 12
Alpine Europe, 135-141
Alps, 105, 135-141, 159-162
Altitudinal (subalpine) treeline, 5, 189
Analysis of variance (ANOVA), 6, 22, 25, 26, 27, 35, 38, 65
Andes, 79, 81, 87, 90, 168, 169
Annual radial increment, 1, 11, 12
    see also ring width
Annual ring, 1, 3, 4, 8, 9, 32, 159
Antarctica, 79-80
Archaeological timbers, 105, 111
Arctic, 105, 107-118
Arctic Ocean, 118
Argentina, 63, 64, 78, 79, 81, 84-90, 159, 168-170, 177
Arid forest margin, see lower forest border
Arid zones, 11
Arid sites, 26, 36, 192
Arizona, 16
Artificial predictability, 53, 54, 56
Asia, 105, 151, 155-158, 193, 196
Assumptions of normality, 174
Assumptions of independence of predictor variables, 174
Atlantic, 67-77, 79-81, 105, 117, 142, 151, 168
Atmospheric CO2 levels, 20, 21
Atmospheric water vapour, 17
Australia, 4, 62, 78-81, 90-96, 103, 170, 184
Australasia, 90-104, 159
Austria, 135, 137
Autocorrelation, 5, 22, 23, 25, 27, 28, 35, 43, 44, 49, 56, 62, 92, 96, 99, 103, 167, 174, 175, 176, 182, 189, 192
Autoregressive time series, 32, 51, 56, 163, 189
Autoregressive moving average (ARMA, ARIMA), 61, 169

Bavaria, 146
β-particles, 8
Beta sources, 12
Benguela current, 82
Belgium, 142
Biennial oscillation, 103
Biological rhythm, 92
Bolivia, 89
Boreal forest, 107
Brazil, 64, 79, 81, 89
British Isles, 38, 39, 45, 142, 143, 147, 148, 149
Bronze Age, 142, 144
Building timbers, 61
    see also Archaeological timbers
Burial of stemwood, 116
Buttressing or fluting, 6
Byelorussia, 150

C-3 plants, 15, 17
Calculated evapotranspiration, 193
Calibration, 12, 32, 51, 52, 167, 174, 185, 186-188, 189, 195
Calibration period or interval, 50, 52, 53, 54, 55, 56, 162, 165, 167, 174, 175, 177, 182, 185, 188
Calibration statistics, 51, 52, 53
Calibrated variance, see R²
Calibration data, 58, 59, 60
Calibration with data at different frequencies, 44, 47
Calibration data, inhomogeneities in, 59
    see also Errors in meteorological data
California, 36, 184, 185-191
Cambial activity, 3, 8, 11

Canada, 9, 108-109, 110, 111, 114, 119, 123, 125, 126, 128
Canonical correlation, 32, 51, 55, 56, 165, 174, 182
Canonical regression, 32, 62, 165, 167, 169, 172, 173, 174, 178, 182, 185
Canonical variate pairs, 165
Cape Province (South Africa), 81, 84
Carbon stable isotopes, 13, 14, 15, 16, 18, 19, 20
Carbon-14, 19, 20, 21
Carbon-dates, 112, 142, 143, 150, 151
Carbon isotope ratios, 1, 13, 14, 15, 16, 33, 111
    see also Isotopes, Stable isotopes
Caucasus, 150
Cell size, 33
Cell types, 1, 8
Cell wall thickness, 8
Cellulose, 13, 14, 15, 16, 17, 18, 19, 20
    α-, 14
    amorphous, 14
    crystalline, 14
    leaf, 16
Central limit theory, 65
Chile, 63, 64, 78, 81, 84-90, 168, 169, 170, 177
China, 155, 156, 157
Chronology analysis, 21-28
Chronology development, 5, 21-28
Chronology variance, 7, 26, 38
Circuit variance, 7, 26, 38
Circuit uniformity/circumferential differences, 3, 6, 19
Classificatory approach to the reconstruction of past climate, 160-162
    see also Discriminant analysis
Climatic cycles, rhythms, see Climatic oscillations
Climatic oscillations, 103, 150, 151
Climatic window, 110, 173
Climate-growth response, 1, 28, 29, 30, 31, 32, 33-38, 45, 50, 51, 56, 57, 62, 63, 165, 177, 193
Climate response characteristics, 6, 35
Climatic responsiveness, 5
Closed-canopy forests, 2, 5, 6, 30, 94, 98, 127, 131, 134, 179
Cluster analysis, 153, 160, 161
Colorado, 36, 48
Competitive status, 3
'Complacent' tree rings, 103
Composite chronologies, 2, 8, 105, 148-149
Composite climate-related variables, 159
Computer processing, 9
Computerised dating, 110
Connecticut, 6, 166-167
Continental climate, 135, 136
Correlation, 7, 16, 20, 27, 29, 34, 35, 36, 38, 40, 48, 50, 57, 140, 157, 167, 169, 173, 175, 184, 185, 188, 190
Corrosion, 116
Cretaceous, 90
'Critical' rings, 5
Crop yields, 61, 190, 193
Cross-correlation, 21, 25
Cross-dating, 2, 4, 5, 7, 25, 29, 47, 84, 85, 88, 90, 96, 99, 101, 102, 104, 110, 112, 128, 148, 160, 168, 192
Cross-sections, 5, 6
Cross-spectral analysis, 152
Cumulative eigenvalues product (PVP or CEP), 40, 42, 46, 47
Curve-fitting, 21-24
Cycles, 150, 155, 156, 157
Czechoslovakia, 137, 148

D/H ratios, 14, 17, 18
Damage by mammals, 97, 105, 135, 152
Data networks, 2, 6, 18, 67-77, 169, 172, 182, 192
Data base, 33, 159, 177
Data network design, 65, 67-77
    spatial density, 67
Dating, 1, 3, 4, 6, 9, 33, 168, 170, 177
Deciduous forest biome, 126-127
Degree days, 112, 193
Dendrochronology, 1, 2, 142
Dendroclimatology, 1, 2, 4, 6, 7, 8, 9, 17, 19, 21, 30, 33, 38, 47, 64, 65, 78, 84, 92, 102, 105, 107, 109, 110, 114, 118, 123, 124, 125, 131, 134, 140, 148, 150, 153, 154, 155, 157, 159, 160, 177, 191-196